Water and Cities in Latin America

Approximately 80 per cent of the population of Latin America is concentrated in urban centres. Pressure on water resources and water management in cities therefore provide major challenges. Despite the importance of the issues, there has been little systematic coverage of the topic in book form.

This work fills a gap in the literature by providing both thematic overviews and case study chapters. It reviews key aspects of why water matters in cities and presents case studies on topics such as groundwater management, green growth and water services, inequalities in water supply, the financing of water services and flood management. Detailed examples are described from Argentina, Brazil, Colombia, Ecuador, Mexico and Peru, and there is also a chapter comparing lessons which might be learnt from US cities. Contributing authors are drawn from both within and outside the region, including from the Inter-American Development Bank, OECD and World Bank to set the issues in a global context.

Ismael Aguilar-Barajas is Professor of Economics and Research Associate, Water Center for Latin America and the Caribbean, Tecnológico de Monterrey, Mexico.

Jürgen Mahlknecht is Executive Director, Water Center for Latin America and the Caribbean, Tecnológico de Monterrey, Mexico.

Jonathan Kaledin is Founder and Principal of the 11th Avenue Strategies consulting firm and a member of the Stockholm International Water Institute's Associate Program.

Marianne Kjellén leads and develops the thematic area of Water Governance and manages the UNDP Water Governance Facility at Stockholm International Water Institute.

Abel Mejía-Betancourt is Senior Water Advisor for the Latin American Development Bank (CAF) and former Water Manager at the World Bank.

Earthscan Studies in Water Resource Management

Contemporary Water Governance in the Global South
Scarcity, marketization and participation
Edited by Leila Harris, Jacqueline Goldin and Christopher Sneddon

Water Governance, Policy and Knowledge Transfer
International studies on contextual water management
Edited by Cheryl de Boer, Joanne Vinke-de Kruijf, Gül Özerol and Hans Th. A. Bressers

Water as a Catalyst for Peace
Transboundary water management and conflict resolution
By Ahmed Abukhater

Sustainable Water and Sanitation Services
The life-cycle approach to planning and management
By Livelihoods & Natural Resource Management Institute, International Water & Sanitation Centre, Centre for Economic and Social Studies, Watershed Support Services & Activities Network

Water for Food Security and Well-being in Latin America and the Caribbean
Social and environmental implications for a globalized economy
Edited by Bárbara A. Willaarts, Alberto Garrido and M. Ramón Llamas

Water Scarcity, Livelihoods and Food Security
Research and innovation for development
Edited by Larry W. Harrington and Myles J. Fisher

Adaptation to Climate Change through Water Resources Management
Capacity, equity and sustainability
Edited by Dominic Stucker and Elena Lopez-Gunn

Hydropower Devleopment in the Mekong Region
Political, socio-economic and environmental perspectives
Edited by Nathanial Matthews and Kim Geheb

Governing Transboundary Waters
Canada, the United States and indigenous communities
By Emma S. Norman

Transboundary Water Politics in the Developing World
By Naho Mirumachi

International Water Law and the Quest for Common Security
By Bjørn-Oliver Magsig

Water, Power and Identity
The cultural politics of water in the Andes
By Rutgerd Boelens

Water and Cities in Latin America
Challenges for sustainable development
Edited by Ismael Aguilar-Barajas, Jürgen Mahlknecht, Jonathan Kaledin and Marianne Kjellén

Catchment and River Basin Management
Integrating science and governance
Edited by Laurence Smith, Keith Porter, Kevin M. Hiscock. Mary Jane Porter and David Benson

For more information and to view forthcoming titles in this series, please visit the Routledge website: http://www.routledge.com/books/series/ECWRM/

Water and Cities in Latin America

Challenges for sustainable development

Edited by Ismael Aguilar-Barajas, Jürgen Mahlknecht, Jonathan Kaledin, Marianne Kjellén and Abel Mejía-Betancourt

Routledge
Taylor & Francis Group
LONDON AND NEW YORK

earthscan
from Routledge

First published 2015 by Routledge

2 Park Square, Milton Park, Abingdon, Oxfordshire OX14 4RN
711 Third Avenue, New York, NY 10017

Routledge is an imprint of the Taylor & Francis Group, an informa business

First issued in paperback 2018

British Library Cataloguing-in-Publication Data
A catalogue record for this book is available from the British Library

Library of Congress Cataloging-in-Publication Data
Water and cities in Latin America : challenges for sustainable development / edited by Ismael Aguilar-Barajas, Jürgen Mahlknecht, Jonathan Kaledin, and Marianne Kjellén.
pages cm. – (Earthscan studies in water resource management)
Includes bibliographical references and index.
1. Water resources development – Latin America. 2. Municipal water supply – Latin America. 3. Water-supply – Latin America.
I. Aguilar Barajas, Ismael, 1956– II. Mahlknecht, Jürgen.
III. Kaledin, Jonathan. IV. Kjellén, Marianne.
TD227.5.W35 2015
333.910098–dc23

2014039807

ISBN: 978-0-415-73097-6 (hbk)
ISBN: 978-1-138-36423-3 (pbk)

Typeset in Bembo
by HWA Text and Data Management, London

Contents

List of figures *viii*
List of tables *x*
List of contributors *xi*
Foreword by Sergio Campos and Vidal Garza *xiv*
Acknowledgements *xv*
List of acronyms and abbreviations *xvii*

Introduction 1
ISMAEL AGUILAR-BARAJAS, JÜRGEN MAHLKNECHT, JONATHAN KALEDIN, MARIANNE KJELLÉN AND ABEL MEJÍA-BETANCOURT

PART I
Water, cities and financing **13**

1 Water, cities and sustainable development: setting the scene 15
ISMAEL AGUILAR-BARAJAS

2 Why does understanding the urban water link matter? 34
ABEL MEJÍA-BETANCOURT

3 Challenges and policy options for financing urban water and sanitation 46
ANTHONY COX AND PETER BÖRKEY

PART II
Managing water and sanitation services **71**

4 Urban water management in São Paulo to achieve universal access to water supply and sanitation services 73
DILMA PENA

5 Water services in Lima, Peru: understanding the linkages between urban development, social inequality and water 89
ANTONIO A. R. IORIS

6 Urban wastewater governance in Latin America: panorama and reflections for a research agenda 102
RAÚL PACHECO-VEGA

7 Issues and challenges for water supply, storm water drainage and wastewater treatment in the Mexico City metropolitan area 109
POLIOPTRO F. MARTÍNEZ AND ERICK R. BANDALA

8 Urban groundwater supply in Latin American cities: panorama and the cases of Mexico City and São Paulo 126
JÜRGEN MAHLKNECHT, RICARDO HIRATA AND ROGELIO LEDESMA-RUIZ

PART III
Climate change, risk management and planning 147

9 Flash floods in the Monterrey metropolitan area, Mexico: lessons from Hurricane Alex 149
NICHOLAS P. SISTO AND ALDO I. RAMÍREZ OROZCO

10 The Río Bogotá environmental recuperation and flood control project 166
CARLOS R. COSTA POSADA, CARLOS FELIPE URAZÁN BONELLS AND EDDER A. VELANDIA DURÁN

11 Protection and environmental restoration of the slopes of Pichincha in Quito, Ecuador 181
XAVIER VIDAL, LUCÍA BURGOS AND OTHÓN ZEVALLOS

12 Water resources management and climate adaptation in the Río Grande Basin, Argentina: a case study introducing the Hydro-BID modelling system 197
FERNANDO MIRALLES-WILHELM, EUGENE BRANTLY, MARCELO BORSELLINO, EDGARDO CASTELLANO, ROBERT DYKES, FEKADU MOREDA, RAÚL MUÑOZ-CASTILLO, JAY RINEER, BILL WHEATON AND ALAN WYATT

PART IV
Green infrastructure development **219**

13 Green growth and water services in Latin American cities 221
GREG BROWDER

14 Water funds as a tool for urban water provision and watershed conservation in Latin America 235
FERNANDO VEIGA, ALEJANDRO CALVACHE, SILVIA BENÍTEZ, JORGE LEÓN AND AURELIO RAMOS

15 Green infrastructure and watershed protection in US cities: Insights for Latin America 256
JONATHAN C. KALEDIN

Index 274

Figures

2.1 Urban water is generally divided into three separate components: drinking water, sanitation/sewerage and drainage 38
2.2 Cycle of water infrastructure projects 40
3.1 Sources of finance for water and sanitation services 50
3.2 Trends in aid to water and sanitation 53
3.3 Investment in public–private partnership projects in developing countries 65
4.1 Hydrographic divisions of São Paulo State 75
4.2 The hydrographic basin Alto Tietê and the Metropolitan Region of São Paulo 75
5.1 The metropolitan region of Lima with built-up areas, main rivers and the three case study locations 91
7.1 Geographical location of RHVM and schematic overview of the water supply sources and drainage outlets in the ZMVM 111
7.2 Water balance in the metropolitan area of the Valley of Mexico 112
8.1 Estimated extent of groundwater use for drinking water in Latin America 132
8.2 Population of and extraction and importation of water to Mexico City, 1900–2010 134
8.3 Location of the Upper Tietê Basin in the state of São Paulo, and municipalities and sub-basin units 137
9.1 Track of Hurricane Alex and location of the Monterrey metropolitan area (MMA) 150
9.2 Isohyetal lines of cumulative precipitation (mm) in the 48 hours between 29 June and 1 July 2010, Nuevo León 151
9.3 Hydrological and hydraulic features in the Monterrey metropolitan area (MMA) and surrounds 152
9.4 Maximum flow recorded during years between 1960 and 2010, Río Santa Catarina, Monterrey metropolitan area 153
9.5 Flash flood in the Río Santa Catarina, Monterrey metropolitan area, following Hurricane Alex 154
9.6 Monthly Industrial Production Index, Nuevo León, January 2009 to December 2011 157

9.7 Narrowing of river channel by road infrastructure, Monterrey metropolitan area, 2007 158
9.8 Narrowing of river channel by residential infrastructure, Monterrey metropolitan area, 2007 158
9.9 Impact of Hurricane Alex on river channel, Monterrey metropolitan area, 2010 159
10.1 The Río Bogotá watershed and the location of municipalities 167
10.2 Areas at risk of flooding in one part of Bogotá 171
10.3 Location of the principal projects of the Hydraulic Upgrade and Environmental Recuperation Project for the Río Bogotá 173
10.4 Recovery works on the Río Salitre, Juan Amarillo, north-western Bogotá 175
11.1 Urban encroachment on the hillsides of central Quito 182
11.2 Urban expansion of Quito, 1535–2009 183
11.3 House at risk of collapse in the lower part of the El Tejar ravine 185
11.4 Bridging and environmental conditioning works, Quebrada El Tejar 190
11.5 Example of environmental conditioning, Quebrada Navarro 193
12.1 Hydro-BID simulation model software flow chart 201
12.2 Jujuy Province, Argentina 203
12.3 Water infrastructure in the Río Grande and upper Río San Francisco basins 205
12.4 Jujuy integrated water resource management approach 206
12.5 Observed and simulated flow, Río San Francisco at Caimancito 209
13.1 The concept of integrated urban water management 231
14.1 General water fund scheme 241
14.2 Water Funds in Latin America and the Caribbean, 2013 242
14.3 Cattle ranching in high *páramo* areas 247
14.4 Governance arrangement of the Bogotá Water Fund 248
14.5 Logotype and slogan of the Bogotá Water Fund 249
15.1 New York City's water supply system 259
15.2 Boston's water supply system 262
15.3 Seattle's water supply system 263

Tables

1.1 The ten global risks of highest concern in 2014 16
1.2 Historic (1990), recent (2014) and projected (2050) urban and rural population data for major regions of the world 20
1.3 Population of the 30 largest urban agglomerations in the world, 1990–2030 22
1.4 Poor populations in urban and rural areas of Latin America 26
2.1 Investments required and targets for closing the infrastructure gap by 2030 in water services in Latin America 42
3.1 Examples of innovative financial mechanisms in the water sector 58
3.2 Typology of contractual arrangements between government and the private sector 64
4.1 Percentage of households with access to water supply and sanitation services in São Paulo, in other states and in Brazil 74
4.2 Deficit in access to sanitation services in São Paulo and in Brazil 77
7.1 Drainage capacity of the Valley of Mexico system 114
7.2 Water governance gaps in Mexico 119
8.1 Comparison of groundwater and surface water 131
9.1 Frequencies of maximum flow, Río Santa Catarina, Monterrey metropolitan area 153
9.2 Reconstruction budget following Hurricane Alex 155
9.3 Manufacturing Production Index (MPI, Nuevo León) and Retail Sales Index (RSI, Monterrey metropolitan area) 156
11.1 Investments in the Pichincha slopes, 1996–2014 189
12.1 Unmet demand and savings for four climate projections and four adaptation options 213
12.2 Cost efficiency of adaptation for urban water supply and irrigation 215
13.1 Regional average and selected water and sanitation hotspots in Latin America and the Caribbean, 2010 223
13.2 Estimated wastewater treatment rates for selected countries 224
14.1 Water fund location and status in Latin America, December 2013 243

Contributors

Ismael Aguilar-Barajas is Professor at the Department of Economics and Research Associate at the Water Center for Latin America and the Caribbean, Tecnológico de Monterrey, Mexico.

Erick Roberto Bandala-González is at the Water Sciences Group, Universidad de las Américas Puebla, Mexico.

Silvia Benítez is at the Nature Conservancy (TNC), Latin America Region.

Peter Börkey is at the Environment Directorate of the Organization for Economic Co-operation and Development (OECD).

Marcelo Borsellino is at the Comisión Regional del Río Bermejo (COREBE), Argentina.

Eugene Brantly is at RTI International Inc., Research Triangle Park, NC, USA.

Greg Browder is Lead Water and Sanitation Specialist at the World Bank on the Latin American and Caribbean Region.

Lucía Burgos Salvador is a Social Management Specialist at the Environmenal Sanitation Programme, Empresa Pública Metropolitana de Agua Potable y Saneamiento de Quito, Ecuador.

Alejandro Calvache is at the Nature Conservancy (TNC), North Andes and Southern Central America Conservation Program.

Sergio Campos heads the Water and Sanitation Division of the Inter-American Development Bank.

Edgardo Castellano is at Comisión Regional del Río Bermejo (COREBE), Argentina.

Carlos Costa Posada is Dean of the Faculty of Engineering, Universidad La Salle, Bogotá, and former Minister for Housing, Environment and Territorial Development, Republic of Colombia.

Anthony A. Cox is Deputy Director in the Environment Directorate of the Organization for Economic Co-operation and Development (OECD).

Robert Dykes is at RTI International Inc., Research Triangle Park, NC, USA.

Vidal Garza leads Fundación FEMSA; a corporate social investment organization created to foster sustainable development of water resources in Latin America and improve the quality of life of its communities.

Ricardo Hirata is Professor of the Institute of Geosciences and Vice-Director of the Groundwater Research Center (CEPAS), University of São Paulo, Brazil.

Antonio A. R. Ioris is Lecturer in Environment and Society, and director of the M.Sc. Programme on Environment and Development at the School of Geosciences, University of Edinburgh.

Jonathan Kaledin is Founder and Principal of the 11th Avenue Strategies Consulting firm and a member of the Stockholm International Water Institute's Associate Programme.

Marianne Kjellén directs the UNDP Water Governance Facility at SIWI, the Stockholm International Water Institute.

Rogelio Ledesma is a water specialist, Water Center for Latin America and the Caribbean, Tecnológico de Monterrey, Mexico.

Jorge León is at the Nature Conservancy (TNC), Latin America Region.

Jürgen Mahlknecht is Executive Director, Water Center for Latin America and the Caribbean, Tecnológico de Monterrey, Mexico.

Polioptro F. Martínez Austria is at the Water Sciences Group, Universidad de las Américas Puebla, Mexico.

Abel Mejía-Betancourt is Senior Water Advisor for the Latin American Development Bank (CAF) and former Water Manager at the World Bank.

Fernando Miralles-Wilhelm is Professor at the Department of Atmospheric and Oceanic Sciences at the University of Maryland, and a Water Resources Advisor at the Inter-American Development Bank, Washington, DC, USA.

Fekadu Moreda is at RTI International Inc., Research Triangle Park, NC, USA.

Raúl Muñoz-Castillo is at the Water and Sanitation Division of the Inter-American Development Bank, Washington, DC, USA.

Raúl Pacheco-Vega is Assistant Professor in the Public Administration Division, Center for Research and Teaching in Economics (CIDE), Mexico.

Dilma Pena is former President of the São Paulo Water Utility Company (SABESP) and former Director of the National Water Agency of Brazil (ANA), Brazil.

Aldo Iván Ramírez Orozco is Professor at the Water Center for Latin America and the Caribbean, Tecnológico de Monterrey, Mexico.

Aurelio Ramos is at The Nature Conservancy (TNC), Latin America Region.

Jay Rineer is at RTI International Inc., Research Triangle Park, NC, USA.

Nicholas P. Sisto is Professor at the Center for Socio-economic Studies, Universidad Autónoma de Coahuila, Mexico.

Carlos Felipe Urazan Bonells is Titular Professor at the Faculty of Engineering, Universidad La Salle, Bogotá, Colombia.

Fernando Veiga is at the Nature Conservancy (TNC), Latin America Region.

Edder, A. Velandía Durán is at the Faculty of Engineering, Civil Engineering Programme, Universidad La Salle, Bogotá, Colombia.

Xavier Vidal Pacurucu is Executive Director of the Environmenal Sanitation Programme, Empresa Pública Metropolitana de Agua Potable y Saneamiento de Quito, Ecuador.

Bill Wheaton is at RTI International Inc., Research Triangle Park, NC, USA.

Alan Wyatt is at RTI International Inc., Research Triangle Park, NC, USA.

Othón Zevallos Moreno is a consultant on water and sanitation, and a former general manager of Empresa Pública Metropolitana de Agua Potable y Saneamiento de Quito, Ecuador.

Foreword

Sergio Campos and Vidal Garza

We are grateful for the opportunity to present a fine book that addresses an array of issues surrounding water and cities in Latin America. The book does this from a development perspective, which sheds light on the challenges deriving from the simultaneous pursuit of economic growth, social inclusion and environmental responsibility. In the 21st century cities will be the key for global sustainability and understanding how they relate to water is essential for achieving a more sustainable development path. This book provides interesting insights and useful policy lessons in this regard.

Understanding cities' potential for coping with the challenges of sustainability is a pressing topic today and for the coming years in Latin America. Latin America is one of the most urbanized regions in the world – its current proportion of urban population (80 per cent) is only second to North America (81 per cent) – and has an historic opportunity to reach higher levels of economic development, reduce inequality and close existing infrastructure gaps by 2030, in particular with respect to water and sanitation. By 2050, almost 90 per cent of the region's population will live in cities, often located close to the sea and exposed to threats intricately tied to water and climate change, such as flooding from rising sea levels and droughts. In this context the knowledge and lessons offered by this book are very useful.

The theme of the 2011 Stockholm International Water Institute's World Water Week was 'Water in an Urbanized World', a truly central topic for our times. There exist however considerable gaps in the literature on this topic, both in terms of theoretical concepts and case studies, and particularly with regards to Latin America. We are convinced that this book will contribute substantially to enriching this literature, as it integrates work from nearly 40 authors from different countries of Latin America and other regions of the world. The book's four parts are well balanced and give the reader sufficient room to discover the power and limitations of planning in connection with nature, water, cities and sustainable development. This book will most likely become essential reading for urban, water and environmental policy planners, as well as for academics, politicians, actors from the private sector and civil society at large, in Latin America and elsewhere.

Acknowledgements

This book could not have been published without the contributions of several institutions and individuals. The peer review of the original book proposal was very useful to better focus the project. Our appreciation is due to the reviewers. The Water Center for Latin America and the Caribbean, at Tecnológico de Monterrey, Mexico, provided the financial resources to fund key tasks: professional translation and editorial services as well as the elaboration and edition of the figures that appear in the book. The generous contribution of the Inter-American Development Bank and Fundación FEMSA in this regard is acknowledged.

The Stockholm International Water Institute's (SIWI) editorial work on this book has been generously supported by the Swedish International Development Cooperation Agency (Sida). Sida contributes both directly to the implementation of SIWI's strategy and supports activities through the UNDP Water Governance Facility at SIWI.

Robert Showman did a superb Spanish to English translation of different chapters. Alice Franek did not only produce a truly excellent edition of the manuscript, but also provided lessons of integrity. Her work was far beyond the call of duty.

The preparation of the manuscript required the collaboration of truly skillful people. Abigail Juárez–Rivero was of great help in making sure the manuscript followed Earthscan publication guidelines. Her careful dedication is reflected throughout the book. Carol Tamez-Meléndez also supported this activity, in addition to partially aiding the editing process. Cynthia Ortega prepared the figures, an essential but difficult task she undertook with great commitment.

The Water Center also supported many administrative activities during the entire process – from the original idea to the final publication of the book. The names of Gloria Catalina Ayala and Adalberto Calzada must be mentioned. The Department of Economics, Tecnológico de Monterrey, also contributed to carry administrative activities through the always diligent work of Reyna Tovar.

We are also largely indebted to Earthscan for the overall encouragement and support to produce this book. The understanding and flexibility of Tim Hardwick and Ashley Wright to address difficult issues were remarkable. Without their help, we would not have completed the book. More particularly,

Ashley had to deal with a great many issues she managed to resolve maintaining a positive attitude throughout. The work done by the firm HWA in the last phase of the production of the book was of superior quality; Holly Knapp dealt with this last part with both energy and excellent attitude.

In the end, but not least, there are our families who had to pay a heavy price for the long hours we had to dedicate to this project. The recognition to them is absolutely necessary.

Acronyms and abbreviations

3Ts	tariffs, taxes, and transfers
ACNUR	Agencia Colombiana de las Naciones Unidas para los Refugiados (Colombian Agency of the United Nations for Refugees)
ADERASA	Asociación de Entes Reguladores de Agua Potable y Saneamiento de las Américas (Association of Regulatory Entities of Water and Sanitation in the Americas)
AFIN	Asociación para el Fomento de la Infraestructura Nacional (Association for the Promotion of National Infrastructure)
AGW-Net	African Groundwater Network
ANA	Agência Nacional de Águas (National Water Agency)
AHD	analytical hydrography data
APPJ	Agua para Pueblos Jóvenes (Water for Young Settlements)
APT	Agua para Todos (Water for All)
ARIMA	auto-regressive integrated moving average
ARSESP	Agência Reguladora de Saneamento e Energia do Estado de São Paulo (Regulatory Agency for Sanitation and Energy of São Paulo State)
ASCE	American Society of Civil Engineers
BAS	Bogotana de Aguas y Saneamiento (Bogota Water and Sanitation)
BIE	Banco de Información Económica (Economic Information Bank)
BOT	build-operate-transfer
BRGM	Bureau de Recherches Géologiques et Minières (Office of Geological and Mining Research)
BRIC	Brazil, Russia, India, and China
BTO	build-operate-transfer
CAF	Corporación Andina de Fomento (Development Bank of Latin America)
CAMAREN	Consorcio de Capacitación para el Manejo de los Recursos Naturales Renovables (Training System for the Management of Natural Renewable Resources Consortium)

CAR	Corporación Autónoma Regional de Cundinamarca (Autonomous Regional Corporation of Cundinamarca)
Cat-Del	Catskill-Delaware
CDHU	Companhia de Desenvolvimento Habitacional e Urbano (Housing and Urban Development Company)
CEAA	Comisión Estatal de Agua y Alcantarillado (State Comission for Water and Sanitation)
CERNL	Consejo Estatal para la Reconstrucción de Nuevo León (State Council for the Reconstruction of Nuevo Leon)
CETESB	Companhia Ambiental do Estado de São Paulo (São Paulo State Environmental Agency)
CFR	Code of Federal Regulations
CIAT	Centro Internacional de Agricultura Tropical (International Centre for Tropical Agriculture)
CIMA	Centro de Investigaciones del Mar y la Atmósfera (Center for Oceanic and Atmospheric Research)
CNA	Comisión Nacional del Agua (National Water Commission)
COMPESA	Companhia Pernambucana de Saneamento (Pernambucana Company of Sanitation)
CONAGUA	Comisión Nacional del Agua (National Water Commission)
CONPES	Consejo Nacional de Política Económica y Social (National Economic and Social Policy Council)
COREBE	Comisión Regional del Río Bermejo (Regional Commission of the Bermejo River)
COSUDE	Swiss Development Cooperation Agency
COVAAP	Comité Vecinal de Administración de Agua Potable (Vigilance Committee for Drinking Water)
CRA	Comisión de Regulación de Agua Potable y Saneamiento Básico (Regulation Commission for Drinking Water and Basic Sanitation)
CSOs	combined sewer overflows
CSR	corporate social responsibility
CWA	Clean Water Act
DAC	Development Assistance Committee
DANE	Departamento Administrativo Nacional de Estadística (National Administrative Department of Statistics)
DC	Distrito Capital (Capital District)
DEM	digital elevation map
DNP	Departamento Nacional de Planeación (National Planning Department)
EAAB	Empresa de Acueducto y Alcantarillado de Bogotá (Bogotá Water Supply and Sewerage Company)
ECLAC	Economic Commission for Latin America and the Caribbean
EEQ	Empresa Eléctrica de Quito (Quito Power Company)

EIA	environmental impact assessments
EMGESA	Empresa de Energía de Bogotá (Bogotá Energy Company)
ENACAL	Empresa Nicaragüense de Acueductos y Alcantarillado (Nicaraguan Company of Water and Sanitation)
EPA	(US)Environmental Protection Agency
EPM	Empresas Públicas de Medellín (Medellín Public Companies)
EPMAPS	Empresa Pública Metropolitana de Agua Potable y Saneamiento (Metropolitan Water and Sewage Company)
EPMAPS	Empresa Pública Metropolitana de Agua Potable y Saneamiento (Water and Sewage Metropolitan Enterprise of Quito)
EPMASEO	Empresa Pública Metropolitana de Aseo (Municipal Garbage Collection Company)
ETr	evapotranspiration
FABHAT	Fundação Agência da Bacia Hidrográfica do Alto Tietê (Agency for the Upper Tietê Hidrographic Basin)
FAD	filtration avoidance determination
FAO	Food and Agriculture Organization of the United Nations
FINDETER	Financiera de Desarrollo Territorial S.A. (Territorial Development Fund)
FONAG	Fondo para la Protección del Agua (Fund for Water Protection)
FOPAE	Fondo de Prevención y Atención de Emergencias de Bogotá (Bogotá Prevention and Emergency Support Fund)
GCM	general circulation model
GDP	gross domestic product
GEF	Global Environment Facility
GIS	geographic information system
GRTB	Grupo Regional de Trabajo de Benchmarking (Benchmarking Regional Working Group)
GWLF	Generalized Watershed Loading Factor
GW-MATE	Ground Water Management Advisory Team (GW-MATE)
GWP	Global Water Partnership
IBNET	International Benchmarking Network for Water and Sanitation Utilities
IDB	Inter-American Development Bank
IFIs	International finance institutions
INEGI	Instituto Nacional de Estadística y Geografía (National Statistics and Geography Institute)
IPCC	Intergovernmental Panel on Climate Change
JMP	Joint Monitoring Programme
km	kilometres
LAC	Latin America and the Caribbean
LAP	Land Acquisition Program

m	metres
m^3/s	cubic metres per second
mm^3	million of cubic metres
mm	millimeter
MDG	Millennium Development Goals
MDMQ	Municipalidad del Distrito Metropolitana de Quito (Municipality of the Metropolitan Government of Quito)
MEA	Millennium Ecosystem Assessment
MMA	Monterrey Metropolitan Area
MMSD	Milwaukee Metropolitan Sewerage District
MPI	manufacturing production index
MWRA	Massachusetts Water Resources Authority
NASA	National Aeronautics and Space Administration
NCEP	National Centers for Environmental Prediction
NGOs	Non-governmental organizations
NHD Plus	United States National Hydrography Dataset Plus
NOAA	National Oceanic and Atmospheric Administration
NRW	non-revenue water
NWC	National Water Commission
NYCDEP	New York City Department of Environmental Protection
ODA	official development assistance
OECD	Organisation for Economic Co-operation and Development
OSE	Obras Sanitarias del Estado (State Sanitary Works)
PES	payment for ecosystem services
PDW	Philadelphia Water Department
PLANASA	Plano Nacional do Saneamento (National Sanitation Plan)
PME	Programa de Modernización de Empresas (Business Modernization Programme)
PNUD	Programa de las Naciones Unidas para el Desarrollo (United Nations Development Programme)
POMCA	Plan de Manejo y Ordenamiento de una Cuenca (River Basin Management Plan)
POT	Plan de ordenamiento territorial (Territorial Development Plan)
PPI	private participation in infraestructure
PPPs	public-private partnerships
PSA	Programa de Protección de Laderas y Saneamiento Ambiental (Slope Protection and Environmental Sanitation Programme)
PSP	private sector participation
RMSP	Metropolitan Region of São Paulo
RHVM	Región hidrológica del valle de México (hydrological administrative region of the Valley of Mexico)
RIO+20	United Nations Conference on Sustainable Development in Rio de Janeiro
RSI	retail sales index

SABESP	Companhia de Saneamento Básico do Estado de São Paulo (Sao Paulo State Basic Sanitation Company)
SACMEX	SACMEX (Sistema de Aguas de la Ciudad de México)
SADM	Servicios de Agua y Drenaje de Monterrey (The Water and Sewerage System of Monterrey)
SDP	Secretaría Distrital de Planeación de Bogotá (District Planning Secretariat of Bogotá)
SDWA	Safe Drinking Water Act
SEDAPAL	Servicio de Agua Potable y Alcantarillado de Lima (Water and Sewerage Services of Lima)
SFP	strategic financial planning
SIRE	Sistema de Información para la Gestión del Riesgo y Atención de Emergencias de Bogotá (Risk Management and Emergency Response Information System of Bogotá)
SISS	Superintendent of Sanitary Services
SPU	Seattle Public Utilities
SRTM	Shuttle Radar Topography Mission
SSPD	Superintendencia de Servicio Público Domiciliarios (Superintendent of Public Home Services)
SUDS	sustainable urban drainage systems
TNC	The Nature Conservancy
TRMM	Tropical Rainfall Measuring Mission
TSS	total suspended solids
UN	United Nations
UNDP	United Nations Development Programme
UNEP	United Nations Environmental Programme
UNESCO	United Nations Educational, Scientific and Cultural Organization
UNFPA	United Nations Population Fund
UN-HABITAT	United Nations Human Settlements Programme
UNICEF	United Nations Children's Fund
UNECLAC	United Nations Commission for Latin America and the Caribbean
US	United States of America
USGS	U.S. Geological Survey
WASH	Water, Sanitation and Hygiene
WEF	World Economic Forum
WHO	World Health Organization
Wisconsin DNR	Wisconsin Department of Natural Resources
WSS	water supply and sanitation
WWAP	World Water Assessment Programme
WWF	World Wildlife Fund
WWTP	wastewater treatment plant
ZMVM	Zona Metropolitana del Valle de México (Metropolitan Zone of Valley of Mexico)

Introduction

Ismael Aguilar-Barajas, Jürgen Mahlknecht, Jonathan Kaledin, Marianne Kjellén and Abel Mejía-Betancourt

Why this book?

There is a lack of literature on water and urban development in Latin America. This book intends to fill the conceptual and operational gap about the interconnection between water, urbanization and sustainable development in the Latin American context. The content is mainly based on the experience and lessons learned from the specific context of Latin America's cities. It has been written with different audiences in mind: principally, policy-makers and planners in the public sector, practitioners working with water and related issues around the world; as well as a larger public including academia, the private sector and the civil society. This book is directed to all who are interested in a comprehensive understanding of policy development in the nexus of water and cities, and how policies in this area have been implemented in countries which are still struggling to achieve higher levels of economic and social performance – including the full service coverage in its cities, while at the same time balancing environmental and climate concerns.

Considering that 80 per cent of the population of Latin America live in urban areas, thus concentrating the municipal and industrial demands and regional pressures on water resources to these areas, there is surprisingly little systematic discussion about how the urban water-related issues are linked from the policy level to specific projects on the ground, and the significant infrastructure and governance gaps to provide adequate water services in most cities. This is the main reason why the Water Center for Latin America and the Caribbean and the Tecnológico de Monterrey in Mexico are taking the lead to fill this analytical gap.

The 2011 World Water Week dedicated to Water in an Urbanizing World provided the impetus to produce this book, and the Stockholm International Water Institute joined the initiative and contributed its policy leadership and a global perspective on these issues. Consequently, to a large extent the book draws on material originally presented at the 2011 World Water Week. It also presents contributions, with an emphasis on water and cities but not restricted to Latin America (there is a chapter containing major lessons from US cities on green infrastructure), by invited authors. The authors of the different chapters have different backgrounds including public management, international organizations, academia and practitioners. This way the book richly reflects on the nexus of water and cities from a range of perspectives.

Outline of the book

Water and Cities in Latin America is structured in four parts. The first part contains three contextual chapters that provide the overarching framework for the book, emphasizing its conceptual and policy relevance. The second part comprises five chapters centred on selected key water and sanitation issues such as universal access to water, wastewater and sanitation governance, groundwater use and social policy. Part III has four chapters that focus on how city planning and urban development can deal with climate change and risk mitigation in constructive ways. The fourth and final part examines, in three chapters, the significant developments in green growth and green infrastructure in Latin American cities, including insights of how cities in the United States have taken these concepts and practices on board.

Chapter summaries are given below.

Part I: Water, cities and financing

These contextual chapters provide the framework of the book by emphasizing its conceptual and policy relevance, as well as the crucial role of financing water infrastructure for cities. Chapter 1 frames the linkages between water, cities and sustainable development based on an updated review of the most recent academic thinking around this nexus. Chapter 2 addresses the question of why urban water matters from the perspective of practitioners, reflecting on integration of policy areas and infrastructure. Chapter 3 brings the perspective of an international organization, and focuses on the pressing issue of financing water and sanitation services.

Aguilar-Barajas argues, in Chapter 1, that for planners and decision-makers it is both essential and complex to link water, cities and development. It requires a robust understanding of the underpinnings of development, so that economic growth, social development and environmental concerns are jointly taken into the policy analysis. Within this framework special attention is given to the issues of poverty, income redistribution and the financing of infrastructure. A part of the chapter is dedicated to connecting water and cities with climate change and risk management, especially in regard to the local dimension, which is where the practicalities of policy implementation really matter.

The author proposes that a more holistic approach is needed, with sound urban water management in line with urban water governance policies and institutions from the national to the local level. The chapter ends with a call for more and better empirical work, statistical information and research on water and cities.

In Chapter 2, Mejía-Betancourt emphasizes a range of physical interconnections in the urban water infrastructure and explains why urban planners and water practitioners need a comprehensive understanding of the linkages between cities and water. The chapter argues that in spite of the many obvious connections between urban planning, water management and development, the institutional links are still largely insufficient, with a minimal integration, at both conceptual and practical levels. There is an underlying assumption that all dimensions of city life – from health to telecommunications, and, of course, water – are somehow addressed within the established framework for urban planning. Contrary to this view, for the author, urban water services are multidimensional and so too should be the nature of policy response.

The chapter has three main messages. First, planners and policy-makers need more analysis and effective integration of issues such as land use, transport, environmental management, housing, and solid waste collection and disposal. Second, putting Latin America's economic growth in a social perspective, it should be possible to close the existing gap and provide universal access to water services. Third, significant reforms to improve policies and institutions are required, which will lead to stronger governance and greater social accountability.

Chapter 3 (Cox and Börkey) starts from the premise that to ensure the adequate financing of water supply and sanitation services is a major policy issue for countries around the world. The chapter takes a close look at the policy options involved in this financing – responding to challenges that are universal as well as those that are particularly pertinent to Latin America. The authors state that many countries, several in Latin America included, tend to elaborate ambitious plans for developing water supply and sanitation infrastructure oriented to economic and social development, but that these are frequently underfunded and struggle to meet not only investment needs, but also those of operation and maintenance. In the absence of 'sustainable financing' for water supply and sanitation, as the authors call it, adverse consequences for users, especially the poorest, arise.

Financial planning is absolutely critical for meeting the objectives of water supply and sanitation service provision. It includes finding the appropriate balance between different sources of financing, thinking over how to face financial challenges, and setting medium- and long-term strategies. Both standard and innovative financing mechanisms are necessary for closing the gap and support the transition towards a sustainable water and sanitation sector. The authors suggest that countries and municipalities should mobilize financing from a variety of sources and increase the contribution from the three Ts (tariffs, taxes and transfers), as well as to look into means for reducing costs and increasing the efficiency of water use. Great challenges lie in the setting of tariff levels to meet financing needs while also taking social issues into account, particularly from political and regulatory perspectives, pointing towards the importance of thoughtful, long-term and integrated planning. As the region's overall economy grows, there is also the increased possibility of making more effective use of private sector financial participation.

Following these overarching framework chapters, the next section concentrates on specific issues with examples from a range of cities in Latin America.

Part II: Managing water and sanitation services

The second part of this book centres on selected key water and sanitation issues such as universal access, wastewater and sanitation governance, groundwater use and social policy. The geographical coverage starts from a panoramic view then focuses on selected cities such as São Paulo and Mexico City. In spite of the differences involved in the case studies, the chapters highlight several similarities, especially in terms of the technical knowledge required for more effective urban and water management, the concern with social policy, and the overall need for more comprehensive planning that enables, for instance, the joint planning of surface water and groundwater resources.

Chapter 4 (Pena) focuses on the overarching policy and main programmes to achieve universal access to water supply and sanitation services in the state of São Paulo, Brazil. Special attention is given to the state water utility (SABESP), the main entity responsible for implementing São Paulo's water and sanitation policy, and the largest water supply and sanitation company in the Americas. The progress towards achieving universal access to water and sanitation that has been made over the past decade has been possible thanks to a combination of targeted investments, political determination and effective planning. In particular, efforts have been undertaken towards getting a more sustainable financing of water and sanitation services. Effective regulation has proven beneficial for both consumers and public utilities. Yet, serious challenges remain, which are illustrated mainly in the conflicts between the state and the federal government concerning the main rivers feeding the water supply system of São Paulo.

The most pressing water challenges in the metropolitan region of São Paulo are water scarcity, pollution of water sources and reservoirs that are surrounded by built-up areas without proper collection of sewage, and urban flooding. Despite clear progress in network coverage, there is still much work needed to improve sewerage services and to expand wastewater treatment and disposal. This is especially evident on the urban periphery, where informal occupation of land by *favelas* (shantytowns), frequently in areas prone to flooding and mudslides, poses serious problems for residents as well as the surrounding water bodies. Thus, greater attention to presently disadvantaged groups is considered to be a must in the public policy of the state. The author recognizes that water supply and sanitation development goals for the coming years are ambitious and will be difficult to achieve in terms of technical complexity, costs, and social and environmental implications. SABESP will need wise regulatory guidance backed by broad civil society participation to be politically and financially viable.

Chapter 5 (Ioris) discusses the political economy of water services development in Lima, Peru. The author's main thesis is that the conventional initiatives of the public sector during the past few decades to keep up with rapid urbanization and water demands have been only partially successful because they treated the different socio-economic classes of the city unfairly. This situation has produced human-made water scarcity, which is felt especially in the low-income periphery of the city – a situation that reinforces the disparities and social inequalities inherited from previous generations. Rural migration to Lima has accentuated the chaotic expansion of the city and exacerbated the water shortages. Yet, the problems are critically linked to poor management practices, characterized by political interference and erratic allocation of resources.

The author advises that in order to understand the current and future water challenges of Lima (and of other Latin American cities), it is necessary both to dig deep into the city's history and to look at the inter-related themes of housing, transportation, employment and economic disparity. Within this more comprehensive perspective, two specific issues are worth highlighting: that grassroots organizations should demand and claim their right to water services, and that Lima is an excellent example to demonstrate that the analysis of water inequalities provides good insight into the complexities of the Latin American city.

Chapter 6 (Pacheco-Vega) reviews the water quality challenges facing urban wastewater management in Latin America. The author argues that analysis of water governance is usually centred on water supply, demand and distribution; that there is far less discussion about sanitation and wastewater. This chapter provides a useful diagnosis as well as suggestions for how governments can improve sanitation services in Latin American cities. In Latin America, the provision of public services is most commonly the responsibility of municipalities. However, state and provincial governments and even federal agencies are increasingly assuming greater roles in the important task of water resources management and service provision. This jurisdictional overlap and lack of a robust institutional framework generates inter-agency conflict and complicates the possibilities for multi-level collaborative initiatives.

Chapter 6 ends with a call for a research agenda for the governance of urban water in Latin America. In drafting this agenda, key issues need to be considered carefully. Wastewater, unlike freshwater, is usually an overlooked area. An issue that emerges from other chapters is also present here: the political dimension of water management. The author concludes that the appropriate policy mix for robust urban wastewater governance should be sought, and the fact that there is no 'magic bullet' in the design and implementation of sanitation policies. The understanding of contextual factors and the causal mechanisms of what works and what does not work becomes essential.

Chapter 7 (Martínez and Bandala) centres upon how urban water services, including water supply, stormwater drainage and wastewater treatment, are managed in the 20 million people global megalopolis of Mexico City. Population and economic growth, especially during the second half of the twentieth century, have always been the main driving forces for new water supply sources and sanitation facilities, as well as protection against droughts and flooding in this closed water basin. Flood control dams cannot work at close to optimal levels because of, among other reasons, informal urban settlements near rivers and ravines. Shortfalls in water supply have been dealt with by over-exploitation of the aquifers in the Valley of Mexico. In turn, the resulting land subsidence is causing great difficulties in providing and maintaining hydraulic infrastructure and sanitation services.

The authors look into large infrastructure projects to address the acute problems linked to stormwater, sewage and wastewater drainage systems, as well as ways for improving solid waste management and the control of non-point sources of pollution. The *Atotonilco* wastewater treatment plant, with its US$1 billion investment, will be the largest in Mexico and one of the largest of its kind in the world. Its construction directly addresses the problem that in spite of the increase in the number of wastewater treatment plants, more than 70 per cent of the wastewater produced by Mexico City is still discharged into the environment without any treatment. One of the main policy issues identified by the authors is the dispersion of responsibilities and the lack of coordination among and between water institutions. Financially, the city has always been in difficult times when it comes to addressing the problems and challenges associated with water supply, stormwater and drainage, derived from a chronic fiscal gap.

Chapter 8 (Mahlknecht, Hirata and Ledesma-Ruiz) provides an overview of groundwater management in urban Latin America. It is illustrated with the two cases of Mexico City and São Paulo, the two largest metropolises in the region. The authors demonstrate that groundwater has a crucial and strategic role as a source of urban water and in sustainable development but that its significance is not well appreciated, not only in Latin America but also in other regions of the world. There is a poor understanding of the complexities of groundwater dynamics. Continuous and comprehensive monitoring of groundwater is rare, even in relatively wealthy countries and well-functioning water authorities.

The cases of Mexico City and São Paulo confirm the complex diagnosis for Latin America, and the need for decision-makers to consider groundwater more seriously. Understanding and managing groundwater as a basic resource is the central premise. The authors of the chapter propose several measures in this direction, among which are the following: collecting better data on water recharge; fostering programmes that protect groundwater resources; tackling the critical issue of unregistered wells; and integrating public networks of groundwater supply with the significant number of private wells. Coordinated management of surface water and groundwater is a long overdue requirement to improve the groundwater situation. The ultimate goal should be the integration of groundwater into overall water planning, leading to more sustainable and efficient long-term groundwater management in Latin America. Mexico City and São Paulo highlight the need to strengthen the institutional frameworks and align appropriate incentives to make this goal achievable.

As highlighted also in other chapters and particularly in the final section on green infrastructure, ground and surface waters, including run-off and rainwater need to be viewed and utilized in a coordinated way in order to emulate more natural solutions in the urban landscape.

Part III: Climate change, risk management and planning

The third part of this book focuses on how cities are managing water risks and climatic uncertainties. The chapters review recent projects to address floods and discuss different tools to aid policy-makers to make better-informed choices, presenting cases from cities in Mexico, Colombia, Ecuador and Argentina. This part addresses the important linkages between geomorphological setting, land and water resources management policies, and the quality of urban growth and design. Climate change heightens the need to address vulnerable urbanization patterns on hillsides and floodplains. The need to understand the perspectives and priorities of different water users and population groups in order to ensure the usefulness and social support of hydraulically sensible urban development is emphasized in this part.

Chapter 9 (Sisto and Ramírez Orozco) explores how flash flooding has been managed in the Monterrey metropolitan area (MMA). The MMA is located in a semi-arid part of Mexico that is particularly prone to flash flooding as hurricanes can cause sudden and heavy precipitation on the slopes of the nearby mountains. This metropolis is Mexico's third most populated centre and an industrial hub of national importance, with a concomitant capacity for adaptation. Hurricanes are not frequent – occurring only twice in the last half-century – but when they do happen, they bring widespread and severe damage. This chapter analyses the unfolding of the passage of remnants of Hurricane Alex in 2010 and how it affected and was dealt with by the authorities in the MMA.

The authors identify the pattern of urban development which is at odds with the MMA's hydrological reality as the root cause of the disaster. More specifically, damage is attributed to the lack of hydraulic capacity of the riverbeds and gullies, which had been narrowed drastically by roads and to some extent also by residential construction in flood prone areas and by recreational use of riverbeds. Reconstruction has been rapid, however, indicating an impressive response capacity, in turn reflecting the relative economic prosperity of the area. Yet, there has been no shift in the planning model towards integrating hydrological risk into the urban design. In order to resiliently face events like Hurricane Alex in the future, a reshaping of the institutional architecture will be needed.

Chapter 10 (Costa-Posada, Urazán Bonells and Velandia Durán) describes the recent Bogotá River environmental recuperation and flood control project aiming to restore water quality and reduce flood impacts. Growing populations and increasing industrial activities, particularly in the middle stretches of the watershed, exert major environmental pressures on the river. The Bogotá River flows along the western perimeter of Colombia's prosperous national capital Bogotá, where several severely polluted tributaries add to the already serious contamination of the Bogotá River. Moreover, urban expansion is compromising the basin's capacity to cope with run-off and, combined with the spread of settlements on the floodplains (without mitigation measures), this has resulted in an increase in the frequency, depth and duration of flooding caused by the river and its tributaries.

The Hydraulic Upgrade and Environmental Recuperation Project for the Bogotá River, one in a series of efforts to address pollution and flooding of the river, focuses primarily on increasing the capacity of existing wastewater treatment plants. It also involves constructing dykes and rehabilitating riparian habitats, including purchasing lands, addressing the issue of illegally occupied areas, and establishing multifunctional zones for recreation as well as for connecting rivers and wetlands. The project is different from previous ones in that sewage treatment and flood control have been designed to be integrated into urban development with a direct impact on the quality of life of the people.

Chapter 11 (Vidal, Burgos and Zevallos) delves into environmental protection and restoration in the Ecuadorean capital Quito, emphasizing the importance of community involvement, institutional coordination, and continuous strengthening of capacity and learning. The eastern flank of the Pichincha-Atacazo massif in Quito presents a typical case where disaster risks have increased due to settlement encroachment into hilly areas that are naturally unstable. In this setting, the frequency of floods and landslides increases further because of the infill of ravines to expand the area available for construction with works that reduce their water carrying capacity. The situation is exacerbated by the disposal of wastewater and the dumping of solid waste in the ravines.

Steps to protect and restore the environment of the slopes have been taken since the mid-1990s. Loans from the Inter-American Development Bank have enabled the construction of engineering works to mitigate the effects of floods and landslides and to create the Slopes Commission to strengthen municipal management relating to planning, land use, community development, relocation of families from high risk zones, and the restoration and replanting of ravines for incorporation into the urban green area. The lead agency for many of these activities, the Metropolitan Water and Sewage Company, has also devoted efforts to institutionalize the activities in order to put them on a permanent footing, aiming to guarantee the sustainability of the overall project.

Chapter 12 (Miralles-Wilhelm, Brantly, Borsellino, Castellano, Dykes, Moreda, Muñoz-Castillo, Rineer, Wheaton and Wyatt) takes the reader through the testing of the Hydro-BID modelling system applied to the Río Grande basin in Jujuy Province, forming part of the upper reaches of the Río Bermejo basin in Northwest Argentina. The Río Grande is the only source of drinking water for the provincial capital San Salvador de Jujuy. Due to high sediment loads in the river, the existing dam has no remaining storage and the city is supplied by run-of-the-river flow, and hence is exposed to seasonal flow variations. Shortages and cut-offs are common during the dry season when the river's flow is very low.

The objective of the Hydro-BID case study was to evaluate the vulnerability and sufficiency of water supply for urban San Salvador de Jujuy, considering also supply for irrigation and other uses, and taking future climate change into account. The application of the Hydro-BID model highlighted the need for data and monitoring of water demand, climate, streamflow and other input parameters. Available data and climate models indicate that there will be a modest and steady increase in average temperature in the study area. While inter-annual variation is projected to remain quite high, the direct impact of climate change on river flow appears to be small. Improving the efficiency of urban water systems will only partially alleviate unmet water demand among urban and industrial users. Fully satisfying San Salvador de Jujuy's water demand will require additional investments in storage or new water sources.

Part IV: Green infrastructure development

Part IV examines developments in green growth and green infrastructure in Latin American cities. It also reviews green growth infrastructure initiatives and programmes in selected cities of the United States and lessons to be drawn from this experience for Latin America. Together, the three chapters provide a well-rounded perspective on urban water issues and the role that green growth and green infrastructure can and should play in urban water development. These chapters show there are opportunities to advance green growth concepts in Latin American cities in the decades ahead. Taking such opportunities will allow for Latin American urban water infrastructure goals to be accomplished in new and adaptive ways, and in a manner that takes into account current thinking and approaches to the economic, social and environmental impacts of water infrastructure.

Chapter 13 (Browder) provides an updated summary of green growth from the perspective of international financial institutions, with a focus on water in urban Latin America. The chapter discusses urban water sector objectives in the context of green growth concepts. Reviewing the accomplishments – and what has not yet been achieved – of the Millennium Development Goals for water, and in particular those goals' successes and failures in urban Latin America, the author establishes that the substantial amount of water infrastructure development still required in urban Latin America can benefit significantly from a green growth approach. Summaries of Chilean, Argentinian and Brazilian accomplishments and challenges regarding water management are presented, and the author asserts the need for relying on green growth approaches in the future.

Chapter 14 (Veiga, Calvache, Benítez, León and Ramos) summarizes the most recent development of water funds in Latin America, an increasingly important instrument to meet water security goals in Latin America's cities by enhancing the role of natural infrastructure within a framework of green and inclusive growth. Since their first appearance in Quito, Ecuador, in the early 2000s, water fund initiatives have emerged throughout Latin America. Water funds are coordinated, multi-stakeholder agreements that establish payment mechanisms for the purpose of investing in green infrastructure approaches to water supply. Water funds pay for programmes such as watershed conservation, the restoration of freshwater ecosystems, and the support of sustainable business activities within a water fund's geographical area of operation.

Chapter 14 first presents the underlying goals of water funds: the provision of water, more often than not to Latin America's urban areas, and the preservation of the water resources being called upon. The chapter then explains the conceptual basis to water funds – how they work – before describing the current status of their implementation in Latin America. Case studies like Bogotá are used to present key aspects of the use of water funds, and as lenses through which lessons and challenges associated with the design and implementation of water funds are discussed. The chapter concludes by asserting that there is a compelling role for the green growth approach of water funds in the future of Latin American urban water systems.

Chapter 15 (Kaledin) discusses the lessons learned from a few examples of green infrastructure approaches from several cities in the United States, and sets forth principles from the United States that can be shared with urban Latin America as its green water infrastructure work takes hold in the years to come. The efforts of New York City, Boston and Seattle in protecting the extensive watershed lands around their water supply systems are discussed. These efforts have been successful, allowing two of these cities (Boston and New York City) to avoid building expensive, conventional water treatment facilities while preserving the water quality of their water resources and protecting substantial amounts of watershed lands from inappropriate development. In essence, several cities in the United States have had a head start on the water funds concept that has taken hold in Latin America over the past decade, and they can be used as examples of green growth water supply development that must be explored fully in the future.

More recently, the cities of Milwaukee and Philadelphia have conceptualized and are now implementing a set of green infrastructure programmes for the purpose of addressing urban stormwater and other water quality issues. Chapter 15 discusses some of the theoretical and practical aspects of these new programmes, including how to fund them. The programmes date back about a decade, so experience with them is similar in length to the Latin American urban water funds experience. Nevertheless, there is a great deal of knowledge sharing already to be gleaned from the Philadelphia and Milwaukee experiences regarding the establishment and running of green infrastructure stormwater programmes, and this information is set forth to conclude the chapter.

Key messages

Policy-making is about the formal and informal processes and interactions across a wide variety of players with multiple and frequently conflicting interests. These processes and interactions are context-specific and dynamic as circumstances change, and so do the nature and outcomes of policies. *Water and Cities in Latin America* presents a variety of examples related to these concerns.

Policy-makers need to thoroughly appreciate the full water and infrastructure cycle to satisfy demands and needs of the urban population and economic activities, while at the same time managing the water resource in the face of extreme events. Latin American cities are on the verge of providing full access to water and sanitation services, while at the same time tackling climate change and escalating resource constraints.

Water supply and sanitation goals for the coming years are ambitious and will not be reached without surmounting some quite severe challenges in terms of technical complexity, financing, and social, environmental and political implications: this is a major lesson from the São Paulo chapter that also applies to other cities in Latin America. Ensuring the adequate financing of water supply and sanitation services is a major policy issue. Significant reforms will be required to close the existing gap in universal access to water services, but it is an achievable goal.

Urbanization is a process of concentration of population and economic activities in a given space. Urban growth inherently leads to greater urban water demands, requiring a more effective integration between urban planning and water management. However, institutional frameworks are still largely insufficient because in most of the cities in the region urban water management in itself is fragmented, with conflicting interests and often overlapping institutional responsibilities. Regulation is of critical importance for the monitoring and adherence to agreed plans and priorities.

Ways to involve variegated sets of stakeholders in deciding on pathways will be of greatest concern in meeting water and sanitation challenges as well as to adapt to climate change and greater weather variability. This book calls not only for improved data, but also a greater integration of economic, social and environmental issues in the analysis, along with the involvement of a broader set of authorities and stakeholders of society in devising the way forward.

Finally, there is a compelling role for the green growth approach in the future of Latin American urban water systems. Investing in the environment can also equate to good business for the economy and benefits for society at large. The US and Latin American experiences suggest that is possible to avoid building expensive, conventional water treatment facilities while preserving the quality of water resources and protecting substantial amounts of watershed lands from inappropriate development. Green growth approaches offer much to Latin American cities in their quest for providing quality services to the whole of its populations while navigating future uncertainties and managing the totality of water resources in a wise and holistic manner.

Part I
Water, cities and financing

1 Water, cities and sustainable development

Setting the scene

Ismael Aguilar-Barajas

Introduction

'Water in an urbanizing world' was the organizing theme of the 2011 World Water Week. Organized by the Stockholm International Water Institute, it brought together public officials, academics and practitioners from all over the world to review the status of the global thinking about the nexus between water and cities. The role of cities in development is becoming an important policy issue for government, international institutions, the private sector and think tanks. For example, *Foreign Policy* dedicated a special issue in 2012 to metropolises. Popular magazines like *National Geographic* also gave wide coverage to the theme of water and cities. Over the past four years, references to the 'water crisis' and the 'urban century' have become common.[1]

This chapter provides a structured discussion about the emerging recognition of the links between water, cities and sustainable development. It looks at the key facts of world urbanization trends and economic geography, as well as infrastructure development and financing. It argues that a path towards sustainable development is also about addressing poverty and income distribution, while considering climate uncertainties and risk management. Lastly, it discusses selected policy challenges and recommendations, and makes an overall appeal for more effective integrated urban water management, including the need for more empirical work, and better statistical information.

Recognizing the relevance of the nexus between water and cities

The marking of the world's population at 7 billion people in 2011 underscored concerns of highly complex economic and social challenges to come. The *OECD Environmental Outlook to 2050* considers water one of the four key areas of concern (Gurria, 2012). The other three – all of which are linked to water – are climate change, biodiversity and the health impacts of environmental pollution.

The conclusions of the 2011 World Water Week recognized that the scale of urbanization poses monumental opportunities and challenges in its path to a prosperous and sustainable future (SIWI, 2011). It advised the global water community to focus their planning and project work on four pillars (SIWI, 2011, pp. 4–6):

- Linking water with urban planning and design;
- Integrating and investing in informal settlements;
- Extending and financing sustainable water and sanitation services;
- Building synergies between water, energy and food.

Water is becoming more relevant as the associated risks with its management are growing. The *Global Risks 2014* report, published by the World Economic Forum, warns of risks that have a great deal of relevance to water and urban management (WEF, 2014). Table 1.1 presents the ten global risks of greatest concern, from which it can be seen that water crises rank as the third highest concern.[2] This concern reflects the growing awareness of managing 'increased competition for already scarce water resources to sustain economic activity and population growth' (WEF, 2014, p. 14). This risk is coupled with two interrelated ones: failure of climate change mitigation and adaptation (fifth in the list), and greater incidence of extreme weather events such as floods and droughts (sixth in the list). The report highlights the additional difficulties that climate change will bring about for countries, especially less developed ones, which are unable to manage effectively even today's climate variability. More severe droughts and floods could worsen the conditions of those countries less able to adapt to climate change, leaving them more vulnerable to crises.

Table 1.1 The ten global risks of highest concern in 2014 according to the World Economic Forum

No.	*Global risk*
1	Fiscal crises in key economies
2	Structurally high unemployment/underemployment
3	Water crises
4	Severe income disparity
5	Failure of climate change mitigation and adaptation
6	Greater incidence of extreme weather events
7	Global governance failure
8	Food crises
9	Failure of a major financial mechanism/institution
10	Profound political and social instability

Note: From a list of 31 risks, survey respondents were asked to identify the five they are most concerned about. The shaded risks (author's shading) highlight direct water-related risks.

Source: WEF (2014, table 1, p. 9, based on the Global Risks Perception Survey 2013–2014)

It is worth noting that three of the ten global risks of greatest concern are water-related. Most of the other top ten risks also have interconnections with water policy and urban management, such as fiscal crises (they reduce infrastructure investments), severe income disparity (which has links to economic and social policy, and is of particular interest to Latin America), and food crises (which are affected by many factors, water crises being one of them). As the *Global Risks 2014* report states, 'It is important to consider the combined implications of these environmental risks on key development and security issues, such as food security, and political and social instability, ranked eighth and tenth respectively' (p. 14). The need for better public information is also stated to be a key requirement given the critical issues at stake. The report stresses the implications for a world that is urbanizing at such a rapid speed and the dangers of mismanaged urbanization. It is estimated that by 2050 the urban population will be twice what it currently is; that is, it will reach approximately 6.4 billion people. As the World Economic Forum states (WEF, 2014, p. 22):

> While urbanization provides important economic and social gains, it substantially increases risks related to ecological disruptions, pollution, climate change and environmental disasters ... Urban flooding has already become the leading form of disaster in the world.

In his commentary on the 30 years of the *World Development Report*, Stiglitz (2009) singles out urbanization as one of the top five themes of development.[3] As the twentieth century ended, the *World Development Report* 1999/2000 focused on urbanization, as one of the megatrends that needed attention (Stiglitz, 2009). Along these lines the Commission on Growth and Development presented the book, *Urbanization and Growth*, claiming that urbanization is central for growth and development (Spence, 2009).

Environmental and social sustainability are at the heart of the nexus between water and cities. The World Economic Forum and its *Global Competitiveness Report* illustrate well the relevance of this interface: the core message is that sustainability – through its environmental, economic and social dimensions – is an essential component in assessing the competitiveness of nations (Bilbao-Osorio *et al.*, 2013).[4] Social sustainability has to do with inclusion, equity, cohesion and the resilience of social systems. Unstable social systems can lead to socio-political instability, which ultimately harms the growth process. Adequate access to drinking water and sanitation and a less unequal income distribution can positively affect the economic welfare of a nation. Environmental degradation can negatively affect human health; unsafe drinking water can be a driver of this degradation. Unplanned urbanization can aggravate all these adverse effects.

In their report on the sustainability of development in Latin America, UNECLAC and UNEP (2002) consider that water management and urban management are two of the five great challenges for sustainable development in the region. Their report also mentions three other areas that have direct connections with urban water in Latin America: protection and sustainable use of natural ecosystems and biodiversity, as well as access to genetic resources; vulnerability; and energy management. Some of these same concerns – sustainable development related to water, energy, health, agriculture and biodiversity – were addressed at the Johannesburg Environmental Summit the same year in which the report was published (Bhargava, 2006).

Urbanization trends and water services

Because economic growth and urbanization are inseparable, the world is more urbanized and this is shown by the latest world urbanization trends presented in Box 1.1 (United Nations, 2014, pp. 4–5).[5] Among the important conclusions that can be derived from Box 1.1 is the need for better understanding of the urbanization trends in order to aid a more coherent policy design.

Table 1.2 presents figures of urban and rural population for 1990, 2014 and 2050 in all major regions of the world. While the pace of urbanization has not been uniform, more than half of the world's population in 2014 was considered urban. North America has 81 per cent, Latin America and the Caribbean (LAC) 80 per cent. Without doubt, LAC is one of the most urbanized regions of the developing world. By 2050, this proportion of city dwellers will increase to 86 per cent (versus 87 per cent in North America). In absolute terms for LAC, this means an increase of urbanites from 496 million people in 2014 to 674 million in 2050.

Examining urbanization trends in Latin America shows a number of interrelated challenges for the nexus between water and cities. The region has a large population and economic concentration in urban areas.[6] Urbanization, however, has not had a uniform pattern in the region but has varied between and within countries. Half of the urban population (about 222 million people) lives in cities of less than 500,000 inhabitants and 14 per cent (65 million people) in megacities. Medium-sized cities are growing at the fastest rate of all cities. Rural to urban migration has lost relevance in the majority of countries in the region, but migration flows are now more complex.

Though megacities seem to lose dynamism, their study is of great relevance for development in general and water management in particular. Megacities are urban centres with more than 10 million people. By 2010, megacities had 5.1 per cent of the world's population, although most had had slow population growth during the 1990s and the first decade of the twenty-first century. Medium-sized cities are growing faster than large cities and they are attracting significant amounts of investment. Most megacities are losing their dynamism in terms of population and economic activity. In all regions of the world, most of the urban population lives in urban areas with less than 1 million inhabitants. But still, in absolute terms, megacities like Mexico City and São Paulo, for example, require good management to deal with their urban problems (Satterthwaite, 2014a), and water is a case in point.

A recent report by the United Nations confirms the above trends (Table 1.3). In 2014, only eight of the 30 largest cities in the world were located in high

Box 1.1 World urbanization trends 2014: key facts

- In 2014 around 54 per cent of the world's population was residing in urban areas. In 1950 the figure was 30 per cent. The projected figure for 2050 is 66 per cent.
- The most urbanized regions are the following: North America, 82 per cent; Latin American and the Caribbean, 80 per cent; and Europe, 73 per cent. More than half of the population in Africa and Asia is still rural. But these regions are urbanizing faster than the former regions. By 2050 they will have 56 per cent and 64 per cent of their populations, respectively, residing in urban areas.
- Asia concentrates 53 per cent of the world's urban population; Europe follows with 14 per cent and then Latin America with 13 per cent.
- In population terms urbanization has expanded rapidly since 1950, from 746 million to 3.9 billion in 2014. Out of the projected 2.5 billion added to the world's urban population by 2050, nearly 90 per cent will be concentrated in Asia and Africa.
- Close to half of the world's urban population resides in cities of less than 500,000 inhabitants. In contrast, one in eight lives in the 28 megacities with more than 10 million people.
- Tokyo is the largest urban agglomeration in the world, with a population of 38 million. Following Tokyo are Delhi, with 25 million inhabitants; Shanghai with 23 million; and Mexico City, Mumbai and São Paulo, each with an estimated population of 21 million people. It is projected that in 2030 the world will have 41 megacities. Tokyo will still be the world's largest agglomeration, with 37 million inhabitants, followed by Delhi with 36 million. At present, large cities are concentrated in the developing south. 'The fastest growing cities are medium-sized centres and settlements with less than one million inhabitants located in Asia and Africa'.
- Some cities have experienced population decline in recent years. Economic contraction and natural disasters have contributed to population losses as well.
- As the world continues to urbanize, sustainable development challenges will be increasingly concentrated in cities, particularly in the lower-middle-income countries where the pace of urbanization is fastest. Integrated policies to improve the lives of both urban and rural dwellers are needed.

Source: Based mainly on United Nations (2014, p. 1)

Table 1.2 Historic (1990), recent (2014) and projected (2050) urban and rural population data for major regions of the world (in thousands of people)

Major region	*Urban 1990*	*Urban 2014*	*Urban 2050*	*Rural 1990*	*Rural 2014*	*Rural 2050*	*Proportion urban (%) 1990*	*Proportion urban (%) 2014*	*Proportion urban (%) 2050*	*Average annual rate of change (%) 2010–2015*
WORLD	2,285,031	3,880,128	6,338,611	3,035,786	363,656	3,212,233	43	54	66	0.9
Africa	196,923	455,345	1,338,566	433,064	682,885	1,054,609	31	40	56	1.1
Asia	1,036,247	2,064,211	3,313,424	2,176,877	2,278,044	1,850,638	32	48	64	1.5
Europe	505,991	545,382	581,113	217,257	197,431	127,954	70	73	82	0.3
Latin America and the Caribbean	313,876	495,857	673,631	131,327	127,565	107,935	71	80	86	0.3
North America	212,935	291,860	390,070	69,351	66,376	56,130	75	81	87	0.2
Oceania	19,059	27,473	41,807	7,911	11,356	15,067	71	71	74	0.0

Source: Adapted from United Nations (2014, table I, pp. 20–25)

income countries, and by 2030 this number will be reduced to four cities: Tokyo and Osaka in Japan and New York and Los Angeles in the United States. Only Tokyo will be in the top ten largest cities, with an estimated population of 37.2 million people (down from the current 37.8 million, but still the largest city in the world). This contrasts with its ranking in 1950 (22), the first year for which the United Nations has data. Five of the world's largest cities today are in Latin America: Mexico City, São Paulo, Buenos Aires, Rio de Janeiro and Lima.

Given the high concentration of population and economic activity in urban areas, urban sustainability is of paramount importance for the region's economic development (BID, 2011). The latest UN-Habitat (2012) report on the state of cities in Latin America and the Caribbean, estimates that two-thirds of GDP is generated in urban areas. Moreover, the report estimates that the major 40 cities produce 30 per cent of total GDP, and that half of this is generated by only four metropolises: São Paulo, Mexico City, Buenos Aires and Rio de Janeiro. These economic facts as well as the impacts of urbanization on the environment, the high vulnerability of LAC cities to climate change and natural disasters, and the financial limitations to addressing all these issues make it necessary to reflect on the sustainability of cities in the region.

Urbanization in developing countries has increased rapidly, but unlike in developed countries, correlation of the rate of urbanization with economic growth and development has been less evident. There is, however, consensus about both the contribution and the potential role of cities in economic growth and development (Clarke Annez and Buckley, 2009; Todaro and Smith, 2012; United Nations, 2013). Urbanization has benefits and costs.[7] There are advantages to be derived from the concentration of population and economic activity in cities. Infrastructure, basic services – like water and sanitation – and economic opportunities are more easily found in urban areas. Cities also tend to concentrate government functions, have high-level education and health services, and have innovation capabilities. In sum, these advantages are referred to as agglomeration economies, which ultimately reflect higher standards of living and well-being.

But urban areas, particularly unplanned or mismanaged urban areas in developing countries, also present costs. Millions of people live in cities with inadequate conditions. Uncontrolled urban expansion increases pressure for natural resources – like water – and energy considerably. In most low and middle-income countries, Latin America included, the explosive urbanization that has taken place has not been accompanied by the needed expansion of services and facilities essential to a healthy urban environment (Satterthwaite, 2014b). The provision of water, sanitation, drainage and solid waste collection are cases in point. Water and sanitation services tend to be of far lower quality in the poorer areas. The quality of the urban environment has profound implications for development. Environmental degradation and its corresponding negative impacts on the health of urban populations is a vivid reality of this pattern of rapid urbanization. Although, overall, cities could be more efficient, this does not mean that financing their infrastructure requirements is trouble-free.

Table 1.3 Population of the 30 largest urban agglomerations in the world, 1990–2030 (millions of inhabitants, ranked by 2014 population size)

Urban agglomeration	*Country*	*1990*	*2014*	*2030*	*Average annual rate of change (%) 2010–2015*
1. Tokyo	Japan	32.5	37.8	37.2	0.6
2. Delhi	India	9.7	25.0	36.1	3.2
3. Shanghai	China	7.8	23.0	30.8	3.4
4. Mexico City	**Mexico**	**15.6**	**20.8**	**23.9**	**0.8**
5. São Paulo	**Brazil**	**14.8**	**20.8**	**23.4**	**1.4**
6. Mumbai	India	12.4	20.7	27.8	1.6
7. Kinki (Osaka)	Japan	18.4	20.1	20.0	0.8
8. Beijing	China	6.8	19.5	27.7	4.6
9. New York-Newark	United States	16.1	18.6	19.9	0.2
10. Al-Qahirah (Cairo)	Egypt	9.9	18.4	24.5	2.1
11. Dhaka	Bangladesh	6.6	17.0	27.4	3.6
12. Karachi	Pakistan	7.1	16.1	24.5	3.3
13. Buenos Aires	**Argentina**	**10.5**	**15.0**	**17.0**	**1.3**
14. Kolkata	India	10.9	14.8	19.1	0.8
15. Istanbul	Turkey	6.6	14.0	16.7	2.2
16. Chongoing	China	4.0	12.9	17.4	3.4
17. Rio de Janeiro	**Brazil**	**9.7**	**12.8**	**14.2**	**0.8**
18. Manila	Philippines	8.0	12.8	16.8	1.7
19. Lagos	Nigeria	4.8	12.6	16.7	3.9
20. Los Angeles-Long Beach-Santa Ana	United States	10.9	12.3	13.3	0.2
21. Moscow*	Russia	9.0	12.1	12.2	1.2
22. Guangzhou, Guangdong	China	3.1	11.8	17.6	5.2
23. Kinshasa	Democratic Republic of the Congo	3.7	11.1	20.0	4.2
24. Tianjin	China	4.6	10.9	14.7	3.4
25. Paris*	France	9.3	10.8	11.8	0.7

Table 1.3 continued

Urban agglomeration	*Country*	*1990*	*2014*	*2030*	*Average annual rate of change (%) 2010–2015*
26. Shenzhen	China	0.9	10.7	12.7	1.0
27. London*	United Kingdom	8.1	10.2	11.5	1.2
28. Jakarta	Indonesia	8.2	10.2	13.8	1.4
29. Seoul*	Republic of Korea (South Korea)	10.5	9.8	10.0	0.0
30. Lima	**Peru**	**5.8**	**9.7**	**12.2**	**2.0**

Note: Agglomerations in bold type are in Latin America. * These cities no longer appear in the projected 2030 ranking. Their places are occupied by Bangalore (14.8 million), Chennai (13.9), Lahore (13.0) and Hyderabad (12.8).

Source: Adapted from United Nations (2014, table II, pp. 26–27)

The nexus between water and cities is better understood within a systemic spatial perspective. Geography is regarded a major driver of development (Galor, 2011; Sachs, 2012). A significant recognition of the role of geography in development came from the fact that the *World Development Report 2009* was dedicated to the theme 'Reshaping economic geography' (World Bank, 2009). More specifically, the provision of water to cities is strongly related to and framed within a spatial/geographical setting. Urbanization is the expression of the localization of population and economic activities in a given space. This location has profound implications on the demand and supply of water and sanitation services. A related major issue is the consideration of water demand and supply in the context of the water cycle and climate variability. Urbanization patterns are often in conflict with their settings, and to a large extent, the problems derived from pressures on water resources are a direct reflection of both the geographical location of cities and the urbanization patterns themselves.

Financing water infrastructure in Latin America presents a major challenge (CAF, 2013).[8] Progress in covering the significant investments required in infrastructure has been modest. It is estimated that in 2010 about 3 per cent of the region's GDP was invested in infrastructure in general. The latest data for 2011 and 2012 show little change in that figure. In comparison, investments in China and India, as a proportion of their GDP, are far greater. The investments in water infrastructure are far lower. It is estimated that during the period 2008 and 2011, investments in drinking water and sanitation were on average 0.2 per cent of GDP. This figure is extremely small compared with infrastructure investments in the transport sector (1.2–1.4 per cent GDP). Almost all investment in the water sector is public. In the telecommunications sector, in contrast, most investment is private. In the case of water services, although privatization has gained a certain acceptance, there has been limited success in attracting such investments (Wade, 2012; UN-Habitat, 2012). The investment gap to achieve universal coverage of water and sanitation services can be closed, but in addition to more investment, better planning will be needed (UN-habitat, 2012; CAF, 2013).

It is appropriate to end this section on financing infrastructure with the conclusions of Bird (2001, p. 122) on the local context of Latin American municipal and inter-governmental local finance:

> The financing of infrastructure is a serious problem when available resources for local capital expenditure are scarce. Notwithstanding this fact, there exist solutions … It is rather little that can be achieved unless countries incorporate in their systems of local government, institutions that encourage the exercise of prevision, imagination, hard-work and prudential – which are the ingredients of good government everywhere.

Water and cities within a comprehensive development perspective

The study of water and cities and their challenges for sustainable development requires a comprehensive approach. It has long been recognized that development is not reduced to economic growth, and that more broad objectives centred on well-being and quality of life are needed (Knight, 1991; Dasgupta, 1993; Meier, 1998; Giraud, 2002). Responsible growth (Hamilton, 2004), inclusive development (World Bank, 2008), inclusive growth (De Mello and Dutz, 2012), green growth and inclusive green growth (OECD, 2012a; b) and urban green growth (OECD, 2013) are all concepts showing that economic growth cannot and should not be the sole indicator or consideration for development. A recent report by UN-Habitat supports the call for a more comprehensive framework to understand development (UN-Habitat, 2013, p. 34):

> There is a need for a shift in attention around the world in favour of a more robust notion of development – one that looks beyond the narrow domain of economic growth that has dominated ill-balanced policy agendas over recent decades.

As well as economic growth, environmental and social considerations must be woven into development objectives and they are aspects too of the analysis of water and cities in relation to development. Urban green growth intends to foster economic growth and development through urban activities that decrease impacts on the environment (OECD, 2013). In this regard, urban green growth is considered as part of sustainable development. The complex interactions between water and urban agendas show very clearly how economic growth and environmental concerns are linked to development (Holloway, 2006, p. 165):

> Development policy that pays attention to environmental quality fuels, not limits, real income growth. The problem is that investments that bring about these positive net benefits come in a "non- market form" – that is they do not have a price in the market ... The alternative view of the development-environment linkage is that development itself depends upon environmental quality in certain critical ways ... Development requires the environment to be improved.

In the case of Latin America, Lagos and Rosales (2013) highlight that income distribution is the great Achilles' heel of sustainable development. They argue that in fact growth and redistribution should and can go together. This situation has long been recognized (Williamson, 2003). If inequality is regarded as an obstacle to economic growth and social development, there is therefore economic rationality behind policies oriented to greater social cohesion (Ocampo, 2006; Emmerij, 2008). Therefore, poverty and income distribution are two issues that deserve the greatest attention in the design of water and urban policies. The access and quality of water services are largely determined by these issues. Poor people are more vulnerable than rich people to the impacts of extreme weather events, basically because of both their location in riskier areas and limited resources. The latest data indicate that about 28 per cent of the Latin American population is regarded as poor (Table 1.4) (ECLAC, 2013).[9]

Poverty, in relative terms, is less marked in urban areas but it is still a major concern. It is estimated that 23 per cent of the urban population across the region is poor. Estimates for individual countries show that, although on a smaller scale than overall poverty, urban poverty is high.[10] As expected, poverty rates are much higher in rural areas. In six out of the 15 countries with available data, more than half of their rural populations are poor. In 2013, it was estimated that almost a quarter of urban residents in LAC live in slums (CAF, 2013). Of course, the provision of basic services, such as clean water and sanitation, in these slums leaves a lot to be desired. In spite of the progress that has been made in achieving universal access to water and sanitation in the region, a quarter of the population does not have access to water services or receive them in inadequate conditions (CAF, 2013).

A more comprehensive understanding of water issues could be obtained if such issues are in fact framed within the development perspective (Biswas, 2009).[11] According to this view, water has to be seen in connection with other development sectors such as housing, agriculture, health, energy, infrastructure, environment and, of course, cities. Water problems and their policy responses can no longer be addressed exclusively and successfully from the water perspective alone. Water policy should go hand in hand with spatial planning. Numerous examples in Latin America and elsewhere show this, and the case studies presented in this book confirm this premise.

Table 1.4 Poor populations in urban and rural areas of Latin America, most recent year available (in percentage of total population)

Country	*Year*	*National*	*Total urban areas*	*Total rural areas*
Argentina	2012	NA	4.3	NA
Bolivia (Plurinational State of)	2011	36.3	16.2	32.1
Brazil	2012	18.6	16.2	32.1
Chile	2011	11.0	11.3	8.7
Colombia	2012	32.9	28.5	46.9
Costa Rica	2012	17.8	17.1	19.0
Ecuador	2012	32.2	28.8	38.9
El Salvador	2012	45.3	41.0	52.6
Guatemala	2006	54.8	42.0	66.5
Honduras	2010	67.4	56.3	76.5
Mexico	2012	37.1	33.2	43.5
Nicaragua	2009	58.3	52.9	65.4
Panama	2011	25.3	15.5	43.6
Paraguay	2011	49.6	42.9	59.3
Peru	2012	23.7	14.5	50.8
Uruguay	2012	5.9	6.1	2.6
Venezuela (Bolivarian Republic of)+	2012	23.9	NA	NA
Latin America (simple average)++	2012	28.2	23.2	48.6

Note: + Since 1998, the survey's sample design does not allow for rural–urban breakdown. Thus, the data correspond to the national total. ++ Estimate based on the above 17 countries plus the Dominican Republic and Haiti. NA, data are not available or cannot be disaggregated. The shaded column (author's shading) highlights the urban dimension.

Source: Adapted from ECLAC (2013, table 1.6.1, p. 75)

Climate change and risk management

There is a wide consensus that climate change will adversely affect Latin America in general and its cities in particular. The region is particularly vulnerable to the observed and projected scenarios of climate change, in part because of its location and the distribution of population and infrastructure (Vergara *et al.*, 2013), and also because of its socio-economic conditions, including poverty and inequality (Rubin and Rossing, 2012). The existing models predict both more intensive flooding and extended dry periods. Although it is difficult to fully estimate the effects of climate change, the damage is expected to be substantial.

It is clear that the cost of inaction will be expensive and that the urban poor population will be most affected (World Bank, 2013a).

However, it is also argued that an appropriate contextual framework of climate change and risk management is needed. It is becoming clear that the links between population dynamics and climate change are more complex than commonly understood (Satterthwaite, 2014b; Guzmán *et al.*, 2009). Focusing more specifically on climate change adaptation in the water sector, Ludwig *et al.* (2009, p. 6) confirm what is emerging from other studies too: the local dimension of the solutions required for adaptation.

> [T]echnical opportunities should be combined with institutional changes to adapt to climate change … [D]espite climate change being a global problem – the solutions required for adaptation are mostly local.

Hurricanes, frequent in Latin America and the Caribbean, can have devastating effects on cities. Although hurricanes have always occurred, the general understanding is that climate change increases the frequency and intensity of extreme weather events. However, these effects have partly to do with unplanned urban growth, where infrastructure is not suitable to cope with extreme weather events. Robust planning must consider the proper design of prevention and mitigation strategies, with the firm intention to reduce the vulnerability and to increase the resilience of those people in the population most at risk (Holloway, 2006). Such design must consider the economic, social and environmental vulnerabilities that contribute to adverse impacts and hardship.

Consideration of the regional and local scale is of great significance in the construction of climate change scenarios (Jacob and van den Hurk, 2009). Down-scaling global scenarios to the local level is a highly complex process. Recent research shows that the costs and benefits of adaptation to climate change are, in fact, context-specific (Marchal *et al.*, 2012). This variation is, of course, climatological and hydrological, but also related to economic, social, legal and political conditions that affect water management at the local level. Proper consultation and communication with stakeholders is essential in this regard. The public needs to be included in, informed of and educated about the policy implications of these scenarios. There is also a psychological dimension to the way in which society perceives and deals with the risks associated with floods and droughts (Shaw, 2005). Short-run focus explains the strong attention these events receive during their occurrence and immediately afterwards, and the back seat they take to other priorities once the events have passed.

Challenges and policy conclusions

Water and cities in Latin America do face a great many challenges that need to be addressed. These challenges will only increase in the future, thus the demand for a more coherent policy design and implementation. As argued by Bhari (2011), the twin engines of urbanization and demand for water resources need an effective integrated urban water management, moving away from the traditional disconnect between urban water issues seen in broader urban planning processes. Up to now, isolation, as opposed to a coordinated effort, has been the rule. Therefore, basin-level management should have a role to take into account the conditions imposed by the logic of the hydrological cycle. From this perspective, it is not rapid urbanization per se but the lack of appropriately functioning governance structure as well as economic stagnation that fuels most urban problems, including the provision of water and sanitation (Satterthwaite, 2006).

A more holistic approach is needed in which sound urban water management is perfectly in line with strong urban water governance at both the national and local level (Lall and Deichmann, 2009; Bahri, 2011; UN-Habitat, 2013). In this revisited urban water management, institutions are properly aligned with coherent legislation and policy frameworks. As emphasized at the sixth session of the World Urban Forum, 'planned urbanization requires robust political and institutional capacity to manage differences, land use disputes and conflicts of interest. Political legitimacy, trust and the rule of law are needed' (UN-Habitat, 2013, p. 12). Effective green growth policies depend on political leadership, public acceptance and, of course, financing, including for rehabilitation and maintenance of infrastructure (OECD, 2012b; Lall and Deichmann, 2009).

An analysis of water, cities and development leads directly to the issues of income distribution and social policy. This recognition comes from urban water dimensions but also from economic growth and development perspectives. Long-overdue policy-making should recognize that investing in development is necessary to generate more economic growth. The reconciliation of income distribution and growth, in particular, has been mentioned as the greatest challenge for the twenty-first century (Kholi, 2006). The connection between these concerns and social development and water policy are straightforward. And so is the relevance of urban policy in the light of Latin America's very high level of urbanization. Coherent urban management and comprehensive water policies are good for sustainable development. This is the set of issues that are addressed in this book.

A more proactive approach to confronting climate change challenges is needed. Prevention could be a cost-effective strategy to deal with the impacts of climate change (UN-Habitat, 2012). To a large extent, cities' vulnerability to climate change depends on their internal capacity to address it, but policy design requires effective coordination between and within government levels and non-governmental actors (UN-Habitat, 2012). Adaptation could best be dealt with through a strategy based on community participation (and its learning) along with more appropriate institutional arrangements (Rubin and Rossing, 2012; World Bank, 2013b). However, only a handful of governments have designed an integrated approach to disaster risk management; this shortcoming is illustrated well in coastal cities, which are prone to unique risks (World Bank, 2013b).

Down-scaling national models of climate change to basin and local levels is needed, as these are the scales at which impacts are felt the most (Rubin and Rossing, 2012).

It is essential to frame public policies regarding water and cities in Latin America within the context in which these policies are formulated and implemented. Policy-making involves a wide variety of actors that interact in different scenarios. Contexts and circumstances change, and so do the nature and outcomes of policies. As argued by Stein and Tommasi (2006) for Latin America, the political process is a major component of the policy design process; to ignore this link can lead, as has been in fact the case, to inadequate reforms and disappointments. Decision-making about the nexus of water and cities in Latin America does present a variety of examples linked to these concerns.

Economic development is not generated exclusively by free market forces. The public sector and the private sector can work together and reinforce each other's work to achieve economic prosperity. This is a sentiment found to be true – to different degrees – everywhere. Latin America is a region that shows this is indeed the case (Sachs, 2013). This recognition has profound implications for water and urban policies, such as financing water infrastructure in cities. The concern of Rojas (1989) expressed long ago regarding the difficulties for Latin American cities to finance the provision of infrastructure in general and for water services in particular, without a dynamic urban economy, is still valid today. This concern is reinforced by the latest UN-Habitat report (2013), which stresses the urgent attention that should be given to the urban economy.

As it has been expressed for the development discourse (Stiglitz, 2001; Coyle, 2002), a greater degree of humility is called for in addressing the urban water agenda in Latin America. We do not have all the answers for the challenges that the complex multi-dimensional nexus between water and cities and their links to sustainable development in Latin America raises. There are no recipes for success. In addition to long-standing problems, new global trends are under way, with profound implications for water, cities and development. Global warming, increasing inequality in the distribution of income, growing prices for food and energy, changing demography, ageing and migration are among them (World Bank, 2008).

That more empirical work and better statistical information is required is one of the recurrent messages in the literature on water and cities. This book provides a rich selection of cases and empirical observations which also support the need for better data. Another major challenge has to do with human skills to address the pressing knowledge agenda about water and cities. Climate change is a case in point, but the list of issues that require professional scientific involvement is large, for which concerted effort across universities, governments, the private sector and civil society is required. Interdisciplinary studies must be fostered, along with strong requirements for specialization. Research on water and cities in Latin America needs to be supported. The research networks that exist in the region could ease the drafting of a Latin American research agenda on water and cities. This book intends to jump start the analytical thinking in these directions.

Notes

1 See the special issues *National Geographic* (April 2010 and January 2012).
2 The definition of this risk, according to the World Economic Forum (WEF, 2014, p. 53) is 'A significant decline in the quality and quantity of fresh water combines with increased competition among resource-intensive systems, such as food and energy production'.
3 The other four themes also have connections with the management of water and cities: knowledge for development; corruption; poverty; and institutions.
4 For the World Economic Forum, sustainable competitiveness is defined as 'the set of institutions, policies and factors that make a nation remain productive over the longer term while ensuring social and environmental sustainability'(Bilbao-Osorio *et al.*, 2013, p. 55).
5 The report underlines that the estimates used are based on national statistics and there is no common accepted definition of what is 'urban', not even within countries (over time). Adjustments have been made to the national estimates. The concept of 'urban agglomeration' is used.
6 This paragraph is based on UN-Habitat (2012).
7 There is a substantive literature that supports this point. See, for example, Manders *et al.* (2012) for the international arena, and Satterthwaite (2014a, b) for low and medium-income countries. The latest *World Urbanization Prospects* revision (United Nations, 2014) provides an updated account of urbanization trends and highlights the pros and cons of urban concentration.
8 The following discussion on this topic is based on this report.
9 The countries with the largest proportions of impoverished inhabitants are Honduras (67.4 per cent), Nicaragua (58.3 per cent), Guatemala (54.8 per cent), Paraguay (49.6 per cent) and El Salvador (45.3 per cent). At the other end of the spectrum, faring much better, are Uruguay (5.9 per cent) and Chile (11 per cent). Most other nations have figures of about 20 per cent to 70 per cent of their populations regarded as poor.
10 Percentages vary from 56 per cent in Honduras to 41 per cent in El Salvador. Mexico has 33 per cent of its urban population living in poverty, and Colombia and Ecuador each have just fewer than 30 per cent.
11 Most of this paragraph is based on this author.

References

Bahri, A. (2011) *Towards Integrated Urban Water Management,* Perspectives Paper, Global Water Partnership, Stockholm.

Bhargava, V. (ed.) (2006) *Global Issues for Global Citizens. An Introduction to Key Development Challenges*, The World Bank, Washington, DC.

BID (Banco Interamericano de Desarrollo) (2011) *Sostenibilidad urbana en América Latina y el Caribe*, Banco Interamericano de Desarrollo, Washington, DC, 2011.

Bilbao-Osorio, B., Blanke, J., Campanella, E., Crotti, R., Drzeniek-Hanouz, M. and Serin, C. (2013) 'Assessing the Sustainable Competitiveness of Nations' in *The Global Competitiveness Report 2013–2014*, World Economic Forum, Geneva.

Bird, R.M. (2001) 'Presentación del escenario: finanzas municipales e intergubernamentales', in M. Freire and R. Stren (eds), *Los retos del gobierno urbano*, Alfaomega for the World Bank, México, DF.

Biswas, A.K. (2009) 'Introduction: Integrated Water Resources Management in Latin America', in A.K. Biswas, B.P.F. Braga, C. Tortajada and M. Palerm (eds), *Integrated Water Resources Management in Latin America*, Routledge, London and New York.

CAF (2013) *Infraestructura en el desarrollo integral de América Latina. IDeAL 2013*, Banco de Desarrollo de América Latina, Caracas.
CEPAL/ECLAC and UNEP (Comisión Económica para América Latina y el Caribe/ Economic Commission for Latin American and the Caribbean and United Nations Environmental Programme) (2002) *The Sustainability of Development in Latin America and the Caribbean: Challenges and Opportunities*, CEPAL, Santiago, Chile.
Clarke Annez, P. and Buckley, R.M. (2009) 'Urbanization and Growth: Setting the Context', in M. Spence, P. Clarke Annez and R.M. Buckley (eds), *Urbanization and Growth*, The World Bank for the Commission on Growth and Development, Washington, DC.
Coyle, D. (2002) *Sex, Drugs and Economics – An Unconventional Introduction to Economics*, Texere, New York and London.
Dasgupta, P. (1993) *An Inquiry into Well-Being and Destitution*, Oxford University Press, New York.
De Mello, L. and Dutz, M.A. (eds) (2012) Promoting Inclusive Growth: Challenges and Policies, OECD Publishing, http:/7dx.doi.org/10.1787/9789264168305 in Economic Commission for Latin America and the Caribbean (ECLAC), *Statistical Yearbook for Latin America and the Caribbean, 2013* (LC/G.2582-P), Santiago, Chile, 2013, United Nations publication, Sales No. E/S.14.II.G.1, accessed 20 August 2014.
Emmerij, L. (2008) 'Teoría y práctica del desarrollo: ensayo introductorio y conclusiones de política', in L. Emmerij and J. Núñez del Arco (eds), *El desarrollo económico y social en los umbrales del siglo XXI*, Banco Interamericano de Desarrollo, Washington, DC.
Galor, O. (2011) *Unified Growth Theory*, Princeton University Press, Princeton and Oxford.
Giraud, P-N. (2002) '¿Qué hemos aprendido?', *Istor – Revista de Historia Internacional*, no. 10, Autumn, pp. 10–24.
Gurria, A. (2012) 'Foreword', in OECD, *OECD Environmental Outlook to 2050: The Consequences of Innaction*, OECD Publishing.
Guzmán, J.M., Martine, G., McGranahan, G., Schensul, D. and Tacoli, C. (eds) (2009) *Population Dynamics and Climate Change*, United Nations Fund for Population Activities and International Institute for Environment and Development, New York and London.
Hamilton, K.E. (2004) 'Responsible Growth for the New Millennium', in World Bank (ed.), *Perspectives on Development*, The World Bank, Washington, DC.
Holloway, A. (2006) 'Disaster Mitigation', in D.A. Clark (ed.), *The Elgar Companion to Development Studies*, Edward Elgar, Cheltenham, UK and Northampton, NH.
Jacob, D. and van den Hurk, B. (2009) 'Climate Change Scenarios at the Global and Local Scales', in F. Ludwig, P. Kabat, H. van Schaik and M. van der Valk (eds), *Climate Change Adaptation in the Water Sector*, Earthscan, London and Washington, DC.
Kholi, A. (2006) 'State and Development', in D.A. Clark (ed.), *The Elgar Companion to Development Studies*, Edward Elgar, Cheltenham, UK and Northampton, MA.
Knight, J.B. (1991) 'The Evolution of Development Economics', in V.N. Balasubramanyam and S. Lall (eds), *Current Issues in Development Economics*, St. Martin's Press, New York.
Lagos, R. and Rosales, O. (2013) 'Globalization and Development Options', *The Brown Journal of World Affairs*, vol. XIX, no. 2, pp. 283–301.
Lall, S.V. and Deichmann, U. (2009) 'Density and Disasters – Economics of Urban Hazard Risk', *World Bank Policy Research Working Paper* 5161, December, pp. 48, WPS5161.
Ludwig, F., Droogers, P., van der Valk, M., van Schaik, H. and Kabat, P. (2009) 'Introduction', in F. Ludwig, P. Kabat, H. van Schaik and M. van der Valk (eds), *Climate Change Adaptation in the Water Sector*, Earthscan, London and Washington, DC.

Manders, T., Chateau, J., Magné, B., van Vuuren, D., Gerdien Prins, A. and Dellink, R. (2012) 'Socio-economic Developments', in OECD, *OECD Environmental Outlook to 2050: The Consequences of Innaction*, OECD Publishing.

Marchal, V., Dellink, R., van Vuuren, D., Clapp, Ch., Chateau, J., Magné, B., Lanzi, E. and van Vliet, J. (2012) 'Climate Change', in OECD, *OECD Environmental Outlook to 2050: The Consequences of Innaction*, OECD Publishing.

Meier, G. (1998) 'Introduction: Ideas for Development', in Gerald Meier y Joseph Stiglitz (eds), *Frontiers of Development Economics. The Future in Perspective*, Oxford University Press for The World Bank, Washington, DC.

Nafziger, E.W. (1997) *The Economics of Developing Countries*, Prentice Hall, Englewood Cliffs, NJ.

Ocampo, J.A. (2006) 'Más allá del consenso de Washington', *Economía UNAM*, vol. 3, no. 7, pp. 7–25.

OECD (2012a) *Inclusive Green Growth: For the Future We Want. OECD Work of the Relevance to Rio +20*, OECD, Paris.

OECD (2012b) 'Executive Summary', in OECD, *OECD Environmental Outlook to 2050: The Consequences of Innaction*, OECD, Paris.

OECD (Organisation for Economic Co-operation and Development) (2013) *Green Growth and Cities*, OECD Green Growth Studies, OECD, Paris.

Rojas, E. (1989) 'El financiamiento del desarrollo urbano en América Latina: estrategias posibles frente a la crísis externa', *Revista EURE*, vol. 45, pp. 49–86.

Rubin, O. and Rossing, T. (2012) 'National and Local Vulnerability to Climate-Related Disasters in Latin America: The Role of Social Asset-Based Adaptation', *Bulletin of Latin American Research*, vol. 31, no. 1, pp. 19–35.

Sachs, J. (2012) 'Government, Geography, and Growth. The True Drivers of Development', *Foreign Affairs*, vol. 91, no. 5, pp. 142–150.

Sachs, J. (2013) *Economía para una planeta abarrotada*, Editorial Debate, Mexico City.

Satterthwaite, D. (2006) 'Urbanisation and Third World Cities', in D.A. Clark (ed.), *The Elgar Companion to Development Studies*, Edward Elgar, Cheltenham and Northampton, NH.

Satterthwaite, D. (2014a) 'Urbanization in Low- and Middle-income Nations in Africa, Asia and Latin America', in V. Desai and R.B. Potter (eds), *The Companion to Development Studies*, third edition, Routledge, London and New York.

Satterthwaite, D. (2014b) 'Urbanization and Environment in Low- and Middle-income Nations', in V. Desai and R.B. Potter (eds), *The Companion to Development Studies*, third edition, Routledge, London and New York.

Shaw, W.D. (2005) *Water Resource Economics and Policy: An Introduction*, Edward Elgar, Cheltenham and Northampton.

SIWI (Stockholm International Water Institute) (2011) *Overarching Conclusions. World Water Week 2011. Responding to Global Changes: Water in an Urbanising World*, Stockholm International Water Institute, Stockholm.

Spence, M. (2009) 'Preface', in M. Spence, P. Clarke Annez and R.M. Buckley (eds), *Urbanization and Growth*, The World Bank for the Commission on Growth and Development, Washington, DC.

Stein, E. and Tommasi, M. (2006) 'La política de las políticas públicas', *Política y Gobierno*, vol. XIII, no. 2, pp. 393–416.

Stiglitz, J. (2001) 'More Instruments and Broader Goals: Moving Toward the Post-Washington Consensus', in *Joseph Stiglitz and the World Bank. The Rebel Within. Edited with a Commentary by Ha-Joong Chang*, Anthem, London.

Stiglitz, J (2009) 'Commentary. The World Development Report: Development Theory and Policy', in S. Yusuf, A. Deaton, K. Dervis, W. Easterly, T. Ito and J. Stiglitz (eds), *Development Economics Through the Decades. A Critical Look at 30 Years of the World Development Report*, The World Bank, Washington, DC.

Todaro, M. and Smith, C. (2012) *Economic Development*, Addison Wesley, Boston.

UN (United Nations) Department of Economic and Social Affairs, Population Division (2014) *World Urbanization Prospects: The 2014 Revision, Highlights* (ST/ESA/SER.A/352).

UN (2013) *World Economic and Social Surveys: Sustainable Development Challenges* (E/2013/50/Rev. 1 ST/ESA7344), cited by United Nations (2014, p. 9).

UN-Habitat (2012) *El estado de las ciudades de América Latina y el Caribe 2012. Rumbo a una nueva transición urbana*, Programa de las Naciones Unidas para los Asentamientos Humanos, Nairobi.

UN-Habitat (2013) *The Urban Future, Report of the Sixth Session of the World Urban Forum*, Naples, Italy 1–7 September 2012, HSP/WUF/6/3, April 2013.

Vergara, W., Rios, A.R., Galindo, L.M., Gutman, P., Isbell, P., Suding, H. and Samaniego, J.L. (2013) *The Climate and Development Challenge for Latin America and the Caribbean. Options for Climate-resilient, Low Carbon Development*, Inter-American Development Bank, Economic Commission for Latin America and World Wildlife Fund, Washington, DC.

Wade, J.S. (2012) The Future of Urban Water Services in Latin America, *Bulletin of Latin American Research*, vol. 31, no. 2, pp. 207–221.

WEF (World Economic Forum) (2014) *Global Risks 2014, Ninth Edition*, World Economic Forum, Geneva.

Williamson, J. (2003) 'No hay consenso en el significado. Reseña sobre el consenso de Washington y sugerencias sobre los pasos a dar', *Finanzas y Desarrollo*, September, pp. 10–13.

Winston, C. (1994) 'A Commentary in Roundtable Discussion. Critical Issues in Infrastructure in Developing Countries', in *World Bank Proceedings of the World Bank Annual Conference on Development Economics 1993*, The World Bank, Washington, DC.

World Bank (2008) *The Growth Report. Strategies for Sustained Growth and Inclusive Development*, The World Bank for the Commission on Growth and Development, Washington, DC.

World Bank (2009) *World Development Report 2009: Reshaping Economic Geography*, World Bank, Washington, DC

World Bank (2013a) *World Development Report 2014: Risk and Opportunity – Managing Risk for Development*, World Bank, Washington, DC.

World Bank (2013b) *Resilient Coastal Cities. The Economic, Social and Environmental Dimensions of Risk*, World Bank LAC Working Paper 78607, Washington, DC.

2 Why does understanding the urban water link matter?

Abel Mejía-Betancourt

Introduction: the nexus between water and cities

More than half of the world's population currently resides in cities. Cities are engines of social progress and economic growth, with most of the global economic output generated within them. Gaining a comprehensive understanding of the connections between the development of cities and water is crucial for all, but particularly so for urban planners and water sector practitioners. It is not only crucial for policy-makers in government agencies and legislatures, but also for decision-makers in industry, which is heavily dependent on water in the industrial process and which generates significant amounts of wastewater.

Water is essential for life and an essential component of livelihood. These fundamental facts help explain why and where cities exist. Historically cities have been established near water sources that could support social and economic activities. Historically, city growth was controlled, among other factors, by the quantity and quality of nearby water. Aqueducts and sewers were built during the Roman Empire, as well as hydraulic infrastructure constructed by ancient civilizations in Asia, the Middle East and America. Yet, the basic need for a means to supply clean water to cities and remove dirty water from them has de facto changed little for millennia.

Understanding the linkages between cities and water is becoming increasingly more important in the light of climate change and ecosystem transformation. It is not yet possible to make fine-tuned, accurate predictions of how these processes will affect individual cities and specific locations on the planet, but a granular image is emerging from global efforts at modelling temperature increases in the upper atmosphere. Science can explain how temperature increases translate into water vapour and rainfall, and it is now known that the effects of climate variability will play out through changes in the hydrological cycle. The good news is that there is some evidence that urban areas – even large and dense cities – can cope better with a changing environment than can more dispersed settlements (World Bank, 2009).

Humans have now mastered the technology to pump and transport water for hundreds of kilometres, which allows supply to old, thirsty cities like Beijing, Tripoli and Baku, or even to entirely new ones like Las Vegas, built in an extremely arid area. Moreover, since the second half of the nineteenth century, there has been continuous technological progress in purifying water for drinking and in treating wastewater for human health and ecosystem integrity. Recently, new mapping technologies have allowed an unprecedented ability to capture, analyse and monitor data and acquire information on water processes both in the natural environment and through the services made possible by engineered infrastructure.

Although there is an obvious connection between urban planning and water management and development, the level of institutional connection is still largely insufficient (CAF, 2013). Urban planners and water managers and practitioners follow similar but basically parallel thinking, and there is minimal integration, neither at conceptual nor practical levels. Urban development plans, programmes and projects are mostly dedicated to land use and transport infrastructure. There is an underlying assumption in such work that social services like health, education and security and also urban infrastructure for the provision of electricity, gas, telecommunications, water, sanitation and drainage will somehow take place within the framework established by urban planning. Similarly, although water planners are used to making master plans for infrastructure investments over time horizons of 25 years or more – based on demographic assumptions, usage indicators and aggregated water demand projections – this work is often poorly coordinated with urban planning and often does not take into account land use change dynamics. This situation has long been recognized (Harberg, 1997). More importantly, it does not consider in full the complexities of the way that water moves and is transformed throughout a city

Urban growth to 2030

Global urban population growth estimates for 2011–2030 are coupled with deep societal changes in overall population, food and energy consumption patterns, hygiene and personal comfort standards, waste generation and communications. Together these factors produce a demand that is pressuring water resources significantly – in both quantity and quality – and that requires expansion and improvement in performance of water infrastructure to meet. According to UN-HABITAT (2009), the planet's urban population of 3,632 million in 2011 is growing by over 70 million people per year, and is projected to reach nearly 5 billion in 2030. Distribution of this generational growth in urban population across the world is uneven: an increase of 808 million people in Asia, 330 in Africa, 113 in Latin America, 58 in North America and as little as 34 million in Europe.

Urban land use is also changing quite dramatically at a rate that is even faster than population growth. The World Bank presents scenarios of land use for urban areas in the developed countries of North America and Europe as well as in Japan and Australia, increasing by 300,000km^2 over the 1990–2030 period, adding to the existing 200,000km^2 of built-up areas. In the rest of the world, the increase is expected to be even greater, estimating that 400,000km^2 will be added to the existing 200,000km^2 (Shlomo *et al.*, 2005). This projected global enlargement of cities is taking place in a period only slightly longer than one generation, and it will be accompanied by the demands for urban services, including health, education, housing, transport and electricity, along with water supply, sanitation, stormwater management and flood control.

Latin America is a highly urbanized region. In 2007, 260 million people lived in 198 Latin American settlements with more than 200,000 people, which together generated a Gross Domestic Product (GDP) of US$3.6 trillion. The top ten cities in size have a population of 95 million with an average GDP per capita of US$18,000 and total GDP of US$1.7 trillion, or 30 per cent of the region's total (McKinsey Global Institute, 2011).

According to the urban performance index developed by McKinsey to measure economic performance, social conditions, sustainable resource use, finance and governance, Latin American cities perform poorly in comparison to international benchmarks (McKinsey Global Institute, 2011). There is a soaring need for improved infrastructure and basic services, and the report notes the urgent need for action, concluding that Latin American cities have fallen substantially short of providing coordinated urban planning. Sufficient infrastructure and services – with water and sanitation services being prominent – are needed for efficient and liveable cities.

In 2005 about 25 per cent of Latin American city dwellers lived in substandard housing conditions, with inadequate urban services (UN-HABITAT, 2009). How, then, will it be possible for Latin American cities to become home for another 200 million people while closing the infrastructure gap for those living in substandard conditions already, especially in regard to water services and sanitation, but also in housing and other urban services like solid waste disposal and collection? This chapter claims that it is viable for Latin America to accommodate the incremental population within existing cities and close the infrastructure gap for those living in slums.This presumption is based on projections of sustained economic growth, macroeconomic stability, and deeply rooted and successful democratic processes.

The Latin America region will also enjoy a demographic bonus or dividend in relation to the dependency ratio, implying that the proportion of working age population will increase, and is estimated to peak by 2040 (World Bank, 2011). All in all, Latin America has an historic opportunity to reach higher levels of economic development while reducing inequality and closing existing infrastructure gaps by 2030.

When examining spatial transformation and urban growth in Latin America from a historic perspective, it is found to have been generally messy (World Bank, 2009). For example, as Latin American cities have grown, so too has the 'informal occupation' of urban areas (UN-HABITAT, 2003). It can be expected that such informal occupation will grow to accommodate the demand of workers and economic activity, while capacity of structured land markets is weak to respond to increasing demand for housing and infrastructure. Most likely, it will be beyond the capacity of governments, the private sector and society at large. Simplistically, it would not be possible to respond to these challenges on time with existing levels of investments, institutional capacity and social governance.

Empirical evidence shows that the somewhat haphazard consequences of accelerated urban growth in Latin America are likely to be unavoidable (World Bank, 2009). As in the rest of the world, urbanization trends in Latin America are already in motion and cannot, without radical intervention by government, be halted. With existing levels of investments, institutional capacity and social governance it will most likely be beyond the capabilities of government, the private sector and society at large to meet housing and infrastructure needs in a planned and timely way. Therefore, a primary test for those in charge of public policy is the planning and implementation of processes to formalize urban land use, build decent housing, and provide critical household services like electricity, water and sanitation.

The real challenge for Latin America involves formulating and implementing policies that make these processes as swift, effective and economic as possible in order to minimize human suffering. In their next generation of work on providing water services to their citizens, as they reach levels of economic growth and production on par with other parts of the world, Latin American countries must create sustainable and transparent institutional governance for water, and use their financial resources wisely as they invest in water.

There must also be recognition of the powerful force of agglomeration economies that move people to cities – the pursuit of economic opportunities and sustained urban advantages, in spite of the high cost of living, a degraded environment and the threat of violence. Attempts to control urban growth through government policy have in most cases failed (World Bank, 2009). Empirical research based on data from thousands of cities has resorted to mathematical laws to explain the benefits of people concentrating in one place: economic activity is enhanced, returns on investment in infrastructure services are higher and there is a general progress of social vitality (Bettencourt *et al.*, 2007). More importantly, the findings on productivity and decreasing costs are fairly consistent across countries with different levels of economic development, technology and wealth.

Managing the urban water cycle

Urban water services are multidimensional. They are susceptible to and must take into account issues in many sectors: the 'informal occupation' of land and low quality housing; the poor quality and administration of urban public services, particularly water and sanitation; the degradation of the urban environment, particularly through water pollution; and the defective management of solid waste. Urban water services' relationship with these is sometimes direct and obvious, sometimes indirect and opaque. Moreover, urban water management is

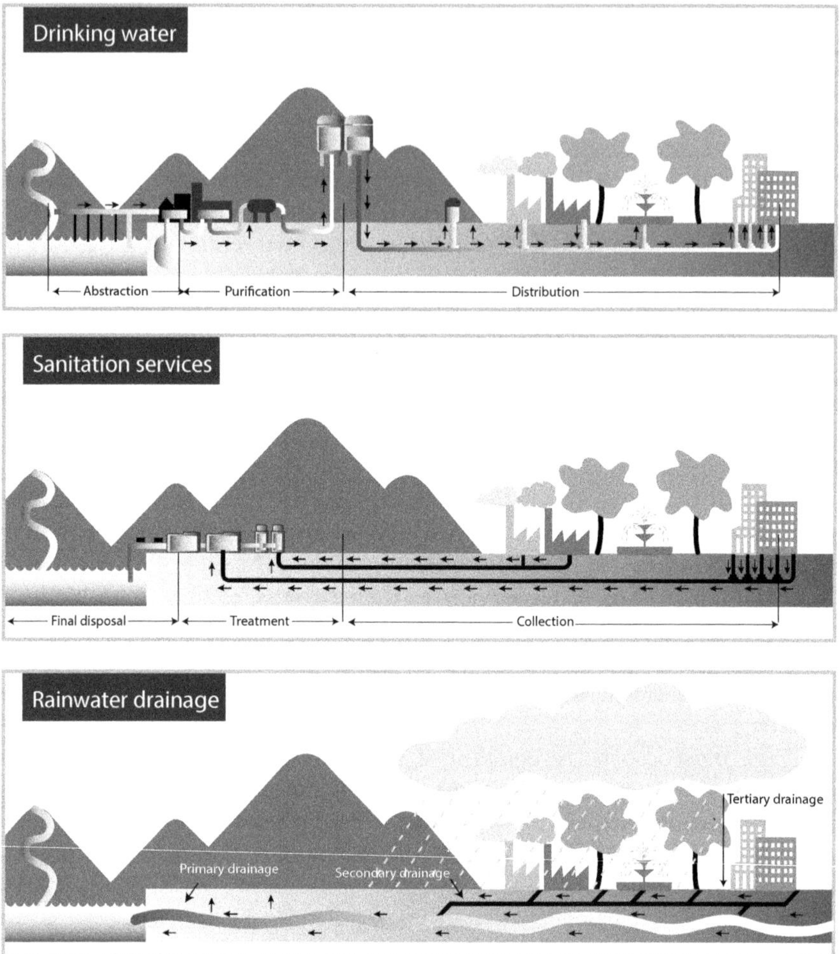

Figure 2.1 Urban water is generally divided into three separate components: drinking water, sanitation/sewerage and drainage

Source: CAF (2013, p. 34)

complex but is generally handled in a rather segmented way; a way that separates it into three distinct components: water, sanitation and drainage (Figure 2.1). Natural hydrological systems of course do not have such distinctions. The natural water cycle is intimately connected to forests, mountains and aquifers that regulate and release water; to land and soil, where run-off takes place and water percolates. It is even connected to urban processes, and to sediments generated by both human activity and natural processes.

In establishing services based on the urban water cycle, it has been the convention that they be designed to satisfy final consumption demand with a reserve capacity of about 15 per cent. Water supply should respond to (peak) demand of 1.5 times average capacity or more, at specific locations, at the right time and with predetermined levels of service in quality, volume, continuity and pressure (Mays, 2000). Water sanitation services involve the collection of wastewater from domestic and commercial activities, and its transportation, treatment and disposal of sewage flows back into the environment (Harberg, 1997). Drainage, also referred to as storm water management, is about the natural and engineered infrastructure needed to manage peak rainfall events with a predetermined return period associated with flood risks, which are generally estimated for 5–25 years. It involves managing storage and imperviousness in cities and during times of flooding, collecting and transporting peak discharges of water that otherwise compromise human life as well as social and economic activities (Tucci and Bertoni, 2003).

These three components of the urban water cycle – water supply, sanitation/sewerage and drainage – are often found under the functional responsibility of several government jurisdictions. In most cities of Latin America, water supply and sanitation are managed by a single public utility; though in some cases, as in Medellín, Colombia, these concerns are managed by a multi-utility in charge also of energy and telecommunications development. In Chile, water utilities are owned by the private sector. Urban drainage is usually under the direct responsibility of municipal authorities; however, they often do not have the territorial jurisdiction or capacity to manage urban flood infrastructure in a way that cooperates with river basins and other natural or engineered contributors to drainage issues (IDB, 1999).

Many if not most water sources for Latin American cities are now affected by uncontrolled pollution processes; and traditional water sources, those nearest to the cities, are insufficient in quantity and quality to satisfy demand from accelerated urban growth (IDB, 2001). For several decades now, Mexico City, São Paulo, Rio de Janeiro and Caracas have relied on expensive water infrastructure systems that extract water from remote river basins and deliver it to these cities. Development of such projects can take decades from planning to operation, due to the complexity of water infrastructure development, including data collection, planning, engineering and construction, and operation (Figure 2.2). Nevertheless, some such projects seem certain in the future for Latin America to meet its urban water supply needs.

The problems of insufficient water supply are exacerbated by a lack of effective measures to control demand. On average, physical losses in Latin American urban systems are close to 40 per cent (GRTB, 2011). In Buenos Aires, for instance, intake and water treatment capacity are designed to supply about 600 litres per person per day, which is almost three times the parameter used in São Paulo and Bogotá, and almost four times that of the most water efficient cities in Europe.

Addressing urban water issues from a single sector perspective is unwise and ineffective. To satisfy the twenty-first century needs of growing Latin American cities, a number of infrastructure considerations (land use, transportation, and solid waste management among others) will need to be integrated with

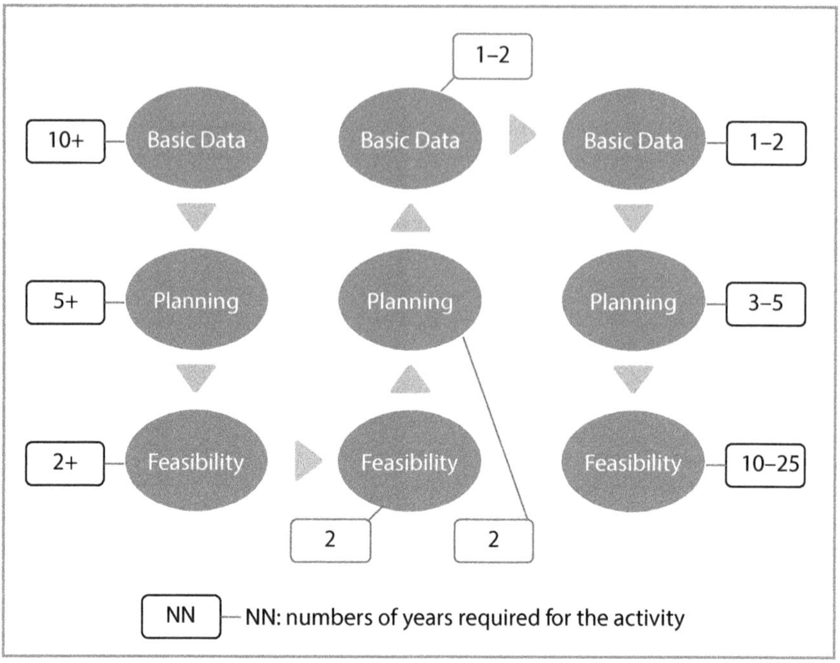

Figure 2.2 Cycle of water infrastructure projects

Source: Adapted from CAF (2011b)

water management to produce a coordinated response. This is perhaps most critical in regard to housing. Addressing the needs of the population living in substandard conditions (in shanty towns, essentially) is at least claimed to be at the top of the political agenda in almost all Latin American countries. But creating better living conditions for people who live in slums requires not only a new generation of housing infrastructure that 'works'; it requires also a new generation of water infrastructure – for water supply and sanitation – that works as well, for without it new housing is almost irrelevant. A coordinated solution to the various urban growth and infrastructure issues, within an integrated policy framework, is required. The solution is really beyond the capacity of a purely sectorial approach.

Similarly, mitigating unacceptable levels of environmental degradation in Latin America's growing cities requires an urban policy approach that closely links water services with the protection and preservation of environmental assets. While environmental conservation is closely related to rising levels of income that facilitate social mobilization which forces governments and polluters to act responsibly, an improved, healthy environment is now also connected to certain services obtained from nature, including water supply and natural wastewater and drainage infrastructure. This is in essence a bottom-up approach and it is emerging in the most advanced cities of Latin America. In some cases this approach is fostering a revision of key technological paradigms and traditional forms of infrastructure in the planning and implementation of development projects. For instance, Latin American cities are beginning to experience hard decisions about water infrastructure, namely, centralized trunk systems versus distributed solutions; the choices made will be based on the technology to build on dense urban areas, life-cycle economic assessment, and availability of long term financing (CAF, 2011a).

Protecting water sources for cities requires looking beyond the water services sector alone. Land use control mechanisms are critical for watershed protection, as well as for protecting groundwater recharge areas and for managing forestry and agricultural activities that threaten water bodies: cattle and dairy farming, and poultry and swine production. Over the years, Latin American water managers, legislators and government officials have had limited success in addressing watershed protection, forestry and agriculture issues with mostly command-and-control approaches, and in many parts of Latin America, pollution of water sources is increasing.

In some cases command-and-control policies have in fact had negative consequences, causing loss of economic opportunities in watershed areas near the expanding cities (World Bank, 1998). The experience of cities such as São Paulo and Caracas shows that command-and-control policies alone cannot effectively manage pollution levels affecting water sources (World Bank, 2007). Cities like Bogotá and Quito – see Chapters 10 and 11, respectively – are initiating promising experiments of payment for environmental services in relation to conservation and land management objectives that directly affect the water sources of these cities.

Closing the infrastructure gap by 2030

Closing the water infrastructure gap by 2030 is a realistic goal for Latin America's cities. A recent assessment by the Andean Finance Corporation (CAF, 2011a, p. 13) shows that annual investments equivalent to 0.31 per cent of 2010 GDP would suffice to close the gap of water and sewerage services, make significant advances in providing and improving these services for people who live in slums, expand water source and drainage infrastructure, and process almost two-thirds of sewage. As shown in Table 2.1, these investments would total US$250 billion for the 2010–2030 period (in 2010 dollars).

The CAF study found that the operational costs of water and sanitation services would be about 0.5 per cent of regional GDP (CAF, 2011a). This estimate, equivalent to about US$20 billion per year, is based on statistical information from countries with the best sectorial data in Latin America: Brazil, Chile, Peru and Colombia. Required investments in water supply, sewerage and wastewater treatment were estimated at 0.2 per cent of GDP per year (2010); they include rehabilitation, renovation and expansion of existing facilities. Annual required investments in urban drainage, aiming to reach 80 per cent of urban built-up areas, and the expansion of water supply sources for

Table 2.1 Investments required and targets for closing the infrastructure gap by 2030 in water services in Latin America

Service	*US$ billion 2010–2030*	*US$ billion Average*	*Target by 2030*
Water supply[a]	45.4	2.27	100% coverage
Sewerage[a]	79.4	3.97	94% coverage
Wastewater[a]	33.2	1.66	64% wastewater
Drainage[a]	33.6	1.68	85% of urban area
Water source[b]	27.1	1.35	100% of incremental demand
Formalization of water supply and sewerage connections	30.5	1.52	50% gap reduction 20 millions of connections
Total	249.2	12.45	

Notes: a = expansion, rehabilitation and renewal; b = new water sources

Source: CAF (2011b, p. 28)

nearly 100 million additional urban dwellers, were estimated at 0.07 per cent of GDP.

Finally, remedying substandard water and sewerage connections in slums was estimated at 0.05 per cent of GDP per year to 2030, with the aim of reducing the deficit in such services by 50 per cent by 2030. During the same period, investment in affordable housing solutions for low-income populations requires an investment estimate of US$200 billion, above and beyond what is needed to provide water services to these populations. This highlights again the interconnectedness of sectorial planning and investment.

Unfortunately, a large and predictable flow of investment into Latin American urban water infrastructure will not on its own be sufficient to close the water services gap by 2030. Most countries of the region still have relatively weak water governance institutions and are in need of substantive policy reforms to improve the administration and efficiency of water services. A rough estimate of the hidden cost of inefficiencies in water and sanitation services in Latin America amounted to US$5.8 billion in 2010 (CAF, 2011a). This estimate is based on normalized comparisons with efficient utilities: those with water losses (unaccounted-for water) of 20 per cent, collection rates of 100 per cent and labour productivity of two staff per 1,000 clients. See also Chapter 3 for a discussion about the issue of balancing tariffs, taxes and transfers.

Latin America does, however, have cities that have water utilities comparable to the best in the world: Monterrey (Mexico), Santiago (Chile) and Medellín (Colombia). There are also countries that have made exemplary progress in improving information, regulating tariffs and service quality, and being accountable to the communities they serve (e.g. Colombia and Chile). But overall, the region needs to reach a societal consensus on a long-term strategic vision for its cities – what quality of water services can be realistically and reasonably provided to all, but in particular to the 25 per cent of urban Latin Americans living in slums.

Conclusions

This chapter has focused on three key messages: integration, growth and reforms. While urban water services need to integrate all components of the urban water cycle, it is equally important that approaches to resolving urban water issues in Latin America cut across other sectors: land use, transport, environmental management, housing, solid waste collection and disposal, and others. Concerning growth, Latin America is at an important historic juncture from an economic and a social perspective. This juncture should allow the gap in urban infrastructure services to be closed by 2030, and provide universal access to water services. Finally, to capitalize fully on the objectives encapsulated in the first two messages, significant reforms to improve policies and institutions are required, which will lead to stronger governance and greater social accountability.

Moving the overall agenda of equitable, economically viable and sustainable water services for growing Latin American cities in the forthcoming decades will require the following. Cities need to engage in collaborative design and decision-making so that the water cycle and the urban environment are viewed as a holistic system. Thus, investment frameworks can be developed valuing sustainability and reliability as much as anything else. Such work can accomplish sustained synergy by coordinating and aligning the actions of key stakeholders, and can enable cities to realize the benefits of integration by planning, designing and managing the whole urban system. This perspective places a value on capital assets (social, natural and engineered) and incorporates a more comprehensive risk assessment in both policy design and decision-making. Finally, not only an unbiased review of technology options but also a championing of innovative water technology is required to meet future urban water demands in Latin America.

Green and inclusive growth, as adopted by international institutions at the United Nations Conference on Sustainable Development in Rio de Janeiro (Rio+20) in 2012 needs to be mainstreamed into all work on water (see Chapter 13 of this book). 'Green and inclusive' recognizes that global paths of economic growth are not sustainable, despite significant reductions in poverty over the past 20 years. The failure to internalize environmental costs is now clearly compromising economic and social progress. The central tenets of 'green and inclusive' are not original: they are about economic growth and poverty reduction, while reducing expensive and irreversible environmental damage.

The crucial question, however, remains: How can cities in Latin America change the existing strategy of 'grow now and clean later'? This question is particularly relevant to climate change and the equitable distribution of the damage caused by its consequences. Wealthy countries have polluted the atmosphere for more than a century, yet the consequences are global and unevenly distributed in poor and rich countries. The dilemma of how to uncouple growth from environment degradation unfortunately remains.

It is important to recognize that human development, poverty reduction and urbanization transcend the explanatory capacity of traditional indicators of income, consumption and welfare. Comprehensive and sustainable solutions to urban water issues have to be centred on the basics of human development and poverty. The debate about sustainability and human economic and social development is crucial for Latin America's cities of today and tomorrow. How to advance the human condition through providing water services in a way that ensures their availability for future generations is the core issue, and poses a truly significant challenge for Latin America's growing urban areas. Some solutions to these challenges already exist – but will require changes in the economics, governance and use of water in Latin American cities – and other solutions have yet to be found.

References

Bettencourt, L. M., Lobo, J., Herbing, D., Kühnert, C. and West, G. B. (2007) Growth, innovation, scaling, and the pace of life in cities. *Proceedings of the National Academy of Sciences*, 104(17); 7301–7306.

CAF (Corporación Andina de Fomento) (2011a) *Diagnostico estratégico para una agenda prioritaria: agua potable y saneamiento.* CAF, Caracas.

CAF (2011b) *La infraestructura en el desarrollo integral de América Latina. Diagnóstico estratégico y propuestas para una agenda prioritaria*. CAF, Caracas.

CAF (2013) *Equidad e inclusion social en América Latina: acceso universal al agua y saneamiento.* CAF, Caracas.

GRTB (Grupo Regional de Trabajo de Benchmarking) (2011) *Informe Anual 2010.* ADERASA.

Harberg, R. J. (1997) *Planning and Managing Reliable Urban Water Systems*. American Water Works Association (AWWA), Denver.

IDB (1999) *Can Privatization Deliver? Infrastructure For Latin America*. IDB, Washington, DC.

IDB (Inter-American Development Bank) (2001) *Investing in Water Quality*. IDB, Washington, DC.

Mays, L. W. (2000) *Water Distribution Systems Handbook*. McGraw Hill, New York.

McKinsey Global Institute (2011) *Building Globally Competitive Cities: The Key to Latin America Growth*. McKinsey, São Paulo.

Shlomo, A., Sheppard, S. C. and Daniel, L. C. (2005) *The Dynamics of Global Urban Expansion*.The World Bank, Washington, DC.

Tucci, C. E. and Bertoni, J. C. (2003). *Inundações urbanas na Amêrica do Sul.* ABRH, Porto Alegre.

UN-HABITAT (2003) *The Challenge of Slums*. UN-HABITAT, Nairobi.

UN-HABITAT (2009) *Global Report on Human Settlement*. Earthscan, London.

World Bank (1998) *Brazil. Managing Pollution Problems. The Brown Environmental Agenda.* The World Bank, Washington, DC.

World Bank (2007) *Water Quality and Pollution Control Project. Project Performance Assessment Report.* The World Bank, Washington, DC.

World Bank (2009) *Reshaping Economic Geography.* The World Bank, Washington, DC.

World Bank (2011) *Population Aging. Is Latin America Ready?* The World Bank, Washington, DC.

3 Challenges and policy options for financing urban water and sanitation

Anthony Cox and Peter Börkey

Introduction

Urban water supply and sanitation services generate substantial benefits for human health, the economy, and the environment; therefore, ensuring the adequate financing of these services is a major policy issue for countries around the world.[1] Significant financial resources are needed for investment in water and sanitation infrastructure, and to operate and maintain it, including expanding its coverage and upgrading service delivery to meet increasing social and environmental expectations. In 2006, the Organization for Economic Co-operation and Development (OECD) projected that the annual capital cost of maintaining and developing water supply and sanitation (WSS) infrastructure in OECD countries and the BRIC[2] countries could amount to US$1,037 billion by 2030 (OECD, 2006). This is far higher than comparable estimates for road infrastructure (US$292 billion per year by 2030) and electricity transmission and distribution (around US$241 billion per year by 2030).

In 2008, and at the half-way point regarding the Millennium Development Goal (MDG) time frame, it was estimated that the investment needed to meet MDG goals regarding access to water and sanitation in developing countries was around US$180 billion a year (Hutton and Bartram, 2008). In the case of Latin America, it was estimated that investments of US$18 billion were required to reach the MDGs, and that if universal coverage for water and sanitation was to be achieved in Latin America, an investment of more than US$46 billion annually needed to be made (IDB, 2010). Notably, none of these estimates included the cost of operating and maintaining new and existing infrastructure.

Most urban water systems, including those throughout Latin America, are perpetually underfunded and struggle to ensure that adequate financial resources are available to meet investment, maintenance, and operation needs. This has been made all the more challenging following the global financial crisis of 2008–2009 and as a result of national fiscal consolidation packages, which have dramatically affected the finances of sub-national governments, including those of cities. Reduced transfers from central governments, coupled with deteriorating borrowing conditions, have severely constrained the ability of sub-national governments to undertake public investment; many are simply seeking to preserve their current spending on welfare, health, and education (Vammalle and Hulbert, 2013).

A lack of what can be characterized as 'sustainable financing' for WSS has significant consequences for water and sanitation users, especially the poorest in urban areas throughout the world, and without exception in Latin America. Access to clean drinking water and sanitation reduces health risks and frees-up time for education and other productive activities, thereby increasing the productivity of the labour force. Safe wastewater collection and disposal helps to improve the quality of surface waters with benefits for the environment (such as healthier ecosystems and greater biodiversity) as well as for economic sectors that depend on water as a resource (fishing, agriculture, tourism, and water-dependent industry). The World Health Organization (WHO) has estimated that in developing countries, meeting the water and sanitation MDGs could generate a benefit to cost ratio of 7 to 1. Three quarters of the benefits stem from productivity gains, while the rest would be driven by reductions in incidences of water-related diseases such as diarrhoea, malaria or dengue fever (Haller *et al.*, 2007).

This chapter takes a close look at the challenges and policy options to financing urban water and sanitation services – challenges that are universal and that are pertinent to Latin America in particular as the most urbanized region in the world. The challenges in particular require governments to take steps to 'close the financing gap' between the costs of investment, operation, and maintenance of water infrastructure on the one hand and the stream of revenues from a range of sources on the other. Where the gap cannot be closed, governments need to employ a wide range of financial mechanisms to bridge the financing gap. The private sector has a potentially important role to play in the WSS financing challenge, and the experiences already observed of private sector involvement in financing WSS can provide valuable lessons in how to improve financial sustainability.

Closing the financing gap

Closing the financing gap requires countries to mobilize financing from a variety of sources. It may include reducing costs through efficiency gains or the choice of cheaper service options, increasing the basic sources of finance that can fill the financing gap, i.e. tariffs, taxes, and transfers (commonly referred to as the '3Ts'), and mobilizing repayable finance, including from the market or from public sources. These potential sources are shown in Figure 3.1. It is important to note that closing the financing gap requires governments to make efforts both on the 'demand side' and the 'supply side' of the sustainable financing equation; both of these dimensions of the financing challenge are examined in this section.

Closing the gap by improving the efficiency of operations

On the demand side, the costs of providing WSS services can be reduced through improved operating efficiencies – a crucial way of generating financial resources, as well as saving on physical resources, particularly in areas of water scarcity. Improved investment planning can also generate substantial savings and better define more realistic investment programmes that, for example, have cheaper and more locally-appropriate investment options, or by adapting service levels to local conditions and development strategies.

Inefficiencies are responsible for significant losses of funds within the WSS sector. Operational inefficiencies can include poor revenue collection, distribution losses (or non-revenue water – NRW), labour inefficiencies, and corruption large and small. Reducing NRW, for example, can significantly reduce operating costs by saving treatment and transport costs as well as the actual amount of water used, which is then available for other uses. Accumulated inefficiencies and deferred maintenance also result in higher costs, especially over time. Operational inefficiencies in the WSS sector can be reduced in a number of different ways: by increasing user charges closer to cost-recovery levels in order to provide more efficient price signals and to help capture lost revenue; by reducing utility operating inefficiencies to prevent the waste of significant resources, support healthier utilities, and improve service quality; and by improving budget planning and execution work.

Similarly, there are many potential ways to increase efficiency in the WSS sector. Incentives for improved efficiency can be introduced with a number of tools, such as price regulation, assignment of risks and rewards, competitive tendering, penalties, and benchmarking. As a first step, benchmarking tools such as the International Benchmarking Network for Water and Sanitation Utilities (IBNET) can be used to compare the performance between various utilities and identify areas of potential inefficiencies within particular systems. As an example of benchmarking, in England and Wales, the water and sewerage companies provide the economic regulator Ofwat with indicators of service performance covering water supply, sewerage services, customer service, and environmental impact, which Ofwat then publishes annually.

Choices regarding hardware and technology can also make a big difference regarding WSS costs. In OECD countries and elsewhere, regulatory regimes in place sometimes influence in unintended ways the selection of WSS financing options, linked to the set of incentives that these investment options seem to prefer. While a rate-of-return regulatory regime may give an incentive to select higher cost options so as to earn a higher return – what is sometimes referred to as 'gold plating'– incentive-based regulatory regimes, such as price cap regimes, introduce incentives to invest at least cost.

For many developing and Latin American countries, a range of options to deliver improved WSS at a range of costs can be achieved through what is commonly referred to as a 'service ladder'. For example, Hutton and Bartram

(2008) demonstrated that the per capita cost of a household connection is more than three times higher than a stand post in Latin America. According to their estimates, the total global costs of attaining the water and sanitation MDGs could be reduced by going from higher technology to lower technology options, from US$327 billion to US$135 billion, equivalent to an annual saving of US$190 billion worldwide. What their work signifies for achieving WSS MDGs in Latin America after 2015, and at significantly reduced costs, is unclear. Cutting down on investment costs may also be achieved by lowering service standards to levels that a country or city can afford; for example, many developing countries have adopted developed country standards without tailoring them to their own circumstances, sometimes resulting in unnecessarily high WSS infrastructure investment costs.

Closing the gap with tariffs, taxes, and transfers – the '3Ts'

On the supply side, revenue can be mobilized through the 3Ts (tariffs, taxes, and transfers) from overseas development assistance or philanthropic donations – and from repayable sources by making the case for investment in WSS, improving the allocation of resources, and reducing risks to attract private investors. Developing the right balance between these sources of revenue calls for governments at all levels to undertake strategic financial planning, so as to evaluate the potential for mobilizing financing from each source of revenue, including demand-side revenue.

It is now recognised that the 3Ts are the ultimate sources of finance for water and sanitation services with transfers still being of particular importance to developing countries (OECD, 2009a). The 3Ts can be used to leverage, and eventually repay or compensate, other funding sources, principally loans, bonds, and equity. Most countries have used public transfers (either from their own government or from external sources) to fund the development of WSS, particularly for capital expenditures. As countries develop and urbanize, and as WSS become more mature, there tends to be a shift toward more use of commercial finance paid for by growing cash flow from user charges – the tariff component of the 3Ts (Figure 3.1).

Tariffs

Conventional economic wisdom calls for setting WSS tariffs at full cost-recovery levels. Very few water and sewer service providers, however, and whether in developed or developing countries, and including throughout Latin America, recover full costs through tariffs. This is true even when only capital, operations, and maintenance costs are in the equation; it is even more difficult when attempting to recover environmental and social costs via tariffs. The concept of sustainable cost recovery has become the central concept for determining tariffs, and is based on the simultaneous application of three principles:

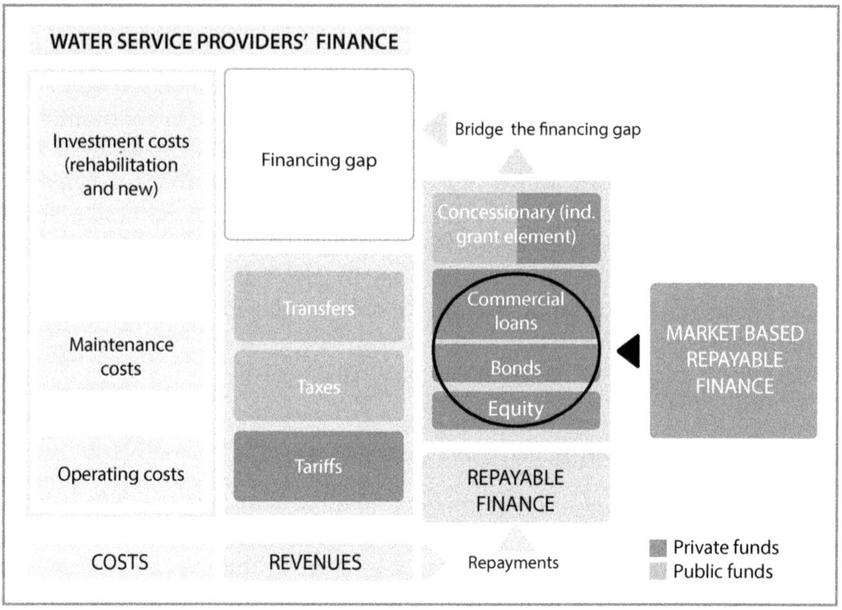

Figure 3.1 Sources of finance for water and sanitation services

Source: OECD (2010b)

1 An appropriate mix of the 3Ts to finance recurrent and capital costs, and to leverage forms of financing other than tariffs;
2 Predictability of public subsidies to facilitate investment planning; and
3 Tariff policy that makes WSS affordable to all, including the poorest, while ensuring the financial sustainability of service providers (OECD, 2009a).

Tariff setting is usually driven by a combination of factors, many of which go beyond the immediate needs of the service. Politicians sometimes insist on keeping tariffs low (i.e. below cost-covering levels) since water is an essential good, for which charging can be politically and socially sensitive. The 'willingness-to-charge' may therefore be lower than the 'willingness-to-pay' due to ulterior motivations. From an economic perspective, the setting of tariffs needs to reconcile a series of potentially conflicting objectives, including economic efficiency, cost recovery (or financial sustainability), and social concerns (or affordability).

In most OECD countries, WSS operating costs are by and large covered by tariffs, but approaches to covering capital costs vary substantially. Prices vary by a factor of more than 10, ranging from 0.49 US$/m^3 in Mexico to 6.7 USD/m^3 in Denmark (Denmark's pricing involves an attempt to incorporate environmental costs into WSS pricing) (OECD, 2010a). Overall, WSS tariffs represent only a small share of average household incomes, ranging from 0.2 per cent in Korea to 1.2 per cent in Poland. These average figures hide some areas of 'water poverty', however, with WSS tariffs representing up to 4.2 per cent and 7.9 per cent of household income for the poorest decile in Mexico and Poland respectively.

Cost-covering tariffs are much less prevalent in developing countries. Charges for WSS services in developing countries have been increasing over the last decade from unusually low levels (OECD, 2010a). Some countries in Latin America now have tariffs above 1 USD/m^3. In most cases, though, tariffs provide little incentive to use water efficiently and do not cover costs. While operating costs are not always covered, capital expenditure for large investments is almost always financed via public funds, either from government taxes or international transfers.

The experience in Latin American countries varies considerably. Countries such as Chile and Brazil have relatively high levels of cost recovery and, as a result, are able to rely on some degree of commercial credit financing for their utilities. In other Latin American countries, cost-recovery levels are low and WSS utilities are dependent on transfers from central governments for financing of capital expenditures and sometimes operating costs. In Colombia, for example, municipalities are legally entitled to receive transfers from the central government calculated through a formula based on their costs and poverty levels.

In virtually all Latin American countries, generating additional revenues via tariff reform (including changes to tariff levels and tariff structures) requires taking into account affordability constraints for the poor, especially in Latin America's burgeoning urban areas. The trade-off between WSS financial sustainability and affordability can be addressed via careful tariff design. Affordability can be assessed at two levels: for society as a whole and for the most vulnerable groups (referred to as micro-affordability). A number of countries have adopted increasing block tariffs with a first 'subsistence' block provided at zero or at an extremely low price, thus enabling poor households to have access to a basic level of water services for free or at low cost. This approach allows for cost recovery by providing a cross-subsidy from larger and wealthier water users, while providing an incentive to conserve water by charging more when more water is used – an increasing block tariff approach.

But actual experience has shown that increasing block tariffs can be regressive in countries with incomplete networks, where the poor are generally not connected to WSS systems and therefore do not benefit from either an underwriting or consumption subsidy. Apart from modifying tariff structures, other solutions to the issue of WSS affordability include providing income support to compensate poor households for increases in the prices of services of vital public interest, and establishing payment systems that help the poor manage their budgets better, for example by paying WSS bills at short intervals.

Taxes

Throughout Latin America, allocations from public budgets still represent an important share of revenue for the WSS sector, and are likely to play a significant role for the foreseeable future. The allocation of public funds to WSS can be justified in a number of ways, including promoting the consumption of merit goods (such as household sanitation and hygiene, whose value consumers may not fully realize) or to compensate for market failures by rewarding WSS providers for supplying 'public goods' such as public health and unpolluted water sources. Public funds may also be used to allow service providers to provide services at tariffs that are below cost for vulnerable consumer groups.

In order to be efficient and effective, tax subsidies should be transparent, targeted, and ideally should taper off over time. The most widespread form is capital expenditure subsidies provided in the form of grants, below market loans, or guarantees. In OECD countries, for example, most of the heavy initial investment that was made in the late nineteenth and early twentieth century (for water supply and sanitation) and since the 1960s (for wastewater treatment) was financed through public funds. When WSS utilities are owned by municipalities, local government budgets are often not sufficient to handle significant capital WSS expenditures, and thus benefit significantly from transfers from the central government.

This is certainly the case in Latin America, where municipalities often struggle to obtain adequate financing from tariffs. The central government therefore makes municipal infrastructure grants to municipal WSS providers to address capital investment backlogs and needs. It is crucial that such transfers are provided in a way that ensures an effective contribution to the long-term sustainable financing of the WSS sector. The goal for intergovernmental transfers should always be to generate stable revenues that can be integrated in medium-term financial strategies of local governments, and the transfers should be limited in time – until the achievement of pre-specified targets. Unfortunately, the experience in Latin American countries has by and large been that intergovernmental transfers have occurred without adequate planning – many Latin American municipal WSS utilities simply receive tax transfers that vary from year to year and without any conditions or goals – making it difficult to increase coverage and improve service quality.

The global financial crisis of the late 2000s demonstrated the potential downside of relying too heavily on tax transfers for investment in water infrastructure. This is particularly significant since during times of crisis there are many competing demands for public funds. Substantial public borrowing is likely to exacerbate the pressure on non-sovereign borrowers – cities and others – through a crowding-out effect, making it harder for them to borrow at acceptable rates. In addition, during periods of economic stress or crisis, governments in developed and developing countries alike are less able to borrow at acceptable rates. As a result, during the recent economic upheaval some were tempted to make temporary cuts in water and wastewater investments and to reallocate those resources to other sectors, with potentially long-term damaging impacts for WSS provision. On the other hand, though, several governments responded to the financial crisis by unveiling substantial stimulus packages, which benefited the WSS sector to a certain extent.

Transfers

International transfers in the form of official development assistance (ODA)[3] and philanthropic donations often play a role in closing the financing gap in transition economies and developing countries. In some countries, ODA subsidizes most investments, in others it plays a more marginal role, and in others still no role at all. ODA has an important role to play both as a source of finance and of capacity development for the provision and financing of WSS services. It can also have a catalysing effect by reducing bottlenecks (particularly capacity constraints in urban areas), ensuring access to the poor, and harmonizing and aligning assistance with national strategies.

After a temporary decline in the 1990s, aid to water and sanitation rose sharply after 2001. In 2010–2011, the total annual average aid commitments to water and sanitation amounted to US$7.6 billion, representing approximately 6 per cent of total sector allocable aid, as shown in Figure 3.2. The poorest countries – least developed countries and lowest income countries – received 33 per cent of the total WSS sector aid. Around 19 per cent of the total aid was directed to water supply activities, 18 per cent to sanitation, 42 per cent to combined water and sanitation activities, and the remaining 21 per cent to capacity development – education and general budgetary support.

In the aftermath of the global financial crisis, it seems likely that transfers in the form of concessional lending rather than grants may emerge as the preferable form of ODA, as public finances come under increasing pressure. This highlights the need for ODA resources to be spent strategically, so as to maximize their

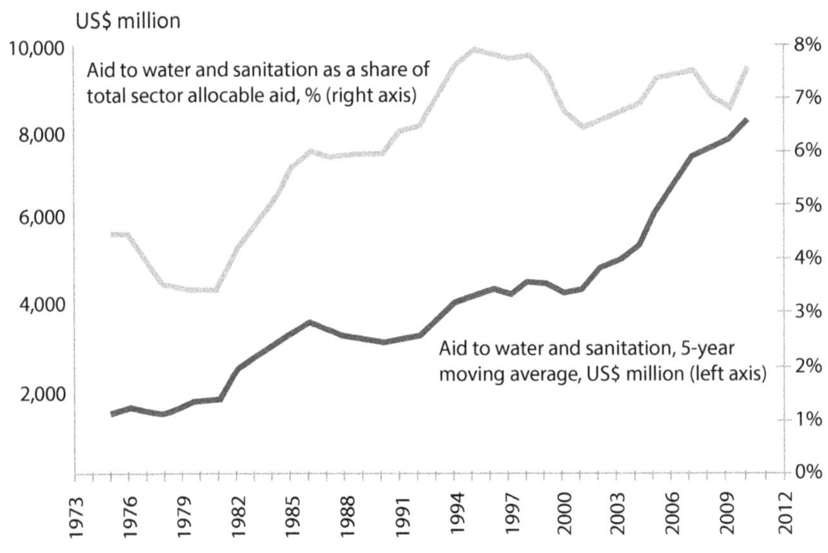

Figure 3.2 Trends in aid to water and sanitation

Source: Adapted from OECD-DAC (2013, Chart 1)

leveraging capacity and effectiveness. Areas where ODA can have a catalysing effect include reducing bottlenecks in the WSS sector, supporting the financial planning process, ensuring access to services by the poor, and supporting the development and use of risk-management mechanisms that help attract private funding, and local private funding in particular.

Bridging the financing gap

While revenues from the 3Ts can close the financing gap for WSS providers, repayable finance is often necessary to bridge the financing gap over the short to medium-term where revenue from the 3Ts is insufficient to cover the full range of costs of WSS service delivery. WSS providers often look to repayable finance in order to finance capital costs associated with repairs, renewals, or expansions of water and sanitation systems, while ongoing operating and ordinary maintenance costs are routinely financed from a mix of the 3Ts. Because repayable finance requires payments in the form of interests or dividends, it can be challenging for WSS providers to access capital markets and to also generate the appropriate level of return on their operations to repay borrowed funds.

Market-based repayable finance refers to a sub-set of repayable finance, where financing is provided through the market by private actors. Sources of market-based repayable finance include: debt finance (loans from commercial banks, bonds issued through capital markets, project finance) and equity finance (from capital markets or private equity funds). Debt financing has been the backbone of most infrastructure investment in developed countries. Depending on the development of local bond markets and the size of the debtor, it has come either in the form of bonds or loans. In developing countries, WSS providers can use bank loans to finance capital investments (although these are usually concessional loans from development institutions). The use of other forms of finance, such as bond finance, project finance or equity finance has so far remained limited in developing countries, but these forms of finance are emerging gradually as ways to complement other methods of WSS infrastructure finance, and may well play an important role in future Latin America WSS development.

Debt finance

Bank loans

Short and medium-term commercial loans are common for financing working capital requirements of WSS providers in developed and developing countries alike. Short and medium-term lending facilities may also need to be used to finance investments in countries where obtaining long-term bank financing to match the long asset life of WSS investments is difficult, as commercial banks are not able or willing to lend over such long periods. In developing countries, commercial banks are usually not familiar with the WSS sector, and often perceive it as a high-risk sector due to difficulties with increasing tariffs, inefficient management, and corruption. Water utility revenues may not be sufficient to reimburse loans. Furthermore, revenues may not be sufficient to cover market-based financing costs, a problem which limits the ability of WSS providers to borrow. Finally, certain types of WSS providers, such as local or small-scale service providers, may not have access to traditional bank financing at all, although they may have the option of accessing credit through microfinance institutions.

The late 2000s global financial crisis made things more difficult, as it led to a dramatic increase in the cost of commercial debt and in a reduction in the availability of overall debt financing, especially for long-term debt; in essence, a severe contraction in bank lending. The crisis also affected the ability of sovereign states to borrow, and reduced the value of sovereign guarantees in some cases. Microfinance institutions suffered as well and may now be less willing to diversify into WSS and away from their more traditional income-generating markets and activities. Indeed, microfinance institutions in many developing countries, including those in Latin America, do not appear to be offering micro-loan/financing for WSS. National development banks – if and where they exist – tend to focus on large WSS projects rather than on small ones. As a result, bigger, richer, and creditworthy cities usually can obtain bank finance, while most small towns and rural areas are neglected.

Bond finance

In developed countries, the WSS sector is considered to have a low risk profile that makes it well suited to the debt market. Bond financing is common in developed markets as it often offers cheaper access to debt finance than loans. The types of bonds issued can include corporate bonds, sovereign bonds, or municipal bonds, depending on the structure and ownership of the water sector. For example, in the United Kingdom, the WSS market is dominated by large private water and sewerage companies which issue corporate bonds.

In the United States of America (US), water companies are often smaller municipally owned utilities; the municipal bond market has provided a major source of finance for water and sanitation investments. The late 2000s global financial crisis somewhat affected the WSS municipal bond market, however, as the credibility of credit rating agencies came into doubt and several monoline insurers – used to enhance the rating of municipal borrowers in exchange for an insurance premium – disappeared. As a result, municipal bonds have somewhat lost their attractiveness for investors, making it a bit more difficult for municipalities to raise funds. In the majority of less developed markets, municipal bonds were not available even before the global financial crisis due to poor creditworthiness, lack of transparency, and other issues inherent in these markets.

Project finance

Project finance is the financing of long-term infrastructure through a 'special purpose entity' set up just for the project, and that can be financed with project debt and equity. A project finance deal typically involves a number of equity investors, known as sponsors, and a syndicate of banks that provide loans to the operation. Following the late 2000s global financial crisis, the feasibility of project finance deals has been severely affected, particularly in countries considered to be risky. New project finance structures are likely to require co-operation with sovereign-backed banks, and will often require bridging loans at less favourable conditions.

Equity finance

Raising equity can be a good way of financing long-term investments in WSS systems, as equity is a source of finance with no deadline for repayment. Equity holders are usually interested in holding their stake over the long term in order to benefit from future dividends and any potential increase in the value of their equity. Equity can also be used as collateral to leverage other forms of private finance, rather than as a way to finance long-term capital investments directly. Use of the private equity market, however, usually results in higher financing costs than the cost of debt finance. Shares are either listed on a stock exchange or held privately, by the founders and managers of the company or institutional investors.

Some Latin American water companies have listed shares on stock exchanges, such as São Paulo State Basic Sanitation Company (SABESP) in Brazil. A key constraint weighing on the ability to raise capital in equity markets, however, is the varying degree of development of local capital markets. Following the late 2000s global financial crisis, equity financing has been more difficult to come by, as the equity risk premium (the return expected by equity investors compared to risk-free investments) has gone up throughout the world, including in Latin America.

The WSS sector is often perceived by potential providers of market-based repayable finance – banks, institutional investors, private equity funds, equity investors, project sponsors – as a high-risk/low return sector. This is so even though its fundamental economics, with relatively stable and almost recession-proof demand for the services and long-life buried assets, would instead seemingly place it in a low risk/low and steady return investment category. The WSS sector's high-risk reputation is frequently linked to difficulties raising tariffs to cover costs, to perceived affordability constraints, or to political issues. As a result, many WSS providers, including in Latin America, constantly find themselves in difficult financial situations, with under-capitalized balance sheets that impede their ability to finance themselves. In the absence of any repayment capacity or history of past lending to the WSS sector, most commercial banks are unlikely to lend to WSS providers, a sector perceived as being somewhat of a credit risk.

In addition, local financial markets may not be able to provide stable, long-term loans with attractive interest rates to local water utilities, which tend to be of medium or small size. There is often a discrepancy between the long-term investments needed in the water sector – to match the life of the assets – and the short-term lending capabilities of local markets: this is referred to as an inappropriate market fit. According to Kariuki and Schwartz (2005), informal WSS operators serve an average of 50 per cent of the population in developing countries, and they have difficulties in accessing credit from the conventional banking sector. Furthermore, in many countries, decentralization of water and sanitation services has transferred large investment needs to local governments and utilities. However, the availability of funds at the local level is restricted: local governments' creditworthiness tends to be low, making it challenging to raise funds on international markets, and the small scale of service of many utilities may result in transaction costs that are too high to make market-based financing viable. Latin America indeed confronts the challenge of how to finance its urban WSS systems, which after all are 'local systems' regardless of size, and for which a wide range of financing mechanisms must be considered and adopted. Given the high rates of urbanization of the region, this matter is of utmost relevance.

Innovative financing mechanisms

In the future, innovative financing can play a major role in attracting market-based repayable finance to the WSS sector. Table 3.1 below outlines examples of critical mismatches, and the types of innovative financial mechanisms that might be used to address these mismatches. In the context of some of these innovations, public funds could be used in order to leverage market-based repayable financing, so as to increase the overall amounts of finance available to the sector.

All of the innovative financial mechanisms set forth in Table 3.1 have been used to a greater or lesser extent in the WSS sector. Country experiences with these mechanisms vary, but are mostly positive. The main challenge lies in ensuring that the general investment conditions in a country are sufficiently stable to encourage investors, and that the economic regulation of the WSS sector provides a solid basis for repayment of the finances. A number of key mechanisms are discussed below.

Blending grants and repayable financing combine concessional financing (either grants or loans with a grant element) with repayable finance in order to support a single project or a comprehensive lending programme. In the water sector, this has been done at the level of specific projects or through the establishment of financing vehicles that aim to combine diverse sources of finance. An example is the Financiera de Desarrollo Territorial S.A. (FINDETER) in Colombia, a public financing entity that rediscounts commercial bank loans for local infrastructure development, including water and sanitation. The main objective of blending is to establish WSS sector funding that would otherwise not be available for a given project, while ensuring that basic public policy goals, such as increasing

Table 3.1 Examples of innovative financial mechanisms in the water sector

Critical mismatch	*Innovative financial mechanisms addressing the mismatch*
Affordability constraints at household level	• Blending grants and repayable financing • Micro-finance • Output-based aid
Limited availability of funds for domestic operators	• Micro-finance • Output-based aid and innovative contract
Risk profile and difficulties in managing certain risks (e.g. political risk, foreign exchange risk)	• Blending grants and repayable financing • Guarantees and risk insurance • Devaluation backstopping facility • Local-currency financing • Revenue agreements in lieu of guarantees
Lack of funds at decentralized level	• Municipal bonds • Pooled funds, revolving funds and bond banks • Instruments to increase sub-sovereign lending
Short tenor of available financing	• Guarantees • Equity contributions
Under-capitalized balance sheets	• Raising equity to strengthen the balance sheet, convertible loans, debt-equity swaps, 'asset-light' expansion models
Lack of understanding by external lenders and investors	• Blending grants and repayable financing • Credit ratings • Project preparation facilities
Lack of 'bankable projects	• Project preparation facilities

Source: Adapted from OECD (2011b)

access and serving the poor, are met. Blended financing structures hold great potential in the water sector, especially in the context of the global financial crisis, given that an element of subsidy is almost always going to be required to make a WSS sector project bankable while reaching the underserved.

Microfinance has been identified as a key way to overcome affordability constraints on providing access to services, particularly for households and small-scale water providers in developing countries. The use of microfinance has so far been limited in the water sector, partly because of limited awareness of its potential, and partly because of a limited understanding by microfinance and water sector professionals of opportunities to work together. A fairly recent review by Mehta (2008), however, makes the case for the strong potential of microfinance in the WSS sector, particularly for loans to households and for community projects such as slum redevelopment projects.

Official Development Assistance can help develop microfinance for WSS by providing seed financing to revolving funds or microfinance institutions, smart subsidies for product development, and guarantees. Donors and international finance institutions (IFIs) can help build awareness of microfinance products, through capacity-building activities and by blending microfinance with other types of financing instruments in the projects they choose to support. For example, they can combine reliance on microfinance, or local commercial banking in the case of small-scale entrepreneurs, with output-based aid – subsidies that are paid to service providers based on effective and measurable results, and are therefore better incentivized.

Although an array of *guarantees and insurance* products is available from donors and international finance institutions, and private institutions, they have not been used on a regular basis or on a large scale in the WSS sector. This is partly a reflection of the changing structure of the market for water services: while international private operators have largely been driven away by adverse conditions, guarantees provided by international institutions for relatively large projects and transactions are less appropriate than they were previously. In addition, ODAs and IFIs usually maintain fairly rigid rules about the use of guarantees – for example, with counter-guarantee requirements or restrictions on the provision of stand-alone guarantees – which means that transaction costs for applicants can be high. The establishment of 'guarantee facilities' at the national level which support all levels of government, and to which ODAs and IFIs can contribute seed financing or underlying guarantees, could facilitate the use of guarantees at the local level which is more in line with the current market structure in the WSS sector. ODAs and IFIs may also need to step in where private entities or governments have become less willing to provide guarantees.

Forming *grouped financing vehicles* can be a helpful way to provide access to financing by a large number of relatively small borrowers, particularly with the combined use of guarantees to improve credit rating. Such groupings are particularly well suited to decentralized WSS situations,, where small and medium-sized service providers struggle to access financing on their own merit. In the WSS sector, the grouped financing approach has mostly been used as a basis for issuing bonds in countries with fairly mature financial markets, such as in the United States, but it has also been used in Mexico. High transaction costs and limited knowledge, once again, partly explain why the grouped financing approach has remained somewhat limited. ODAs and IFIs need to step up their efforts in order to create such structures or help define institutional environments that would be conducive for grouped financing vehicles to be established where appropriate. This may require establishing grouped financing structures directly, such as revolving funds or bond banks, or fostering the adoption of legislation that make such structures more attractive – such as tax-exemptions on bonds issued by such structures, as practised in the US, or requirements that grouped financing vehicles be formed in order to access government financing.

Direct lending to sub-sovereign entities, without the need for a central government guarantee, has been practised with success for some time by some ODAs and IFIs. Many ODAs and IFIs, however, have not been able to lend at the sub-sovereign level, either because their internal rules do not allow them to do so or because they are not willing to take on the risk they perceive exists with such lending. Besides, many sub-sovereign entities, including cities of all sizes, are either too weak financially to borrow or lack the capacity to put together a bankable project eligible for donor financing. Central governments may also be reluctant to let sub-sovereign entities borrow directly, especially when such debt may affect the overall debt burden accumulated at the national level. Sub-sovereign debt may have to be covered by central governments in the event of bankruptcy or repayment failures, even if explicit guarantees have not been provided.

ODAs and IFIs should evaluate how they can relax guarantee requirements for sub-sovereign lending, so as to pave the way for commercial lending to sub-sovereign borrowers. For example, reliance on revenue agreements with the sub-sovereign borrowers to either increase tariffs or intercept central government transfers can provide security to lenders without the need for central government guarantees. Finally, donors can combine these lending instruments with guarantees to commercial lenders so as to broaden the pool of financiers and investors interested in investing in water and sanitation at the local level. Direct lending to entities at the sub-sovereign level, including cities and municipal utilities, can help these borrowers build a credit history and give them access to a broader range of investors, including commercial banks and equity investors.

Raising *equity* can help strengthen the balance sheets of water companies, which are often under-capitalized. Interesting models have been developed in the water sector to mobilize equity via financial markets – such as the Hyflux Water Trust in Singapore – which act to diversify away from mobilizing funding from private water companies (whose ability to bring in equity capital is limited in any case) and using such equity injections to leverage other forms of finance for capital investments. Mobilizing equity through capital markets can strengthen financial discipline and improve transparency, including for companies that are primarily government-owned, including a number of the state water companies in Brazil that are publicly listed.

Credit ratings can also help improve transparency and facilitate access to financial markets for borrowers. Significant progress has been made in establishing credit ratings for municipal governments and water companies, although the use of such ratings remains limited, particularly in markets that are too small to develop a national rating scale. The global financial crisis also significantly affected the credibility of rating agencies, by calling their efficacy into question. ODAs and IFIs can potentially step in to develop shadow credit-rating systems (i.e., based on a one-off exercise) to examine the creditworthiness of particular companies, and make recommendations on how they can improve their creditworthiness.

Other initiatives to improve overall transparency and improve knowledge of the WSS sector for external financiers should also be encouraged. For example, the World Bank's IBNET benchmarking system can and should be strengthened so as to improve the reliability and comparability of the information produced. Finally, it is important to note that significant information is already collected by different lending and research institutions. Coordinated efforts to improve its quality, increase collaboration between sources, and improve dissemination would have a positive impact on raising the profile of the WSS sector with external financiers.

Finally, *project preparation facilities* can also help with the definition and preparation of bankable water projects. A limited number of such facilities have been set up at the international level, and have, on the whole, enabled the preparation of bankable projects in an accelerated manner, thereby improving the effectiveness of lender and donor financing by pooling funds together for support to project preparation. Project preparation facilities have been particularly useful in well-defined geographical areas where they have been set up to accompany well-defined policies. The establishment of such facilities at the national level could also be encouraged, as they can reduce transaction costs and connect more easily to other domestic financial mechanisms outlined earlier. In addition to the establishment of project preparation facilities, local expertise for project preparation should be strengthened, from project conceptualization all the way down to design stage and implementation.

The role of the private sector in water/sewer service sector financing

The private sector is involved in many different ways in the WSS sector. Private actors in the various segments of WSS service provision include the following:[4]

- *Formal private water and sanitation service operators* provide services based on a contract or licence with the public authorities responsible for delivering services in a given region or country.[5] They may be either domestic operators, which tend to focus on a single country, or regional and international operators, which provide services to customers in many countries.
- *Informal private water and sanitation service operators* can be small-scale independent providers, which generally operate informally where no public services are provided or are provided at low standards of quality.[6] They can also be large-scale developers of both residential and commercial property, who often provide and operate water and sanitation services to their own development projects, thereby making a considerable contribution to a city's infrastructure and service coverage. This approach can be found throughout Latin America.
- *Private financial institutions (banks and investment funds)* provide financing from private sources in the form of loans, bonds, or capital in the form of equity. Such private financing can be provided to both public and private operators.
- *Private companies* whose main business is not water but which are heavy water users, such as beverage, mining, energy, and construction companies.

During the 1990s and early 2000s, the introduction of private sector participation (PSP) in the management of WSS in developing countries was construed as a way to bring additional financial resources to the sector and therefore to close the financing gap. But PSP too often was based on the misconception that the private sector would bring financing with it through concession contracts or other similar contracts that contained investment obligations. The early termination of some high-profile concessions – such as in Buenos Aires – following national financial crises, and in which the private operator was exposed to substantial foreign exchange risk on debt incurred to finance WSS programmes, challenged the initial expectations about PSP. Early experiences with PSP also helped highlight the fact that private operators themselves have to source external capital and arrange financing.

Recent experience has allowed governments to gain a better understanding of the ways in which private operators can either directly or indirectly mobilize financing for the WSS sector. Three main areas of experience can be summarized. First, by improving overall sector efficiency, operators can reduce costs (and financial needs), and improve the sector's creditworthiness and ability to attract financing. By reducing costs, PSP can contribute to filling the financing gap. Improved services can contribute to creating a 'virtuous circle': customers are more willing to pay their bills when service improves, more efficient operation increases cash flow from operations, more funds are available for investment, which in turn increases the customer base and the utility's revenues. As creditworthiness improves, a utility can more easily access funding and invest in service expansion.

This has been borne out by the experience of PSP over the last 20 years, which has proven to be a useful tool for improving WSS sector performance and efficiency. For example, Marin (2009) looked back at 15 years of experience with public–private partnerships (PPPs) for urban utilities in countries and evaluated their impact on four dimensions of performance: access (coverage expansion), quality of service, operational efficiency, and tariff levels. He found that many private operators succeeded in reducing water losses, and notably in both Brazil and Colombia. In some cases, private operators reduced NRW to less than 15 per cent, a rate similar to the best-performing utilities in the world. Such efficiency gains have contributed to improving the financial position of water utilities by cutting costs, increasing revenues, and reducing the need for external subsidies to fill funding gaps.

Second, PSP can help the WSS sector by financing investment costs, particularly when the public sector's ability to borrow is limited. Private operators are sometimes brought in because they are deemed more able to mobilize financing, especially from private financial institutions. While PSP can facilitate crucial access to repayable market finance, especially given the need for such funding to cope with huge upfront capital investment costs, it does not per se contribute to closing the financing gap, but rather helps bridge the gap. Private financing ultimately needs to be repaid, after all, through a combination of the 3Ts. To a certain extent, the ability for PSP to mobilize financing can depend on the type of contractual arrangements entered into with the public sector. PSP can occur under a broad variety of contractual arrangements with the public sector, reflecting the ways in which risks have been allocated between the parties. The allocation of responsibilities for investment (and for financing such investment) can vary substantially according to the contractual arrangement in place.

An overview of the range of contractual arrangements entered into between PSPs and the WSS sector is presented in Table 3.2. In the case of concession contracts, schemes like build-operate-transfer (BOTs), divestitures, and some joint ventures, PSPs can be requested to mobilize substantial funding for capital investments directly, which they then usually recoup via tariff revenues or fees. In the case of the other contractual arrangements – although there are always exceptions – responsibility for mobilizing investments rests with the public sector and PSP involvement is brought in largely for its capacity to drive efficiency gains or to mobilize financing indirectly. The PSP also has more control over management in the case of 'higher-powered' forms of private sector participation, such as concessions, BOTs or divestiture, which is usually associated with a greater ability to deliver efficiency gains.

The third way in which PSP can assist the WSS sector is by managing and enabling capital investment programmes of public authorities. Here, the PSP manages an extensive capital investment programme on the behalf of a public authority, and closely coordinates the capital improvements with the ongoing operation and maintenance of the WSS provider. This approach has generated confidence and stability in WSS sector capital investment programmes, which in turn has significantly increased public sector investment in the sector.

Overall, private financing of urban water utilities in developing countries has been limited when compared with other infrastructure sectors – only 5.4 per cent of the total investment commitments in private infrastructure between 1990 and 2000 were in the WSS sector (Marin, 2009). Based on figures from the World Bank's Private Participation in Infrastructure (PPI) Database,[7] Marin found that investment commitments by private operators (made in the year of financial closure) decreased sharply in the aftermath of the Asian financial crisis, from a peak of US$10 billion in 1997 to a low of US$1.5 billion in 2003, and they have not really recovered since then, as shown in Figure 3.3.

In a study by Gassner *et al.* (2008), private investments were examined based on detailed regression analysis of water and electricity PSP contracts, with a data set of more than 1,200 utilities in 71 developing and transition economies. The study was not able to conclude that investment always increases with PSP, despite evidence of actual increases in water connections of about 12 per cent on average. The study also did not find conclusive evidence that investment increased in developing countries as a result of concession contracts. For lease and management contracts – where there is no investment obligation for the private party – the study suggested that the public asset holding company usually did not increase investment even if the PSP brought operational improvements.

This study and others with similar findings have led more and more countries to adopt PPP models and risk sharing arrangements in which investment largely remains in the hands of the public sector while the private partner focuses on improving service and operational efficiency (Marin, 2009; OECD, 2009b).[8] In practice, funding for investment under mixed finance PPP projects comes

Table 3.2 Typology of contractual arrangements between government (G) and the private sector (P)

	Service contract	*Management contract*	*Affermage/ Lease*	*Concession*	*BOT*	*Joint venture*	*Divestiture*
Asset ownership	G	G	G	G	P/G	G/P	P
Capital investment	G	G	G	P	P	G/P	P
Commercial risk	G	G	Shared	P	P	G/P	P
Operations/ maintenance[a]	G/P	P	P	P	P	G/P	P
Contract duration	1–2 yrs	3–5 yrs	8–15 yrs	25–30 yrs	20–30 yrs	Infinite	Infinite
Source of retribution of operator	Municipality	Municipality: fee is fixed or based on performance	Operator collects user fees[b, c]	Users	Municipality	Users	Users
Occurrence 1991–2009 (World Bank PPI Database)	Not part of scope	Together: 111 of 715 projects	278 of 715 projects	294 of 715 projects	Not a separate category	32 of 715 projects	
Examples	Mexico City	Johannesburg	Cartagena	Aguascalientes (Mexico) Jakarta Manilla	Mexico	Cartagena	England Chile

Note: a = Maintenance may lead to considerable amounts of investments on the part of the responsible partner
b = *Affermage*: Revenue shared
c = Lease: Fee paid by municipality

Source: OECD (2009a) updated based on World Bank Private Participation Infrastructure Database. http://ppi.worldbank.org

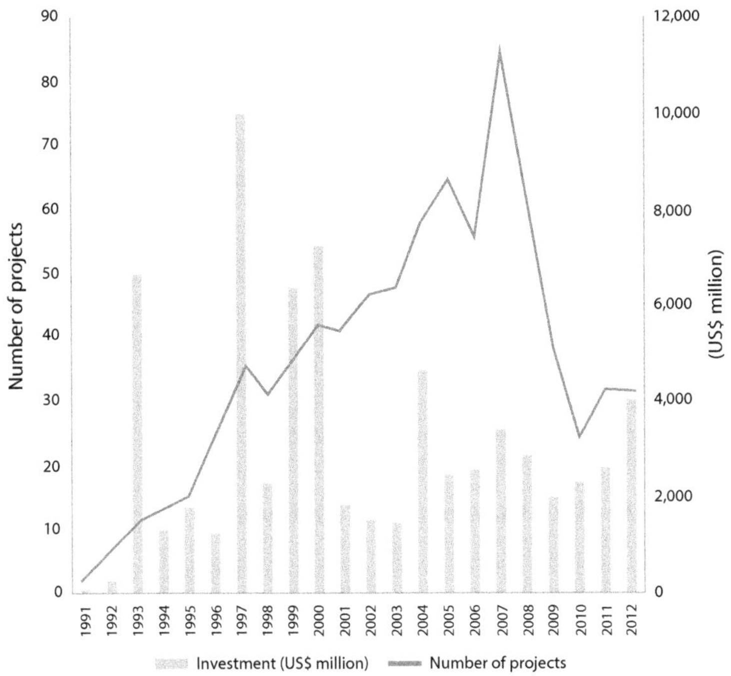

Figure 3.3 Investment in public–private partnership projects in developing countries, 1991–2012

Source: World Bank PPI (2014)

from direct cash flows from revenues plus a variable mix of government and private sources, which tends to make the traditional dichotomy between leases/*affermages* and concessions increasingly obsolete.

Several approaches to combine PSP, so as to benefit from efficiency gains, and a mix of public and private financing have been developed over the past decade, as follows.

- Concessions that rely largely on revenue cash flow for investment, with cross-subsidies from electricity sales or tariff surcharges, or both.
- *Affermages*, as originally applied in West Africa, bolstered by enhanced incentives for operational efficiency, a programme of subsidized connection to expand coverage for the poor, and a gradual move to full cost recovery through tariffs.
- Mixed ownership companies, such as those already operating in Latin America in numerous small, middle, and large municipalities in Colombia, Havana in Cuba, and Saltillo in Mexico.
- Concessions with public grants for investments to spearhead access expansion or rehabilitation while minimizing the impact on tariffs. This approach is typified by the PPPs designed under Colombia's Programa de

Modernización de Empresas (PME, Business Modernization Programme). A similar approach has been adopted in Guayaquil, Ecuador, and in a few concessions in Argentina, such as in Córdoba and Salta.

The above shows that the private sector can contribute to closing the financing gap in urban water and sanitation in several distinct ways. However, the ability of the private sector to contribute requires appropriate institutional and regulatory environments as well as sustainable cost recovery. In order to provide guidance to countries on the allocation of roles, risks, and responsibilities between public and private partners, as well as on the institutional, regulatory, and policy frameworks necessary to improve investment conditions in the water sector, the OECD has developed a tool – the Checklist for Public Action for Private Sector Participation in Water Infrastructure (OECD, 2009b) – and has supported its use in a number of countries, including in Mexico.

Moving forward on the financing challenge

Throughout Latin America, the extent to which a particular type of financing source can meaningfully assist with WSS sector financing challenges will be highly location-specific. The factors at play include the overall enabling environment, and the willingness of government to set realistic objectives and to adopt reforms so as to improve the efficiency and creditworthiness of existing WSS providers. Goals that are set politically and are not matched by real revenue streams result in major financing gaps and unexecuted plans, and the poor often suffer the most through absent or seriously deficient services.

In some cases, donors share some of the responsibility for the lack of realism. For example, the use of best available wastewater treatment technologies may not be affordable or even necessary if scaled up beyond the project level. Strategic financial planning must occur, and in the context of broader WSS sector planning that addresses roles and responsibilities of government agencies, policy priorities, and related legislative and regulatory reforms, in order to ensure that a package of financially realistic measures can be advanced.

As part of a government's setting of realistic objectives for the development of the WSS sector, the objectives must be 'checked' against available resources and 'agreed to' in a multi-stakeholder policy dialogue – together a process termed strategic financial planning, or 'SFP'. SFP has several objectives:

- To provide a structure for a policy dialogue to take place, involving all relevant stakeholders including Ministries of Finance, with the aim of producing a consensus on a feasible future WSS.
- To illustrate the impact of different objectives and targets in a long-term perspective, linking sector policies, programmes and projects.
- To serve the important aim of facilitating external financing, providing clear and transparent data on financing requirements. Such process can be carried out either at national level or at municipal or regional levels.

Ultimately, the WSS sector must use a full range of financing approaches, making the most of potential efficiency gains, adjusting targets and combining funding from both public and private sources, in order to meet its investment needs and successfully maintain and expand service. To achieve this, policy makers and water service providers need to engage in SFP and other strategic financial planning processes so as to identify what needs to be financed, how much additional resources can be generated from existing sources, and how the performance of utilities can be improved to generate efficiency gains and mobilize external financing (OECD, 2011c).

The WSS sector should also make use of all available tools that can help with financial planning. Numerous tools, such as the World Bank's IBNET, FEASIBLE (developed by the OECD), and many others exist and can assist Latin American countries and their WSS providers with a wide range of tasks, including the definition of achievable targets and financially sound planning that takes into account limited public funding. Given the era we are in regarding access to data and tools, all levels of government should take advantage of these tools. For example, the Checklist for Public Action for Private Sector Participation in Water Infrastructure tool seeks to assist policy makers in assessing and managing the implication of PSP in the WSS sector. It identifies key policies needed to bring about greater co-operation, and provides a set of approaches and practices aimed at addressing identified issues, based on country experiences.

Conclusions

The importance of ensuring that WSS systems are sustainably financed cannot be over-emphasized. Countries often develop elaborate and ambitious plans for developing WSS infrastructure so as to benefit their citizens and to help with their development and economic prospects. Such plans, however, are likely to be thwarted unless adequate and stable plans are made for financing both the initial investment in WSS infrastructure and then the operation and maintenance of the infrastructure. Many systems throughout Latin America are underfunded, with serious consequences for water and sanitation users, and especially the poor. Ensuring sustainable financing of WSS systems is an ongoing problem for both OECD and non-OECD countries. The challenge was exacerbated by the late 2000s global financial crisis. This is certainly the case for Latin America.

Closing the financing gap requires countries and their municipalities to mobilize financing from a variety of sources. Doing so requires reducing costs, increasing the efficiency of water use, increasing funding from the 3Ts, and establishing healthy repayable financing mechanisms from both public and market sources. WSS financial planning is absolutely critical, and arguably has been lacking in Latin America and other parts of the world. It is needed to find the right balance between potential sources of WSS financing, to think through and overcome financing challenges, and to set a medium and long-term strategic approach to the financial planning necessary to make the WSS sector a success throughout the region.

The importance of thoughtful, long-range planning is evidenced by the fact that setting tariff levels to meet financing needs while taking into account social issues can be deeply challenging, particularly from a political perspective. Yet an over-reliance on public subsidies through taxes, or through ODA transfers, increases the vulnerability of the WSS sector to exogenous shocks and changes in budgetary priorities by governments and donors. Bridging the financing gap through standard and innovative finance mechanisms is a key to supporting the transition to sustainable financing arrangements for the WSS sector in Latin America. As the region's overall economy grows and becomes more mainstream, there is also the increased possibility of making more effective use of private sector financial participation in Latin America's WSS sector. Underlying all these opportunities, however, is the need to take into account all financing possibilities and evaluate the most promising ones, and a strong and stable regulatory framework that provides a solid foundation for households, businesses, and investors to reap the benefits of sustainable financing for the vitally important WSS sector.

Notes

1 In this chapter water supply and sanitation (WSS) is used interchangeably with the terms water supply and sanitation sector, water and sanitation sector, water and sanitation sector. The first term is more commonly used while the others may be technically and slightly different.
2 Brazil, Russia, India, and China.
3 Term coined by the OECD to measure voluntary transfers (aid) from one country to another.
4 Note that households can be also considered as private actors, either when they invest in building and maintaining water and sanitation facilities in the private domain to enable them to benefit from connection to full public services or when such services are self-supplied (as with the case of on-site sanitation systems for example).
5 The contribution of the formal 'private water and sanitation service operators' constitutes what is usually referred to as Private Sector Participation (PSP) in the sector.
6 Note that, in cases where households or private actors provide services without government instruction in order to fill an 'access deficit' gap, they would tend to finance all necessary investments themselves either from their own funds, borrowing from family or friends, local money-lenders or microfinance institutions where they exist.
7 It should be noted, though, that a weakness of the database is that it does not provide a comprehensive picture of private investment in water infrastructure because it fails to cover some of the deals that involve mainly domestic players as well as some of the re-financing that occurs over the lifespan of PSP contracts. As a result, PPI numbers may fail to capture an important new source of private investment and distort the overall picture.
8 Different PSP contracts have different impacts on operational efficiency and therefore on access to finance. This can be explained by the different risk-sharing and responsibility-sharing arrangements of the various contract types; for example, concessions are responsible for both operations and investments, while lease contracts provide direct incentives to increase operation efficiency through the revenue structure.

References

Gassner, K., Popov, A. and Pushak, N. (2008) *An Assessment of Private Sector Participation in Electricity and Water and Sanitation Services in Developing and Transition Countries*, World Bank, Washington DC.

Haller, L., Hutton, G. and Bartram, J. (2007) 'Estimating the Costs and Health Benefits of Water and Sanitation Improvements at Global Level', *Journal of Water and Health,* vol. 5, no. 4, pp 467–80.

Hutton, G. and Bartram, J. (2008) 'Global Costs of Attaining the Millennium Development Goal for Water supply and Sanitation', *Bulletin of the World Health Organization*, vol. 86, no. 1, pp 13–19.

IDB (Inter-American Development Bank) (2010) *Drinking Water, Sanitation, and the Millenium Development Goals in Latin America and the Caribbean*, Washington, DC.

Kariuki, M. and Schwartz, J. (2005) *Small-Scale Service Providers of Water Supply and Electricity*, World Bank Policy Research Paper 3727, World Bank, Washington, DC.

Marin, P. (2009) *Public-Private Partnerships for Urban Water Utilities. A Review of Experiences in Developing Countries*, PPIAF Trends and Policy Options no. 8, Washington, DC.

Mehta, M. (2008) *Assessing Microfinance for Water and Sanitation: Exploring Opportunities for Sustainable Scaling-Up*, A report to the Bill and Melinda Gates Foundation, Seattle.

OECD (2006) *Infrastructure to 2030: Telecom, Land, Transport, Water and Electricity*, OECD, Paris.

OECD (2009a) *Managing Water for All: An OECD Perspective on Pricing and Financing,* OECD, Paris.

OECD (2009b) *Private Sector Participation in Water Infrastructure*, OECD Checklist for Public Action, OECD, Paris.

OECD (2010a) *Pricing Water Resources and Water and Sanitation Services*, OECD, Paris.

OECD (2010b) *Innovative Finance Mechanisms for the Water Sector*, OECD, Paris.

OECD (2011a) *Benefits of Investing in Water and Sanitation: An OECD Perspective*, OECD, Paris.

OECD (2011b) *Meeting the Challenge of Financing Water and Sanitation: Tools and Approaches,* OECD, Paris.

OECD (2011c) *Ten Years of Water Sector Reforms in Eastern Europe, the Caucasus and Central Asia*, OECD, Paris.

OECD–DAC (Organization for Economic Co-operation and Development – Development Assistance Committee) (2013) *Financing Water and Sanitation in Developing Countries: the Contribution of External Aid*, OECD, Paris.

Vammalle, C. and Hulbert, C. (2013) 'Sub-national Finances and Fiscal Consolidation: Walking on Thin Ice', OECD Regional Development Working Papers, February 2013, OECD Publishing, http://dx.doi.org/10.1787/5k49m8cqkcf3-en, accessed 15 February 2014.

World Bank PPI (Private Participation in Infrastructure) (2014) Database http://ppi.worldbank.org, accessed 15 February 2014.

Part II
Managing water and sanitation services

4 Urban water management in São Paulo to achieve universal access to water supply and sanitation services

Dilma Pena

Introduction

This chapter describes the overarching policy and main programmes of the state of São Paulo in Brazil to achieve universal access to water supply and sanitation services.[1] The guiding principle behind these policies is that access to these services is a fundamental human right. At the same time, sanitation is essential to safeguard environmental sustainability in the state. The chapter describes the socio-economic context and provides information on the availability of water resources and the organizational structure for managing them. It also outlines the principles, stages and institutional framework for managing water and sanitation policy. Special attention is given to the Companhia de Saneamento Básico do Estado de São Paulo (State Water Utility – SABESP), which is the main organization responsible for implementing São Paulo's water and sanitation policy. A few remarks about the learning process over recent decades are made to conclude the chapter.

Background

São Paulo is the most populated state in Brazil, with 43.6 million inhabitants in 2013 (IBGE, 2013). It is home to 21.7 per cent of the country's population, distributed in 645 municipalities. However, its territory of 248,210 km^2 is only 2.9 per cent of Brazil's total. Therefore, São Paulo has a relatively high population density – 176 inhabitants per square kilometre – which is amplified by an urbanization rate of approximately 96 per cent. However, the population of São Paulo State is growing at a smaller rate than that of the country as a whole, a trend which has been seen in the last three census periods. Between 2000 and 2010, population growth was only 1.09 per cent per year, less than the national average of 1.17 per cent (SPDR, 2011). São Paulo State is the country's economic engine, concentrating about one-third of the Gross Domestic Product and exports of Brazil (IISD, 2012). It is the largest industrial centre in the country and home of a growing service economy.

Since the 1970s, the state government has made substantial investments in water supply and sanitation infrastructure, the main outcome of which has

Table 4.1 Percentage of households with access to water supply and sanitation services in São Paulo, in other states and in Brazil for 2010

	Water supply (general network) (per cent with access)	*Sewage collection (general network) (per cent with access)*
São Paulo	96.4	86.8
Other states	85.0	46.5
Brazil	87.5	55.5

Note: The data are from urban zones (cities and districts); the rural population is not included

Source: Pena (2012, p.145, adapted from IBGE (2010))

been high network coverage of water supply and sewage collection (Table 4.1). Despite advances in network coverage, there is still much work needed to improve sewerage services and to expand wastewater treatment and disposal. This is especially evident at the periphery of cities, which have serious problems derived from the informal occupation of land by *favelas* (shantytowns), frequently in areas prone to flooding and mudslides.

The hydrographic division of Brazil adopted by the Agência Nacional de Águas (National Water Agency (ANA, 2006)) has 12 watersheds. The state of São Paulo is located in two of these watersheds: the Parana River and the Atlantic Southeast. To manage the state's water resources, 22 Unidades Hidrográficas de Gerenciamento de Recursos Hídricos (State Water Resource Management Units) were created (Figure 4.1). Stream water available to the state of São Paulo is estimated at 9,800 m^3/s considering both local water and water imported from upstream states. Municipal water supply represents the greatest share of total water demand in São Paulo (39 per cent), followed by industrial use (28 per cent) and irrigation (27 per cent). Other uses amount to 6 per cent (ANA, 2005).

With surface water availability of 2,209 m^3 per person per year, São Paulo State is classified as 'poor' according to the Falkenmark Water Stress Indicator (Falkenmark and Lindh, 1976), but scarcity is not evenly distributed (CERH, 2006). The most serious water stress is found in the Metropolitan Region of São Paulo (RMSP) (Piracicaba and Alto Tietê river basins, Figure 4.1) where 60 per cent of the state's population lives with a water availability of less than 200 m^3 per person per year (Figure 4.2).

The concentration of population at the headwaters of the Piracicaba and Alto Tietê rivers has a direct impact on the quality and quantity of water in downstream areas of these two rivers, which cross the state from east to west. The Tietê River, however, undergoes a significant self purification[2] process which improves river water quality downstream. The Alto Tietê basin is divided into six sub-basins – Billings-Tamanduatei, Cotia-Guarapiranga, Tietê-Cabeceiras, Juqueri-Cantareira, Penha-Pinheiros and Pinheiros-Pirapora – all within the RMSP (Figure 4.1). In these sub-basins, water availability is only 200 m^3 per person per year. This level of availability is lower than that of Northeast

Figure 4.1 Hydrographic divisions of São Paulo State

Source: (Pena, 2012)

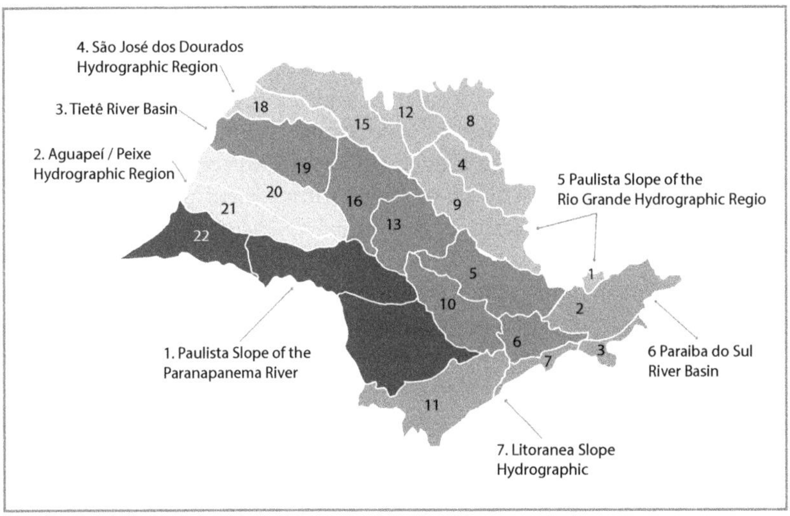

Figure 4.2 The hydrographic basin Alto Tietê and the Metropolitan Region of São Paulo

Source: Courtesy SABESP

Brazil, considered the country's driest region. These limitations present a great challenge for the state of São Paulo.

The most pressing water challenges in the RMSP are water scarcity; pollution of water sources and reservoirs that are surrounded by built-up areas without proper collection of sewage; and urban flooding. In response to these challenges, the public policies of the state government have been designed to guarantee water supply to the population; collect, treat and dispose of sewage; and collect and dispose of solid waste. The overarching environmental priorities are the reduction of water and air pollution, as well as preservation of environmentally sensitive areas near water sources.

Water and sanitation policy in the state of São Paulo

The main water policy objective in the state of São Paulo is to provide universal access to water supply and sanitation services by 2020. In order to achieve this objective, the state's Secretaria de Saneamento e Recursos Hìdricos (Secretary of Sanitation and Water Resources) relies on Companhia de Saneamento Básico do Estado de São Paulo (State Water Utility – SABESP). While this arrangement follows federal and state sanitation law, the provision of water supply and sanitation services is a municipal responsibility according to the Constitution. Under this framework, São Paulo State government policy is to promote different municipal and private operating options, including SABESP as a state company with mixed capital.

The Secretaria de Saneamento e Energia (State Secretary for Sanitation and Energy) coordinates the state sanitation policy. This policy separates planning, regulatory and supervisory responsibilities, which are managed by state agencies, from the actual delivery of services, which are provided either by the state government (largely by SABESP under a concession arrangement with individual municipalities) or directly by municipal service companies and private companies. The primary source of financing for capital investment for and operation and maintenance of services is revenue generated by water supply and sanitation tariffs charged to customers. However, in recent years, to accelerate the implementation of an ambitious investment programme, there has been increasing and increasingly diversified financing from national and international sources.

São Paulo State's sanitation policy assigns the highest importance to planning, which is intended to provide a guiding framework to all water agencies and stakeholders. Therefore, state and local governments formulate local and regional sanitation plans by river basin covering water supply, sewage collection and treatment, urban drainage, and collection and disposal of solid waste. An important advance in planning for water supply and sanitation was the decision to integrate the entire metro area of São Paulo (the Paulista Macrometropolis) in comprehensive investment planning.

Universal water supply is oriented towards matching population growth, while maintaining the high network coverage already achieved; increasing efficiency throughout the system; augmenting service to areas suffering from deficient service; and providing service to areas lacking it altogether. These latter two areas include isolated communities and risky environments that have been built up or occupied by squatters, such as *favelas*, which need comprehensive urbanization reforms. In recent years, most of the investments have gone towards expanding the collection and treatment of sewage; nevertheless, it is expected that the Paulista Macrometropolis will achieve universal drinking water coverage before 2020.

The state has developed a comprehensive pollution control and water quality management strategy to ensure the protection and recovery of water supply sources. These efforts are supported by legislation and special investment programmes. The overall strategy includes several programmes to reduce intolerable increasing levels of pollution in rivers, reservoirs and groundwater. The efforts are coordinated across the entire water cycle, including urban drainage in cities and the installation of sewage collection and treatment facilities.

According to the latest census (2010), the number of households in São Paulo still lacking access to water and sanitation services is lower than the average number for Brazil (Table 4.2) (IBGE, 2010). The state government's policies to improve living conditions for the poorest 40 per cent of the population (slum upgrading) have produced tangible results, as the share of this population without access to water and sewer networks is smaller than in the rest of the country. In São Paulo, the deficit is about 170,000 households without connection to a water network (5.4 per cent of the state total) and 630,000 households without connection to a sewer network (20.2 per cent of the state total). For Brazil as a whole, there is a deficit of more than 4.5 million (22 per cent) and 12 million (59 per cent) households without network connections to water supply and sewerage, respectively.

In achieving universal coverage of sanitation services, SABESP has encountered some resistance to voluntary connection to existing sewerage networks because of the incremental investment cost that low-income households incur to reconfigure the internal plumbing of their houses to discharge into the public network. To mitigate the financial burden, the state government created a programme targeting poor families – *Liga na Rede* (Connect to the Net) – which provides an investment subsidy and manages the incremental investment cost.

Table 4.2 Deficit in access to sanitation services in São Paulo and in Brazil (in the poorest 40 per cent of the population)

Region	*Water network*		*Sewer network*	
	Deficit (households)	*Per cent*	*Deficit (households)*	*Per cent*
São Paulo	169,487	5.4	629,947	20.2
Brazil	4,534,043	22.0	12,102,977	58.6

Source: (Pena, 2012, p.150, adapted from IBGE, 2010)

Policy and institutional framework

The Federal Constitution of 1988 included the policy principle of the right to essential public services for all citizens and the obligation of the federal government to guarantee that this right is fulfilled. In the case of water supply and sanitation services, the Constitution assigns the federal government the exclusive right to 'create guidelines for urban development, including sanitation' (Article 22); to define areas of common attributions for the three administrative jurisdictions of the country (federal, state and municipal) to promote improved sanitation conditions (Article 23); and to combat the causes of poverty by integrating the underprivileged sectors (Article 23). The existing legal framework establishes areas of competence for the three levels of government. It also includes the obligatory nature of the local sanitation plan and the establishment of regulatory bodies, both of which are under the state government.

The financing of operations and investments of water and sanitation companies in São Paulo State is supported by tariff revenue generated by services associated with water supply and with collection and treatment of sewage. Tariffs are set by a semi-autonomous state regulator (law no. 11.445/97). There are also subsidies that can be used to reduce charges to low-income customers. In addition, direct transfers of government funds may be used to ensure the financial viability of the service provider. Regulation and supervision of water supply and sanitation services has been strengthened following the positive incentives created by encouraging private sector participation in the water supply and sanitation sector in the late 1980s.

There are two federal laws that have had a large impact on the regulation of water supply and sanitation services. The first is law no. 11.107 of 2005, which refers to the constitution of public consortia and cooperative agreements, and voids contracts that transfer the responsibilities of planning, regulation and supervision of services to water utilities. The second is law no. 11.445 of 2007, which established national guidelines for water supply and sanitation services. This law states that contracts for water supply and sanitation services must abide to regulatory norms, and regulatory agencies should be in place with independence in decision-making and administrative autonomy.

This innovative legislation has been important because it allows for (i) greater integration between water supply and sanitation companies and other service providers; (ii) the adaptation of companies to the regulatory norms; and (iii) public disclosure of information about company results and performance to regulators and customers. The changes in the legal framework have caused an adjustment in SABESP. They have stimulated for-the-market competition of water services in different municipalities, generating new responsibilities and obligations but also new business opportunities for SABESP.

In São Paulo, these laws came into effect in 2007 with the approval of the complementary law no. 1.025/07, which created the Agência Reguladora de Saneamento e Energia do Estado de São Paulo (Regulatory Agency for Sanitation and Energy of São Paulo State – ARSESP). This agency has planning, regulatory and supervisory powers over the water supply and sanitation, and electricity sector, creating an orderly and predictable environment for developing the sector in the state.

There are several reasons to justify a joint regulatory mandate for water supply and sanitation and electricity services under a single agency. These include economies of scale, the scope to combine administrative and financial activities, and the sharing of institutional knowledge and cross-sectoral learning. Electricity and water supply and sanitation effectively face similar conceptual and operational problems. Both sectors share the characteristic of natural monopoly, mainly in transmission and distribution grids. To reach individual customers, both require infrastructure networks whose duplication would be very expensive. A regulator therefore faces similar conceptual issues relating to the efficiency, management quality and operation of both of these networks.

ARSESP has sectoral divisions and operating councils (in which municipalities participate) for electricity and for water supply and sanitation. Its legal framework establishes a fixed mandate, stable management and independence in decision-making, as well as administrative and financial autonomy. ARSESP is expected to guarantee legal stability with respect to the implementation of norms and to provide legal security for concession contracts signed by municipalities with service providers, as well as to supervise service quality as defined by the service contracts. More importantly, ARSESP is responsible for setting service tariffs to be charged to consumers as well as for processing consumer complaints. ARSESP's budget is financed by 0.5 per cent of the tariff revenue for electricity and water supply and sanitation and by state budget allocations to cover staff costs.

Brazil's experience with the regulation of public utilities is that both consumers and service companies benefit, particularly because the provision of infrastructure services such as water supply and sanitation and electricity requires assets that are amortized over a long period of time. These services are dependent on decisions of a technical nature, and should be less susceptible to interference from short-term political decision-making. The regulatory approach of ARSESP is guided by three main principles: (i) *efficiency* – rates should reflect the efficiency of each service to each client (type, connection, geographic location, volume, socio-economic status); (ii) *equality* – rates should be determined according to the capacity of each client category to pay; and (iii) *simplicity* – the categories used to classify clients should be simple and easy to explain to everyone.

Water allocation is guided by the Constitution of 1988 and the Water Law approved in 1997 after several years of negotiation. The Water Law adopted modern management principles and instruments, among them: (i) water is a public good and a natural resource with an economic value; (ii) in conditions of scarcity, priority should be assigned to human and animal consumption; (iii) water resources should be managed by river basins integrating multiple uses; and (iv) water management should be decentralized following a principle of subsidiarity and it should be participatory. The law created the National Water

Agency (ANA) with a mandate to implement it, including: (i) responsibility for the administration of water rights and water rates in federal rivers; (ii) classification of rivers according to use; (iii) strategic water resources planning; and (iv) management of the water resources information system. States have their own water resources laws, which in the case of São Paulo, was approved several years before the federal law. Apart from the ANA, the institutional framework for water resources management comprises the National Water Council, water resources councils at state level, river basin committees and river basin agencies. The main rivers feeding the water supply system of São Paulo require federal intervention to allocate water rights. This condition presents a number of issues and potential conflicts because of competing demands from different users in different river basins, cities and municipalities. Interstate rivers are controlled at the federal level by ANA.

Water quality in São Paulo State is monitored by the Companhia de Tecnologia de Saneamento Ambiental (São Paulo State Environmental Agency – CETESB), which operates a surface water monitoring network created in 1974 as a result of state law no. 118 of 1973 (CETESB-Infoaguas, n.d.). Water quality monitoring programmes should take into consideration population and industrial growth, including specialization and diversification of industry and agroindustry. The goal is to provide incentives to water pollution control programmes, especially those targeting water quality at the source. Since 1978, CETESB has released water quality data in its annual reports (CETESB-Infoaguas, n.d.). These reports have several objectives apart from informing the public about water quality parameters and environmental conditions across the state, among them: (i) help define actions to control pollution and improve water quality in rivers and lakes; (ii) provide information to formulate public policies; and (iii) cooperate in applying the tools outlined in the state's water resource policy. According to the most recent annual reports, the main rivers in the state of São Paulo have reasonable water quality in the upstream part of the basin; however, in the downstream part, from the border of São Paulo to other downstream states, the rivers are unfortunately classified as being of extremely poor quality.

SABESP's role in implementing São Paulo's water and sanitation policy

SABESP is the largest water supply and sanitation company in the Americas. As of 2012, it supplied water to 27.7 million people – 24.2 million directly and 3.5 million on a wholesale basis – about 67 per cent of the urban population of São Paulo State. It operates 214 water treatment plants and 502 sewage treatment stations. The water distribution network and sewerage network are 67,600 km and 45,800 km long, respectively. The number of employees is about 15,000 (SABESP, 2013).

Operation, maintenance and expansion investments for the area covered by SABESP are undertaken under municipal concessions lasting 30–40 years, depending on their economic and financial viability. Under this scheme, SABESP directly operates water supply and sanitation services in 363 out of 645 municipalities in São Paulo State, and supplies bulk water to another six. Of these municipalities, five also receive sewerage services from SABESP (SABESP, 2013).

In municipalities not covered directly by SABESP under concession contracts, the state government supports implementation of policies and actions to improve and expand water supply and sanitation services according to municipal plans. The state also provides technical and financial support, particularly to reduce the deficit in access to sanitation services, to municipalities with less than 50,000 people (about 125 of the 645 municipalities in the state).

Progress to achieving universal access to water and sanitation made over the past decade have been possible thanks to a combination of investments, political determination and effective planning. Over the past ten years, more than 2.8 million people have gained access to potable water in areas covered by SABESP, making a total of 24 million with access in these areas. Now that the challenge of universal water supply has almost been achieved, attention is focused on expanding collection and treatment of sewage. Between 2002 and 2013, the indicator for sewage collection increased by 7 per cent, which translates to an additional 4.8 million people with access to this service. Currently, sewerage services reach around 21.5 million people (86 per cent). The indicator for sewage treatment also grew significantly: from 62 per cent in 2002 to 78 per cent in 2013 in the municipalities where water supply and sewerage services are under the responsibility of SABESP.

Moreover, since 2007, SABESP has begun providing services in a more flexible manner thanks to federal and state laws allowing to: (i) renewing expired concession contracts with municipalities; (ii) engaging in public–private partnerships (PPP); and (iii) financing and leasing assets. State Law no. 1,025 allows SABESP to associate with the private sector (e.g. specific purpose companies) in bidding processes for projects such as new water producing systems, to apply subsidies and to work in the energy and waste sectors. This flexibility is aimed at achieving greater efficiency in the sanitation sector. In 2010, SABESP signed a concession agreement with São Paulo State and the city of São Paulo to provide water supply, sewerage and wastewater treatment services in the city of São Paulo. At the time, this contract represented 55 per cent of the company's total income.

Investments for universal access to water and sanitation services

Investment plans covering urban areas with legal land titles where SABESP operates are structured to complete the infrastructure needed to supply universal sanitation services. Water and sewage treatment services have already achieved universal coverage in 206 municipalities, including the entire RMSP. The objective for universal access in the remaining municipalities under concession contract with SABESP is expected to be achieved by 2020 (SABESP, 2014).

To achieve universal access objectives, SABESP has substantially increased its investments. Average annual investments between 2008 and 2011 were R$2.2 billion (at 2012 prices), more than double the annual average between 2004 and 2007, estimated at R$1.0 billion. In 2012 and 2013, the annual investments were R$2.5 and R$2.7 billion, and in the coming years it is expected that these investments will increase further.[3] SABESP's investment plan covers a portfolio of programmes to meet two development objectives: (i) to guarantee the security of water supply; and (ii) to achieve universal coverage of sanitation services and control pollution with adequate infrastructure for collection, treatment and environmentally safe disposal of sewage (SABESP, 2014).

The improvement of water supply and sanitation services cannot be achieved without reducing pollution levels in reservoirs, rivers and groundwater that supplies water to the RMSP. This is a high priority for which the state government created the *Vida Nova/Mananciais* (New Life/Springs) programme. The programme has been designed to improve the quality of life for the population that lives close to these water sources, and is a typical example of a comprehensive cooperative programme carried out by various governmental agencies. The first stage of the programme provides interventions in sanitation, urbanization, urban infrastructure, and construction of housing, parks and recreational areas in 43 shantytowns and low-income housing areas around the Guarapiranga and Billings reservoirs within the RMSP. The total investment is about R$1.3 billion, which is financed by the federal, state and municipal governments, the Companhia de Desenvolvimento Habitacional e Urbano (Housing and Urban Development Company – CDHU), SABESP and the World Bank. An expansion of the programme is being developed with additional financing for upgrading a greater number of shantytowns.

Finally, the RMSP is being confronted with a serious water shortage. Water allocation from basins that have traditionally supplied the region cannot meet the increasing demands of the population and economic activity. As a result, SABESP must consider new water sources that are subject to complex legal authorizations, agreements with other users and high investment costs. In addition to projects oriented to augment water supply, SABESP is dedicated to control water demand through programmes aimed at reducing water losses, encouraging the rational use of water and promoting water reuse. In the following sections an account is given of major programmes oriented to drinking water and sanitation. The financial considerations are also discussed.

Programmes to guarantee security of water supply

Metropolitan Water Programme

The goal of this programme is to ensure the security of potable water supply for the RMSP. Water demand in the RMSP is about 60 m^3/s, but it exceeds 80 m^3/s

when considering the entire upper Tietê River basin. The programme aims to increase water production, transportation and distribution capacity by 13.2 m^3/s by 2020. Improvement in the quality of treated water is also a major policy goal. An investment of R$1.7 billion was arranged for this end, with financing from the Caixa Econômica Federal (Federal Economic Bank – CEF), which administers the Fundo de Garantia do Tempo de Serviço (Guarantee Fund for Service Time).[4] This programme encompasses the PPP of the Alto Tietê Production System and the São Lourenço Production System. The Alto Tietê PPP project concluded in 2011 and increased water treatment capacity of the Alto Tietê Production System from 10 to 15 m^3/s. The São Lourenço Production System, located 80 km from greater São Paulo, will have a production capacity of 5 m^3/s and will serve almost 1.5 million people. It is expected to become operational in 2018 (SABESP, 2014).

Coastal Water Programme

The goal of this programme is to increase water production, improve water quality, increase water conveyance capacity and improve capacity of reclaimed water in the Baixada Santista metropolitan area. The coastal region of São Paulo (Baixada Santista) has a population of about 1.7 million in nine municipalities, but the population reaches up to 3.6 million during the summer season. It has been estimated that an investment of approximately R$500 million is necessary to increase capacity of water supply by 3.2 m^3/s.

Corporate Programme to Reduce Water Losses

The Corporate Programme to Reduce Water Losses began in 2009 with the initial goal of reducing losses from about 35 per cent to 22 per cent by the year 2019. In 2011, the level of non-revenue water was 35.6 per cent, compared to the Brazilian average of more than 37 per cent (Pena, 2012). SABESP started this programme, which was boosted through an international cooperation agreement with the Japan International Cooperation Agency (JICA) in 2012.

Programme for Rational Use of Clean and Reclaimed Water

The Programme for Rational Use of Clean and Reclaimed Water is aimed at reducing water consumption among the biggest users, especially in the public sector. Reclaimed water is produced from treating sewage and has limited uses apart from domestic use; for example, in industrial and power plant cooling. SABESP has started to expand the output of reclaimed water. One related project, implemented in 2012, is the Aquapolo Ambiental Project which supplies reused water to a petrochemical complex in the City of Mauá. Currently, 450 l/s of reused water is available, with the expectation that this figure could grow to 1,000 l/s by 2020 (SABESP, 2014).

Programmes to achieve universal coverage of sanitation services

Developing infrastructure for collection, transport and treatment of sewage is more challenging than supplying drinking water. The market where SABESP operates is a vast territory in dispersed localities, with an enormous number of small rural cities and intermediate-sized cities, as well as the RMSP itself, with its 20 million inhabitants. Universal collection and treatment of sewage will allow recovery and progressive reduction of contaminated run-off into water bodies. With the Tietê Project, SABESP has been dedicated to reducing pollution in the Tietê River for more than 22 years, and this is the most important and longstanding structural water project in the RMSP, aimed at providing collection services and sewage treatment for 20 million inhabitants. Its actions are complemented by the *Córrego Limpo* (Clean Stream) Programme, with the objective to clean up numerous streams; the *Onda Limpa* (Clean Wave) Programme, which cleans up surface water for the São Paulo coastal areas; the *Mananciais/Vida Nova* (Spring/New Life) and *Liga na Rede* programmes, mentioned above; and the Água *é Vida* (Water is Life) Programme, which provides sewage treatment to low-income communities. The Clean Wave and Clean Stream programmes are discussed below.

The Tietê River Decontamination Programme

This programme began in 1992 to clean up the then biologically dead upper reach of the Tietê River. This is SABESP's largest investment in a sanitation programme. The programme is divided into four stages over 28 years, and the total investment is projected to be US$4.5 billion. When the Tietê Project is finished, the infrastructure being installed will be able to treat the sewage generated by 18 million people and will avoid the discharge of 1,030 t/day of organic waste into the rivers of the RMSP. The first and second stages of the project, which were completed in 2008, were designed to expand the sewerage system and increase sewage treatment capacity. Sewage collection increased from 70 per cent to 84 per cent of the population of the RMSP, while sewage treatment increased from 24 per cent to 70 per cent of collected wastewater. More than 8.5 million people in the metropolitan area now have access to sewage collection and treatment as a result of this project. The infrastructure built during the past 19 years is equivalent to servicing a city the size of London.

The third stage of the project began in 2010, with an investment of US$2 billion, financed by the Inter-American Development Bank (IDB), Banco Nacional de Desenvolvimento Econômico and the CEF. The aim is to provide collection services for more than 1.5 million people and sewage treatment for 3 million new inhabitants living in peripheral areas of the RMSP, which happen to have the highest population growth rates. With all three stages completed, 11.5 million people in the metropolitan area will have comprehensive sanitation services by 2016 (SABESP, 2014).

SABESP is now planning the fourth stage of the Tietê Project, with an estimated investment of US$4 billion. This last stage foresees works to: (i) consider adjustment and expansion of the existing sewage collection and transport system in the central region of RMSP; and (ii) consider the implementation of new networks in the remote, low-income peripheral areas of the RMSP. This stage will be highly complex. It will require large amounts of financial resources for land expropriation and re-urbanization. The goal for the fourth stage is to increase the level of sewage collection to 95 per cent of the population by 2018 and 100 per cent by 2024 in the cities of the metropolitan region where SABESP operates.

Clean Stream Programme

This programme is being developed in conjunction with the city government. The goal is to clean up the city of São Paulo's main rivers (Tietê and Pinheiros), to be achieved through improvement of the sewage system and elimination of discharge of raw sewage into these rivers. The programme also covers cleaning up the shores of the rivers and streams, and removing and relocating people occupying stream margins. Between 2007 and 2013, 146 streams were cleared in the capital, benefiting nearly 2 million people. Through December 2014, an additional 20 larger streams are expected to be cleared. A recent survey showed that river water quality has improved at 49 sampling sites (SAPESP, 2014).

Clean Wave Programme

This programme is centred on cleaning up rivers and canals in the coastal areas of the state, and the improvement of beaches and their surrounding areas. It is expected these works benefit the tourism industry and as a result, the regional economy. On the Santos coast (Baixada Santista) activities started in 2007 and were completed in 2013, with a total investment of R$1.9 billion. SABESP is now working on the second stage of the project valued at R$700 million, which aims at collecting and treating sewage in all the cities of Baixada Santista to achieve a coverage of 88 per cent by 2018. In the state's northern coastal region, the goal of the programme is to increase sewage collection and treatment in the cities of Caraguatatuba, Ilhabela, São Sebastiao and Ubatuba from the current level of 49 per cent to 85 per cent by 2015. This requires an investment of R$500 million. Before the programme was initiated, only 30 per cent of the population had access to sewage collection and treatment. About 600,000 people will benefit from these investments, both residents and tourists.

New Life Programme

This programme was established in 2008 to recover and protect dams supplying water to the RMSP that are surrounded by informal housing areas. It involves interventions in 43 shantytowns in the southernmost sub-basins of the Guarapiranga and Billings reservoirs. The programme will benefit 50,000 families. SABESP's responsibility in this project is to expand and improve water and sewage systems. The programme is expected to be completed by 2015, with a total investment of R$1.6 billion. Funds come from several sources: SABESP; federal, state and local governments; CDHU; and the World Bank.

Financial sustainability

SABESP's performance is closely tied to sound financial management. The company uses its own funds and judicious debt management to finance its investment plan. Most of its debt (60.2 per cent) has been raised with official agencies of local and foreign governments, as well as with multilateral organizations. The lower costs of long-term debt helps to stretch out repayments and minimizes pressure on the company's cash flow (SABESP, 2014). The market has recognized the company's efficiency and performance throughout the years. In 2011, Standard & Poor's raised the global credit rating for SABESP from BB to BB+ and increased its national rating from brAA to brAA+. This means that SABESP's risk rating fell, thus allowing the company to borrow money at lower interest rates. Since then the outlook for both ratings was 'positive' until in May 2014 the risk rating agency revised the outlook from positive to stable as result of the expected financial impact of the prolonged drought affecting the RMSP. Before this adjustment, SABESP and the state government implemented a programme to encourage lower water usage by its customers by giving a 30 per cent discount on the bills of clients who reduced their consumption by 20 per cent. This measure is expected to reduce net revenue for 2014 (Reuters, 2014).

Conclusions

The progress made in Brazil's water supply and sanitation sector over the past 50 years is undisputable, despite any errors made and notwithstanding the challenges yet to be overcome. The achievements are even more impressive when compared to countries in the developed world, which have taken 150 years to achieve similar network coverage levels. During the past five decades, the pressures from population and urban growth have driven an accelerated expansion of infrastructure, the creation of new institutions, the development of human capital, and large investments and spending. Of course, there are still important challenges – the two main ones that São Paulo State faces in the coming years being increasing service to the disadvantaged groups living in urban and rural areas, and reversing the intolerable environmental degradation and water pollution that affect health and quality of life in general.

The Plano Nacional do Saneamento (National Sanitation Plan – PLANASA) implemented during the period of the 'Brazilian economic miracle' from the early 1960s was based on the premise that, in time, state water and sanitation companies would become financially self-sufficient as they increased their scale of operations, became more efficient and charged adequate rates for their services. While the scale of their activities grew, efficiency and rates did not live up to the plan. In reality, these companies had to reduce spending, and service quality thus deteriorated. The termination of the PLANASA model 30 years later following the breakdown of the Brazilian economy in the 1990s was, then, inevitable because of the companies' limited capacity to pay, their mounting debt and the absence of new financing.

After the federal government phased out PLANASA, state governments adopted different strategies: some of them attempted to grant concessions to the private sector, and others returned the operation of services to the municipalities. São Paulo State is among those states that opted for a strong state water company with a mixed capital structure and other forms of private sector participation to guarantee its public mandate, breaking with the false notion that the water and sanitation sector could not use private funds to achieve public policy objectives. This path has had its favourable elements. For example, unlike many other infrastructure projects in Brazil, the Tietê Project kept going through hyperinflation, financial crises and changing political leaders. This shows a solid foundation for success, based on social acceptance and market realities.

Water supply and sanitation development goals for the coming years are ambitious and will be difficult to achieve in terms of technical complexity, costs, and social and environmental implications (SABESP, 2014). It will be necessary to increase the pace of investment in the sector to more than R$2.5 billion per year (R$12.8 billion in 2014–2018). The great challenge for public policy will be to balance water and sanitation tariffs, improve efficiency and optimize investment plans. The company will need wise regulatory guidance backed by broad civil society participation so that it will be politically and financially viable. Finally, the implementation of non-structural measures, such as conservation and water reuse, need to be reinforced.

Sustaining universal services is a critical expression of development that reflects the fundamental aspirations of a society. Universal services are the result of political consensus, which is never easy to attain. For that reason, what is required is a sectoral planning policy supported by the central government that includes clear development objectives, a sound and transparent financial policy and incentives for efficient operation and which is induced by public regulation. In addition, the sector needs well-structured, innovative service companies that look for the best technological solutions for each demand situation and employ the best business practices to manage their operations and assets. All of this must occur in an environment that includes widespread consultation with citizens and their active participation. This is the basis for universal sanitation services in Brazil, and indeed anywhere.

Notes

1 This chapter draws substantially from an essay by the author published in 2012 (Pena, 2012). The editors are grateful to the author and CAF for providing their authorization to reuse the material.

2 Self-purification of natural water systems is a complex process that often involves physical, chemical and biological processes working simultaneously with the result of a better water quality downstream.
3 On 14 March 2014, the rate of exchange was 2.36 Brazilian reals against the United States (US) dollar. One year before, it was 1.96 (*The Economist*, 15 March, p. 84). Between 2010 and 2013 the exchange rate fluctuated between 1.6 to 2.0 Brazilian reals, on average, to 1 US dollar.
4 Public Employees' Pension Fund.

References

ANA (Agência Nacional de Águas) (2005) *Cadernos de recursos hídricos: disponibilidade e demandas de recursos hídricos no Brasil,* SPR, Brazil.

ANA (2006) Plano nacional de recursos hídricos, Brasilia, www2.ana.gov.br/Paginas/servicos/planejamento/planoderecursos/pnrh.aspx, accessed 8 June 2014.

CERH (Conselho Estadual de Recursos Hídricos) (2006) *Plan estadual de recursos* hídricos 2004/2007: *resumo*, DAEE, São Paulo.

CETESB-Infoaguas (n.d.) 'Sistema de Informação Info ÁGUAS', www.cetesb.sp.gov.br/agua/infoaguas/131-sistema-de-informacao-infoaguas, accessed 5 June 2014.

Companhia de Saneamento Básico do Estado de São Paulo (SABESP) (2014) 'Sustainability Report 2013', www.sabesp.com.br/sabesp/filesmng.nsf/376A20693D2B063683257CD70054D2AF/$File/RS_ingles_final.pdf?OpenElement, accessed 7 June 2014.

Falkenmark, M. and Lindh, G. (1976) 'How can we cope with the water resources situation by the year 2015', *Ambio*, vol. 3, pp. 114–22.

IBGE (Instituto Brasileiro de Geografia e Estatística) (2010) *Censo 2010*, http://censo2010.ibge.gov.br/, accessed 31 March 2014.

IBGE (Instituto Brasileiro de Geografia e Estatística) (2013) 'Estimativas populacionais para os municípios brasileiros em 01.07.2013', ftp://ftp.ibge.gov.br/Estimativas_de_Populacao/Estimativas_2013/estimativa_2013_dou.pdf, accessed 7 June 2014.

IISD (2012) *Sustainable Public Procurement in the São Paulo State Government: An In-Depth Case Study: TKN report*, International Institute for Sustainable Development, São Paulo.

Pena, D. (2012) 'The technical and management perspective: social demands, longterm actions an efficient management in São Paulo', in A. Mejía-Betancourt and M. Pardón (eds), *Equity and Social Inclusion in Latin America: Universal Access to Water and Sanitation*, Reflections on Social and Environmental Policies Series, Development Bank of Latin America (CAF), Development Bank of Latin America, Caracas.

Reuters (2014) 'Companhia de Saneamento Básico do Estado de São Paulo', www.reuters.com/finance/stocks/overview?symbol=SBSP3.SA, accessed 8 June 2014.

SABESP (2013) 'Financial statements as of and for the years ended December 31, 2012 and 2011', www.sabesp.com.br/sabesp/filesmng.nsf/AA66690C7304BD5E83257B4F006BC343/$File/Sabesp_ITR_4Q12.pdf, accessed 7 June 2014.

Secretaría de Planejamento e Desenvolvimento Regional (SPDR) (2011) 'Plano Plurianual 2012–2015', www.planejamento.sp.gov.br/noti_anexo/files/PPA_2012-2015_vol-I.pdf, accessed 7 June 2014.

5 Water services in Lima, Peru

Understanding the linkages between urban development, social inequality and water.

Antonio A. R. Ioris

Introduction

Lima's water supply system is one of the most complex and challenging in Latin America, and has been since the early sixteenth century, when the city was founded by the Spanish on the arid Peruvian coast in an area of small rivers and low rainfall. In the twentieth century, the chaotic expansion of the city caused largely by internal migration from rural areas exacerbated Lima's water shortages. In recent decades, the provision of water services in Lima has also been compromised by lack of investment, inadequate maintenance, operational failures, and shifting political commitments resulting in infrastructure 'bottlenecks'.

Since the early 1990s, in tandem with macroeconomic reforms and international support for the modernization of Peru, a series of water and sewerage investment programmes have been put in place by successive national administrations. Despite institutional and operational improvements in Lima's public water services, however, water shortages and socio-spatial disparities persist. Investment initiatives have produced uneven results, and rather than overcoming water management problems, they have accentuated a pattern of inequality and privilege: water services in poor areas of Lima are worse than in rich areas of the city.

According to performance indicators presented by the World Bank (2009), there has been a considerable improvement of water and sewerage services in Lima between 1994 and 2008. Water coverage expanded from serving 79 per cent to 91 per cent of the beneficiary population, sewerage from 71 per cent to 90 per cent (p. 27), and service continuity from 11.5 hours per day in 1995 to 21.5 hours per day in 2008 (p. 10). These figures, however, conceal great disparities in service quality. Nearly half of the population receives unsafe water, and around half a million people, mainly living in the most deprived peri-urban sections of Lima, still have to purchase water from private vendors at a price 12 times higher than those receiving water through public service providers (AFIN, 2013).

For many poorer residents of Lima, water problems represent a daily struggle. They must use creative coping strategies to secure a minimum amount of water for household consumption. Even in areas now served by a public water utility, the quality and reliability of the service is normally lower than in high-income areas of the city. This suggests that long-lasting, embedded conditions of discrimination and partial social integration persist within the water service sector of Lima.

This chapter considers the mutually reinforcing relations between uneven and inequitable urban development, multiple deprivations, and socio-spatial asymmetries in Lima. It briefly reviews urban water challenges in the broader context of the Latin American megacity, to demonstrate that Lima's water situation, caused by multiple factors including uneven urban development and the selectivity of macroeconomic policies, is not unique. The text then discusses the limited achievements of conventional initiatives by the public sector in light of growing population demands. It concludes by investigating how Lima's water problems, particularly in the periphery of the city, have been exacerbated by unequal treatment of the city's different socio-economic classes – a treatment that has produced human-made water scarcity.

Primary and secondary data presented toward the end of the chapter were obtained through fieldwork carried out by the author in 2009, which included ethnographic research in three locations: Pachacútec, Huaycán and Villa El Salvador (Figure 5.1). The author's original research was followed-up in subsequent years with a second fieldwork in Lima carried out in 2013. The fieldwork focused on the interests and behavioural patterns in the geographical locations of Pachacútec, Huaycán and Villa El Salvador; of their stakeholder groups, and of the institutional frameworks within which they operate.

The dynamics of water scarcity

Before dealing with the concrete experience of Lima, it is necessary to briefly revisit the ontological basis of scarcity and abundance. 'Water scarcity' is not a state that is achieved in isolation; it can be reinforced or mitigated by human action. The primary factor determining water scarcity, in addition to actual physical availability, relates to how water is managed when there are multiple and often competing uses for water resources. Societal responses to water scarcity reflect power asymmetries and social inequalities.

Water scarcity should be recognized as a highly politicized phrase that brings together social demand and water resource availability. Much contemporary public policy is actually the offspring of the 'dialectics of scarcity' (Panayotakis, 2003), considering that scarcity serves as justification not only for large infrastructure works, but also for the adoption of market-based environmental conservation and development policy. The standard regulatory instruments generally adopted by water agencies around the world – water licences, user fees, decision-support systems and payments for ecosystem services – can all be rationalized to a certain extent through the lens of scarcity (Ioris, 2010). By limiting the analysis of water management issues to the rather utilitarian balance between supply and demand, mainstream public policy often neglects

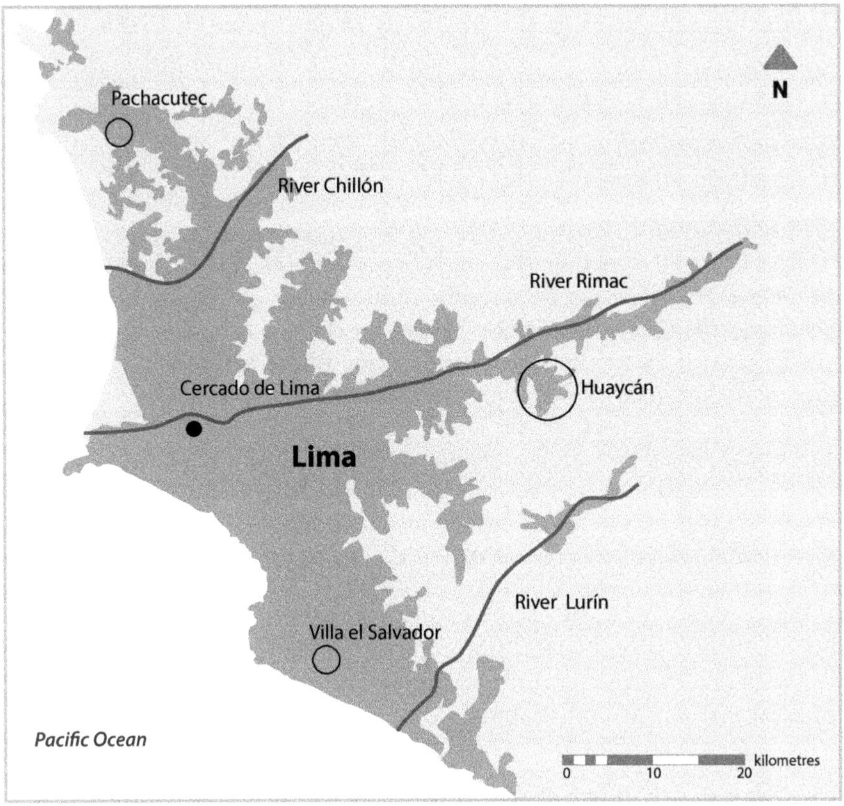

Figure 5.1 The metropolitan region of Lima with built-up areas (shaded), main rivers and the three case study locations

Source: Ioris (2012a, p. 270)

the social construct of water scarcity, i.e. the multiple interactions and disputes between social groups with uneven political forces that result in the appropriation of resources, investments and policies by the stronger sectors of society.

The insufficiencies of Lima's public services still remain unresolved, notwithstanding advances in several parts of the city and in the operation of the public utility. This chapter will show that the discourse of water scarcity continues to be a driving force behind government interventions and public policies, and a powerful impetus for managerial reforms, foreign loans and cooperation projects. Moreover, water scarcity goes beyond the physical insufficiency of resources and reflects the long-term inadequacy of social institutions devoted to the allocation and use of water (Ioris, 2012b).

Escalating scarcities in the unsteady urban space of Lima

During colonial times, access to water in Lima was restricted to the upper classes. For example, the *Juzgado Privativo de Aguas* (Water Tribunal) was established in 1556 to resolve disputes between local nobility, the clergy and artisanal industrialists in regard to water. This system was largely maintained for close to three centuries. In the decades that followed independence in 1824, Lima was a relatively small administrative centre with responsibility for an overwhelmingly rural society. The provision of public water services started in the 1850s, with the sudden but short-lived availability of investment capital produced by the guano bonanza (SEDAPAL, 2003). At the time, Lima was known as the 'garden city' because of the irrigated green areas of the Rímac river valley, which were built on the original water distribution system of the Incas.

With the removal of its city walls in 1872, Lima began to spread into surrounding areas. Early in the twentieth century, it had approximately 100,000 inhabitants, equivalent to 4 per cent of the population of Peru. Gradually, the city attracted more and more people searching for jobs and other economic opportunities. The main period of Lima's expansion was between 1950 and 1980, during which growth occurred at more than 5 per cent per year. Because of the slow pace of investment in affordable housing for poor migrants, an 'informal city' emerged within and around Lima's central urban areas, in large part because these central urban areas already contained some reasonable water infrastructure (Calderón Cockburn, 2005).

The evolution of the informal city

Historically, the majority of new settlers migrating to Lima managed to find accommodation in *tugurios* (slum tenements) established in abandoned town houses in the city centre. From the mid-1950s on, however, the *barriada* became the main option for the incoming population (Barreda and Ramirez Corzo, 2004). The *barriada* (from *barrio* which means neighbourhood) is a form of urbanization by which a plot of land is first obtained usually by illegally occupying the area and then the household is constructed, generally without urban services, such as water, which would be arriving only later. In 1961, 17 per cent of the population of Lima lived in *barriadas* and 30 per cent in *tugurios*; between 1960 and 1968, 111 new *barriadas* emerged, especially in the Northern Cone (Calderón Cockburn, 2005). The distinction between *barriada* and *tugurio* became blurred with the development of *barriadas* in central areas and *tugurios* inside the *barriadas* (Sánchez León *et al.*, 1986).

In 1968, new legislation dispensed with the need for property deeds as a requirement for the provision of water, entitling some *barriadas* to access the public water supply even without deeded land tenure, and serving as an incentive for consolidating and promoting barriadas as a de facto solution for the lack of housing in Lima. Shortly thereafter the Peruvian government changed the name of *barriadas* to *pueblo jóven* (young settlement), and by 1974 Lima had

371 such settlements, which by the end of the 1970s housed some 1.5 million people (Calderón Cockburn, 2005).

Lima has been expanding in three directions known as the Northern, the Southern and Eastern Cones. In the 1980s, Lima's expansion focused on its Eastern Cone (prior expansion had focused to the South), as well as the borders of other cones; most new residents no longer migrated from rural Peru but instead came from older *barriadas* where land availability was becoming increasingly scarce. During the 1990s, the policies adopted as part of the reforms promoted by the Fujimori Administration reinforced the long trend of economic growth with only partial social inclusion that has historically characterized Lima's development. The continuous formation of *barriadas* became tacit urban policy employed to address the city's expansion with limited investment by the government (Calderón Cockburn, 2005). For more than five decades now, Lima's urban perimeter has expanded, with scarcity of urban services, including water and sewerage, serving as the key mediator of social relations (Caria, 2008). Old and new *barriadas* have remained practically the only alternative to many newcomers and second generation residents (Portes and Roberts, 2005).

Water services

In 1962, Lima's public water services were reorganized as a metropolitan utility in charge of overall service management and large-scale investment projects. After 20 years, in 1981, and with the return of democracy to Peru, Lima's public water utility got its current name, SEDAPAL (*Servicio de Agua Potable y Alcantarillado de Lima*; Water and Sewerage Services of Lima). At that time, campaigns for better access to water constituted the highest number of grassroots protests. In the 1980s, around 20 per cent of Lima's residents relied on private water vendors, and of residents with access to public services, over 40 per cent had intermittent supply and as much as 50 per cent of treated water was lost due to leakage and illicit connections (Zolezzi and Calderón, 1987).

SEDAPAL's failings and the outbreak of a cholera epidemic in Peru in 1991 provided the impetus to include SEDAPAL in the list of public utilities to be privatized. SEDAPAL was seen as having inadequate system maintenance, a high level of unaccounted-for water, excess staff, low metering rates and poor water quality – quite an indictment (Corton, 2003). SEDAPAL's preparation for privatization included a large investment programme. The World Bank, together with other agencies, put together a US$600 million financing package to rehabilitate the pipeline network, expand connections, install metering and improve SEDAPAL's efficiency (Alcázar *et al.*, 2000). Three international consortiums prequalified to bid in November 1994, but the privatization process was postponed many times and eventually cancelled in 1997. The main reason for the postponement was that the political price of privatization was not affordable to the president, particularly when his popularity was declining nationwide, and Lima was one of the main political strongholds in his campaign for re-election.

With the decision not to privatize Lima's water services, SEDAPAL expanded a programme of operational recovery and infrastructure expansion during the 1990s, which received considerable public funding. Initially, the programme targeted leak reduction and reduced water demand by installing meters and increasing tariffs to reflect cost of provision. In the more remote areas of Lima, where standard water infrastructure was too costly, alternative projects were adopted with the help of NGOs and with international funding. One of the main initiatives was APPJ (Water for Young Settlements) which was supported by the European Union and aimed to build community systems comprising cisterns filled by water tankers and with distribution from the cisterns through hosepipes. However, even with these community systems and other forms of investments in infrastructure, 70 per cent of the population of Lima remained systematically affected by rationing and intermittent services (ICOM, 2001).

At the end of the Fujimori Administration in 2000, SEDAPAL faced a deteriorating financial situation and disjointed management direction. Under the new Toledo administration, tariffs continued to rise, and frequently became a point of public and regulatory contention (Bonifaz and Malásquez, 2008). The average tariff rose from US$0.39 per m^3 in 2001 to US$0.77 per m^3 in 2008, an increase of almost 100 per cent. Concurrently with increasing tariffs, SEDAPAL massively expanded the installation of domestic water meters, from 6 per cent of households in 1995 to about 70 per cent in 2005 (SEDAPAL, 2005). An additional investment of US$657 million was announced in 2005 through the programme MIAGUA – funded by multilateral foreign loans (Ministerio de Vivienda, Construcción y Saneamiento, 2005).

Water scarcity, however, remained an ordeal for the poor, particularly for those estimated 10 per cent left to buy water from private tankers, who could on average afford only 30 litres per capita per day; in comparison, average consumption in San Isidro, one of the wealthiest neighbourhoods in Lima, was 405 litres per capita per day (Grupo GEA, 2005). While Lima's poorest neighbourhoods are the areas where most of the population was not served by public water services, even in wealthier areas it is possible to find pockets of houses without access, and that depend on alternative sources of water. The severity of the water problem was aggravated by the sustained rate of population increase (around 1.7 per cent per year in 2005) and the rising demand for water after the economy resumed growing during the Toledo administration. Yet, the supply of water did not change to match this demand increase.

In 2007 the programme *Agua para Todos* (APT–Water for All) was launched with an expensive portfolio of new and ongoing projects. Regrettably, APT provides an emblematic example of the interplay between water scarcity, abundance of financial resources and political manipulation. Just a few years after being started, it became evident that APT was serving less laudable agendas. In 2011, a commission was formed by the National Congress and tasked with investigating corruption in the APT programme during the relatively new García administration. One of its specific remits was to scrutinize the apparent mismanagement of SEDAPAL (including the allegation of fraud, incompatible transactions and the abuse of public office in 1,584 engineering works with a total cost of around US$ 1.5 billion). Evidence gathered by the commission revealed that by claiming 'urgency' to the public water services work it was supporting, the APT programme was facilitating adjustments in original infrastructure plans without the need for careful technical justification or without determining the legality of the changes. The commission found that major projects were initiated under the APT programme with only superficial technical plans, and sometimes without even sorting out the ownership of the land where the water infrastructure would be located (Escalante, 2013; Peru21, 2013). Lima's APT programme experience clearly shows that the availability of significant public funds does not necessarily translate into better water services or less water scarcity.

Scarcity and abundance reassessed

Water scarcity in Lima reflects the long-term development of the city since its foundation during the early Spanish colonial period, and continues to be caused by multiple socio-spatial relationships, rather than only by physical or incidental constraints. Despite significant public investment and the expertise of foreign and national operators, water scarcity persists in Lima, and continues to dominate local water policy decisions.

More significantly, the interplay between scarcity and abundance has been used as a political device to manage expectations – downwards – in the impoverished areas of Lima (Ioris, 2012c). Lima, as with virtually all other large Latin American cities, continues to be a 'dual city' when considering the legality of land rights. Some well-known low-income neighbourhoods, where houses were built without legal title several decades back, have now secured a certain level of public services and infrastructure, but generally with a qualitative gap when compared to rich areas of the city. Essentially, a second class of city exists within Lima (Barreda and Ramírez Corzo, 2004). In certain poor neighbourhoods that have existed for several decades but that do not have clarity as to legal title to the land, about half of the households received water for only a few hours per day. Moreover, in the poorest neighbourhoods, even after the installation of pipelines, many residents wait for years before they can afford to connect to the system.

Communication between SEDAPAL and its customers is typically about two things: increases in tariffs and service deficiencies. Because of the tension that these two issues generate, a high level of suspicion toward SEDAPAL exists in Lima's under-serviced neighbourhoods, even in regard to SEDAPAL's initiatives aimed at expanding water and sewerage services to low-income households. This was the case with the introduction of 'condominial' sewerage systems in Lima in 2003, as a low-cost sewerage network technology used in Brazil to reduce significantly investment costs by up to 40 per cent through the joint connection by blocks rather than traditional and more expensive individual connections. Despite the reduced investment and operational costs provided by the condominial approach, the Lima neighbourhoods where it was tried were not satisfied with being offered an alternative solution to their wastewater issues when the central areas of Lima continued to benefit from traditional systems.

In 2006, the condominial approach was essentially abandoned by SEDAPAL. Analysing SEDAPAL's experience with it, it can be compared to what occurred in El Alto, Bolivia, where communities complained about their lack of participation in all aspects of the decision-making around the condominial networks (Laurie and Crespo, 2007). So unfortunately, some of SEDAPAL's important initiatives aimed at helping the poor have inadvertently caused significant resentment and reinforced discrimination in water services between social groups. Although low-cost technologies (Paterson *et al.*, 2007) may act as a normalizer for disadvantaged neighbourhoods, the fact that wealthier neighbourhoods have sewerage networks that use traditional engineering standards is viewed as a social justice issue.

While water scarcity in Lima has largely been caused by the differing economic conditions of the city's neighbourhoods – and exacerbated by the hydrological conditions of limited rainfall and small water courses – the picture is not completely captured by conventional analyses of supply and demand. Multilateral agencies that finance the water systems call for improved economic efficiency, including where possible the adoption of economic incentives, to guide water policy making (Marmanillo, 2006). There is broad agreement among these organizations that reform of the Peruvian water sector, and Lima, seems to be going in the right direction, but that the institutional fragility of national and municipal public agencies and the low cost of financial return on water infrastructure investment are still major obstacles. There are, however, other fundamental problems associated with Lima's unfettered growth, its urban development and accompanying social exclusions, which can be better demonstrated through the following three case studies.

A fluid and sometimes adversarial relationship exists between the water utility, SEDAPAL, and the low-income populations of Lima. Residents of the peri-urban areas of Lima have not been passive when it comes to the troubled operation and expansion of water services. People in these areas have learned to demand water not only as a basic need, but as a legitimate right. Expressing the need for and right to water requires massive mobilization and specific negotiation. Fieldwork undertaken in three relevant areas of Lima, located in different extremes of the metropolitan region, demonstrates how the provision of water and water scarcity has changed through the years, but remains set against a background of urban expansion and politicized socio-spatial relations.

Water scarcity in each of the study areas is closely related to the level of collective organization and the balance of power. Villa El Salvador was founded in southern Lima in 1971 as a *comunidad autogestionaria* (self-administered community). Four decades later, Villa El Salvador still has some symbolic elements of the original organization. However, with time, public participation, in a context of conservative politics and individualist ideologies, has become increasingly bureaucratized and fragmented, which has inevitably affected the struggle for public services such as water. The decline of active popular involvement in Villa El Salvador's public affairs has eroded the possibility of advocating strongly and effectively with the government for essential services such as water services. Interviews with residents demonstrate their belief in the importance of grassroots campaigns to secure and expand water and sanitation in the area. But these same people also expressed frustration with the growing electoral appropriation of popular initiatives.

Pachacútec, in the Northern Cone of Lima and the second study area, is a settlement founded in 2000, when the government transferred a group of people that were then living in Villa El Salvador but could not be permanently settled there. Pachacútec is located in a vast sandy area where the arriving settlers (around 40,000 people) and original residents (10,000) had to organize themselves to demand some level of emergency water supply. The sudden collapse of the Fujimori Administration in the late 1990s left the recently transferred families without the promised infrastructure.

Because of its tenuous situation, Pachacútec soon became a priority area to test alternative systems of water supply and sanitation. In the early 2000s, funds from Germany (US$860,000) were provided for the construction of nine community water systems. A Vigilance Committee for Drinking Water (COVAAP) was required for each community system, so as to coordinate the purchase and distribution of water. Interviews indicated that problems with the management of most COVAAPs existed right from the start; allegations were made of financial mismanagement and clashes with other community organizations. This suggests that the problem of water scarcity is not only reinforced by top-down neglect or competing initiatives, but it is also a problem that reflects community disputes and intra-community political tensions.

The persistence of water-related diseases in Pachacútec prompted another social experiment involving water: a micro-credit scheme, which was introduced in 2009 by national and international agencies and overseen and managed by an American non-governmental organization. The goal was to tackle sanitation problems with the collaboration of the private sector through the creation of a small sanitation market, more specifically oriented to small and medium-sized enterprises (Bendezú, 2006). Local shops were encouraged to sell sanitation equipment and toilet units, while residents were persuaded to commit themselves to loans to purchase such equipment, with these loans being paid back over two to three years (Baskovic, 2008). Instead of advancing the establishment of a uniform level of sanitation service throughout Pachacútec, however, Pachacútec's micro-credit project exacerbated the inequalities within the neighbourhoods of Pachacútec. Residents within certain neighbourhoods complained that the sanitation equipment and technology provided were not appropriate for their wooden houses, and that only the better off in the overall Pachacútec community could really benefit from the scheme.

Huaycán, the third study area, was also created as a *comunidad autogestionaria*, but with a different socio-spatial trajectory and with the weakest level of political mobilization and support of the three study areas. Huaycán is situated on steep slopes – between 500 and 900m above sea level – 20km to the east of central Lima. After the initial occupation of the land, the left-wing administration of Barrantes (the Mayor of Lima between 1984 and 1986) launched a programme to accommodate 30,000 people in Huaycán. The settlement was organized into community household units, but from the beginning received only rudimentary water infrastructure. Lack of water and sanitation has been one of the most significant failures of the settlement of Huaycán.

Self-management of Huaycán was abandoned in 1987, and the Huaycán community programme suffered from serious corruption and opportunism by politicians (Soto *et al.*, 2005), as the basic essential services of the community, including water and sanitation services, went unfulfilled. For years, most of Huaycán's residents had to buy water and store it in containers, while SEDAPAL and the local utility authority demonstrated low levels of commitment to the success of the Huaycán initiative. Precarious water services became a central and pressing problem for the community, reinforced by its constant expansion and high levels of unemployment within the community (Zambrano, 1997). After living with unfulfilled promises for years, Huaycán's residents came to resent what they saw as partial investments motivated by politicians' electoral concerns.

Despite several promising community-based initiatives, the high altitude neighbourhoods of Huaycán, which have continuously expanded over the years, have limited prospects of receiving water. Interviews indicate that the introduction of the APT program has in fact sometimes increased community disorganization and fostered internal community conflict, for example, between those streets likely to and not likely to receive new water infrastructure.

The sagas played out in three areas of Huaycán, Villa El Salvador and Pachacútec show that long-lasting water scarcity in Lima is directly associated with the lack of political will, and a persistent reluctance to address the demands and needs of low-income communities. Ironically, the issue of funding for Lima's water infrastructure in recent decades has not been much of a limiting factor, given the variety of funding sources that the city has been able to tap into. Moreover, the problems identified above have been amplified by internal disputes within the communities and by a lack of sustained mobilization by those Lima neighbourhoods without satisfactory water and sanitation services.

Conclusions

The provision of water services have not kept up with Lima's rapid urbanization. These unsolved water problems reveal an interconnection between water scarcity, political manipulation and ineptitude, and selectivity in how the investment of public funds is directed. While Lima's relatively new but poor peripheral areas suffer from precarious water services, Lima's increasing role in the global marketplace encourages residents in wealthier neighbourhoods to consume and undertake activities that inevitably increase water consumption and exacerbate inequality. The inflow of money and technology to the water sector in the past two decades has failed to provide a solution to Lima's water problems. Rather than lack of finance and absolute water scarcity, the major issues surrounding access to water in Lima can be explained primarily by sustained practices of social and spatial discrimination, whether conscious or not.

Lima's revitalization and expansion, and the accompanying water management that has gone with the boom, have operated within the hegemonic asymmetries that dominate Peru's politics, and unfortunately reinforce the disparities and social inequalities inherited from previous generations. The modernization of water services in Lima has been based on investments that were not sustained, on the commercialized management of the public utility, SEDAPAL, and on alternative water systems (e.g. partial privatization and micro-credit schemes).

With unplanned and haphazard development occurring in Lima, and a long-standing and deeply engrained structural inequality, the lack of consistent, fairly apportioned water supply services throughout the city is not simply the outcome of ill-conceived interventions, but also permeates decision-making regarding the allocation of resources. SEDAPAL's difficult task of covering the huge and expanding area of Lima becomes impossible when confronted with the erratic allocation of resources and constant political interference. Water scarcity in Lima is integrally connected to the lack of decent housing, employment, and social services for many of its residents.

Recent experiences with the expansion of water services in Lima have only been partially successful, and many of the city's disadvantaged residents are required to use their resilience and organizational capacity skills daily to survive. Attempts at providing low-cost services in low-income areas have generated resentments as high-income areas are provided with expensive infrastructure and abundant water. Further, seemingly well-intentioned programmes have been caught with hidden agendas and poor use of funds. Unfortunately, poor management practice has been found at all levels of authority, including the community organizations. Still, while residents interviewed were disillusioned with the treatment by local and national politicians, a firm belief in the importance of grassroots organizations to demand and claim their right to water services remains.

Lima as a thriving large city offers the promise of abundance, but in practice excludes many by failing to provide some of the most essential needs for human life, including water and sanitation. To understand the current and future water challenges of Lima (and of other Latin American cities), it is necessary to both dig deep into the city's history and to look at inter-related issues such as housing, transport and economic disparity as they play out in different parts of the city. The analysis of water inequalities provides true insight into the complexities of the Latin American city, and Lima is an excellent example.

References

AFIN (Asociación para el Fomento de la Infraestructura Nacional) (2013) *AFIN señala graves carencias en los servicios de saneamiento en Lima y Callao. Lima: Asociación para el Fomento de la Infraestructura Nacional*. http://www.afin.org.pe/images/noticias/notas_de_prensa/np_04_04_13_afin_graves_carencias_en_servicios_de_saneamiento_lima_callao.pdf, accessed 20 February 2014.

Alcázar, L., Xu, L. C. and Zuluaga, A. M. (2000) *Institutions, Politics, and Contracts: The Attempt to Privatize the Water and Sanitation Utility in Lima, Peru*, World Bank Group, Washington, DC.

Barreda, J. and Ramírez Corzo, D. (2004) 'Lima: consolidación y expansión de una ciudad popular', *Perú Hoy* 6, 199–218.

Baskovic, M. R. (2008) 'Un nuevo paradigma: el saneamiento como negocio: modelos de mercado inclusivos para los pobres del Perú', *Agua. Revista del Comité Sectorial de Agua y Saneamiento*, 26, pp. 32–37.

Bendezú, C. (2006) 'Agua potable de Alto Huampaní: modelo de autogestión sostenible'. *Agua. Revista del Comité Sectorial de Agua y Saneamiento*, 21, pp. 6–9.

Bonifaz, J. L. and Malásquez, E. (2008) 'Tarifas de SEDAPAL: La tortuosa tarea de regular empresas públicas', *Apuntes de Estudio* 67, pp. 51–68.

Calderón Cockburn, J. (2005) *La ciudad ilegal: Lima en el siglo XX*, UNMSMS, Lima.

Caria, A. S. (2008) *Títulos sin Desarrollo: Los Efectos de la Titulación de Tierras en los Nuevos Barrios de Lima*, DESCO, Lima.

Corton, M. L. (2003) 'Benchmarking in the Latin American water sector: The case of Peru', *Utilities Policy*, 11, pp. 133–142.

Escalante, V. (2013) 'Megacomisión del Congreso dará más detalles de la gran estafa. Más perlas de "Agua para todos" aprista'. *La Primera. Digital*, http://www.diariolaprimeraperu.com/online/politica/mas-perlas-de-agua-para-todos-aprista_131743.html, accessed 20 February 2014.

Grupo GEA (2005) *Perspectivas del medio ambiente urbano: GEO, Lima y Callao*, PNUMA, Mexico.

ICOM (2001) *Investigación Cuali-cuantitativa sobre la Imagen Externa de SEDAPAL*, SEDAPAL, Lima.

Ioris, A. A. R. (2010) 'The political nexus between water and economics in Brazil: a critique of recent policy reforms', *Review of Radical Political Economics*, 42, pp. 231–250.

Ioris, A. A. R. (2012a) 'The neoliberalization of water in Lima, Peru'. *Political Geography*, 31, pp. 266–278.

Ioris, A. A. R. (2012b) 'The geography of multiple scarcities: urban development and water problems in Lima, Peru', *Geoforum*, 43, pp. 612–622.

Ioris, A. A. R. (2012c) 'Scarcity, neoliberalism and the "Water Business" in Lima, Peru'. *Human Geography*, 5, pp. 93–105.

Laurie, N. and Crespo, C. (2007) 'Deconstructing the best case scenario: lessons from water politics in La Paz-El Alto, Bolivia', *Geoforum*, 38, pp. 841–854.

Marmanillo, I. (2006) 'Potable water and sanitation', in M. Giugale, J. L. Newman and V. Fretes-Cibils (eds), *Opportunity for a Different Peru: Prosperous, Equitable, and Governable*. World Bank, Washington, DC.

Ministerio de Vivienda, Construcción y Saneamiento (2005) 'MIAGUA: la reforma silenciosa del sector de saneamiento', *Agua, Revista del Comité Sectorial de Agua y Saneamiento*, 19, pp. 4–11.

Panayotakis, C. (2003) 'Capitalism's "dialectic of scarcity" and the emancipatory project', *Capitalism Nature Socialism*, 14, pp. 88–107.

Paterson, C., Mara, D. and Curtis, T. (2007) 'Pro-poor sanitation technologies', *Geoforum,* 38, pp. 901–907.

Peru21 (2013) 'Denuncian colusión y fraude y agua para todos', *Peru21*, http://cdn.peru21.e3.pe/politica/denuncian-colusion-y-fraude-agua-todos-2119313, accessed 27 February 2013.

Portes, A. and Roberts, B. (2005) 'The free-market city: Latin American urbanization in the years of the neoliberal experiment', *Studies in Comparative International Development* 40, pp. 43–82.

Sánchez León, A., Guerrero, R., Calderón, J. and Olivera, L. (1986) *Tugurización en Lima metropolitana,* DESCO, Lima.

SEDAPAL (Servicio de Agua Potable y Alcantarillado de Lima) (2003) *Historia del abastecimiento de agua potable de Lima 1535–2003,* SEDAPAL, Lima.

SEDAPAL (2005) *Plan Maestro Optimizado,* SEDAPAL, Lima.

Soto, L. Z., Wilches-Chaux, G. and Orrego Ocampo, J. C. (2005) *HUAYCÁN: Construyendo una ciudad segura y saludable,* UNDP, Lima.

World Bank (2009) *Implementation Completion and Results Report (IBRD-38110) on a Loan of the Amount of US$ 150 Million To the Republic of Peru for the Lima Water Rehabilitation and Management Project,* Washington, DC, World Bank Group, http://www-wds.worldbank.org/external/default/WDSContentServer/WDSP/IB/2009/10/16/000333037_20091016004236/Rendered/INDEX/ICR12530P0080510disclosed0101141091.txt, accessed 27 February 2013.

Zambrano, D. (1997) *Agua y saneamiento: experiencia en el Perú,* Intermediate Technology Development Group, ITDG-Perú, Lima.

Zolezzi, M. and Calderón, J. (1987) *Vivienda popular: autoconstrucción y Lucha por al agua.,* DESCO, Lima.

6 Urban wastewater governance in Latin America

Panorama and reflections for a research agenda

Raúl Pacheco-Vega

Introduction

This chapter provides an overview of the challenges facing urban wastewater governance in Latin America. Usually the analysis of water governance is centred on water supply, demand and distribution; there is far less discussion about sanitation and wastewater. While preliminary, this chapter provides a useful diagnosis as well as a list of suggestions on how governments can improve sanitation services in Latin American cities. In spite of the differences within and between countries and cities, there are commonalities around Latin American urban sanitation issues. Still, the author points to the urban heterogeneity as one of the distinctive challenges of wastewater governance in comparison with freshwater provision.

The chapter addresses a research agenda on the politics of urban water governance in Latin America. Drafting this agenda requires careful consideration of key issues. First, wastewater, unlike freshwater, is usually an overlooked subject area. Second, urban heterogeneity affects the way in which water policy can be designed. Income inequalities aid in the explanation of divergences in sanitation coverage and performance indicators. The chapter concludes with a major question about the appropriate policy mix for robust urban wastewater governance, and the fact that there is no 'magic bullet' in the design and implementation of sanitation policies. The understanding of contextual factors and the causal mechanisms of what works and what does not becomes essential.

Urban wastewater governance: technical issues or social morass?

While recognizing that the generation of waste through bodily functions has been a traditionally shunned and taboo topic (George, 2008; Jewitt, 2011b; Movik & Mehta, 2009), it is undeniable that the governance of urban wastewater is a very important topic, currently both deserving of academic research and understudied. Two and a half billion people lack access to water and sanitation services and more than 940 million people still defecate in the open (UNICEF/WHO/JMP, 2012).

The hype surrounding achievement of some of the United Nations' Millennium Development Goals (MDGs) in the water and sanitation sector is unwarranted and very premature (Pacheco-Vega, 2012b). Income disparity, population heterogeneity, budgetary constraints and lack of political will are some of the negative compounding factors that affect population access to sanitation services (Pacheco-Vega & Vega, 2008; Pacheco-Vega, 2009, 2012a, 2013a). These factors interact to hinder vulnerable communities' development. Achievement of a quantitative figure that summarizes the current state of sanitation, sewerage and wastewater treatment infrastructure for an entire geographical region does not guarantee that the entire population will be served adequately.

There are substantive global challenges facing urban wastewater governance, even in the face of increased access to drinking water and sanitation infrastructure. Moreover, income disparities even within the same city can often exacerbate the difficulties facing vulnerable communities with regards to access to improved sanitation facilities. Even more importantly, aging infrastructure, social opposition to privatization of water provision and sewerage and a generalized disinterested population unaffected by the disparities in sanitation coverage are all compounding factors that generate not only bitter disagreements among public service providers but also hinder further development of a solid foundation for wastewater treatment and reuse.

Wastewater, sewerage and sanitation have all been the centre of attention in the engineering and natural sciences literature, with very little if any examination from the social sciences perspectives (Grigg, 2012). Sanitation and wastewater in the social sciences (and in particular in the human geography literature) have been mostly examined around topics such as implementation of cleaner toilet technologies, access to sewerage, and development of pipeline and infrastructure of effluent distribution (Kar, 2012). Most scholarship focused on wastewater governance has examined empirical cases in developing countries in Africa and Asia. India, specifically Mumbai (Gandy, 2008; McFarlane, 2008a, 2008b), has been one of the geographical regions whose sanitation processes have been studied in more depth. Particular attention has been paid for access to appropriate toilet technologies (Burra, 2003) and innovations around reducing open-air defecation (Chaplin, 1999).

Wastewater governance is not only an issue of technology, though. It is (and has been for many years) a complicated tale of countervailing forces. On the one hand, pressure on local governments to provide better water and sanitation services have placed undue stress and strong pressure on the public utility to deliver water of high quality and provide timely and robust sewerage. At the same time, local politicians and bureaucrats recognize that to be able to offer these services, sometimes a balance must be struck between partnering with the private sector (public/private partnerships), service co-production and private sector service delivery.

Sanitation, wastewater treatment and sewerage all suffer from many of the issues mentioned in the literature on water privatization. Water sector privatization has been seen by many as a necessary evil, whereas others perceive it to be a good method for public service provision. Bakker suggests that a middle ground is necessary. As her research argues, 'conventional models of both government and private provision have serious flaws' (Bakker, 2010, p. 6). Matter-of-factly, Bakker is right in her assertion. There is no right or wrong answer in regards to whether water sector privatization will work or not. Contextual factors will determine whether a privatization project succeeds. Urban water supply, as well as urban wastewater treatment, are both environmental issues, but at the same time, economic and social ones (Bakker, 2010; Pacheco-Vega, 2013a).

The technical component is important but it is not the only one. In fact, in Latin America, we have some of the best environmental engineering research. The technical capacity is there, as well as the human capital. However, we need to look beyond just technocratic approaches to sanitation and understand the human elements underlying wastewater governance.

Water has evolved from being a natural resource to becoming a political one – to many scholars' surprise (and dismay). Sharing transboundary resources has quickly given rise to inter-state and international conflicts (Pacheco-Vega, 2013b). Yet, it is usually water in its purest form that has become a political resource rather than wastewater. This phenomenon is most likely the result of years of shunning bodily functions as though they do not exist, and as though clean water resources were not wasted in the transportation and disposal of human waste (Franceys & Gerlach, 2008; Jewitt, 2011a; Krause, 2009). Latin America offers a perfect laboratory for the study of water governance, and in recent years, for the analysis of wastewater and sanitation policy.

Towards a research agenda on urban wastewater governance

Urban wastewater governance in Latin America offers interesting challenges that its freshwater provision counterpart does not. For one, sanitation policy is often an overlooked subject area. As indicated by Jaimie Benidickson, the culture of flushing is so prevalent that we refuse to acknowledge that there is a global sanitation crisis (Benidickson, 2004). While rural regions are deeply affected (Mehta, 2011), cities face major challenges that often are kept out of mind because the issue at stake is out of sight. For two, urban heterogeneity affects the ways in which water policy can be designed. Even within the same municipality, we will often find communities that are more profoundly affected by improper sanitation, whereas in other neighbourhoods we may find no visible trace of negative effects (Pacheco-Vega, 2011).

In outlining a research agenda on the politics of urban water governance in Latin America, it is important to distinguish the heterogeneities and specificities of each Latin American country and city. Sub-national analyses require an in-depth examination of the institutional architecture of wastewater policy. Without a clear map of jurisdictional responsibilities, Latin American cities will continue to face almost unsurmountable challenges. An insight gained from this chapter that goes well beyond what the traditional literature on water and sanitation privatization points out, Latin America presents specific and unique issues that are also derived from a lack of rule of law and poor accountability on the part of bureaucrats and politicians.

It is important to point out that the water and sanitation crisis is not only an issue in developing nations, but also in developed ones. Privatization of public service delivery is not the 'magic bullet' that many would like to purport it to be. Solomon (2011) demonstrates with cases in the United States (Atlanta and Indianapolis) that privatization transactions can have diverging effects. In the case of Atlanta, privatization in 1999 was followed by the reacquisition of the water sector responsibilities by the local government, as a result of poor performance, lacking water quality and general malaise about the system. In the case of Indianapolis, the public/private partnership in the water and sanitation sector achieved great results (Solomon, 2011). This variation in performance shows that there are contextual factors beyond pure technical advances or managerial effectiveness. Economic context variables can be also used to explain divergences in sanitation coverage and performance indicators. Poorer cities can be expected to perform less aptly in public services delivery, particularly drinking-water provision and sanitation coverage.

Attitudes towards human waste use and reuse have been slowly shifting, although wastewater, sanitation and shit are still very much taboo topics. The challenge remains to offer infrastructure improvements without resorting to its use as political weapons. The provision of adequate sanitation infrastructure as a political tool has evolved much like water itself. Water has become more than a natural resource, a political one. This phenomenon can be seen from a policy implementation perspective, where the political environment conditions are such that access to sanitation is reserved for the well-off, privileged members of society (much like Chaplin observes in her 1999 study of sanitation in India), or from an agenda-denial perspective, similar to what Pacheco-Vega establishes in his study of the Mexican Lerma-Chapala river basin council and sanitation (Pacheco-Vega, 2005a, 2005b, 2006), where even given the technical conditions for improved sanitation provision, wastewater remains out of the list of highly relevant topics in Latin American water policy agendas (Chaplin, 1999).

Conclusions

Clearly, there is much heterogeneity in the way in which municipalities, small towns and mega-cities all deal with urban wastewater governance challenges. However, it is clear that contextual factors matter, a lot. Cultural practices, financing mechanisms, social norms and expectations around what the appropriate level of wastewater treatment should be, biases and perceptions around effluent treatment and human capital capabilities to design and implement liquid effluent treatment processes are all important variables that need to be taken into account when designing urban wastewater governance processes. Contrary to common belief, just increasing the levels of effluent treatment is not enough to sustain a robust system of sewerage and sanitation. There are a number of additional factors that need to be taken into account when developing policy solutions within the sanitation sector.

What is the appropriate policy mix for robust urban wastewater governance still remains one of the biggest policy puzzles of the twenty-first century. There is no 'magic bullet' to design and implement sanitation policies. There is the need to understand the contextual factors and the causal mechanisms of what works and what does not.

Acknowledgements

I am grateful to Alberto Hernández, Fernando Basurto, Alejandro Ramírez, Carla Villegas, Luis Esparza and Fernando Aquino for superb research assistance. All errors remaining are my sole responsibility. This chapter is a component of my larger research agenda on the politics of waste and wastewater in Latin American countries (Brazil, Argentina and Mexico).

References

Bakker, K. (2010). *Privatizing water: Governance failure and the world's urban water crisis.* Ithaca, N: Cornell University Press.

Benidickson, J. (2004). *The culture of flushing: A social and legal history of sewage.* Vancouver, BC: The University of British Columbia Press (UBC Press).

Burra, S. (2003). Community-designed, built and managed toilet blocks in Indian cities. *Environment and Urbanization*, 15(2), 11–32. doi:10.1177/095624780301500202.

Chaplin, S. E. (1999). Cities, sewers and poverty: India's politics of sanitation. *Environment and Urbanization*, 11(1), 145–158.

Franceys, R., & Gerlach, E. (Eds.). (2008). *Regulating water and sanitation for the poor: Economic regulation for public and private partnerships.* London, UK: Earthscan.

Gandy, M. (2008). Landscapes of disaster: water, modernity, and urban fragmentation in Mumbai. *Environment and Planning A*, 40(1), 108–130. doi:10.1068/a3994.

George, R. (2008). *The big necessity: The unmentionable world of human waste and why it matters.* New York: Macmillan.

Grigg, N. S. (2012). *Water, wastewater and stormwater infrastructure management.* Boca Raton, FL: CRC Press.

Jewitt, S. (2011a). Geographies of shit: Spatial and temporal variations in attitudes towards human waste. *Progress in Human Geography,* 35(5), 608–626. doi:10.1177/0309132510394704.

Jewitt, S. (2011b). Poo gurus ? Researching the threats and opportunities presented by human waste. *Applied Geography,* 31(2), 761–769. doi:10.1016/j.apgeog.2010.08.003.

Kar, K. (2012). Why not basics for all? Scopes and challenges of community-led total sanitation. *IDS Bulletin,* 43(2), 93–96. doi:10.1111/j.1759-5436.2012.00312.x.

Krause, M. (2009). *The political economy of water and sanitation.* New York: Routledge.

McFarlane, C. (2008a). Governing the contaminated city: infrastructure and sanitation in colonial and post-colonial Bombay. *International Journal of Urban and Regional Research,* 32(2), 415–435. doi:10.1111/j.1468-2427.2008.00793.x.

McFarlane, C. (2008b). Sanitation in Mumbai's informal settlements: State, 'slum' and infrastructure. *Environment and Planning A,* 40(1), 88–107. http://dx.doi.org/10.1068/a39221.

Mehta, L. (2011). Why shit matters: Community-led total sanitation and the sanitation challenge for the 21st century. In L. Mehta & S. Movik (Eds.), *Shit matters: The potential of community-led total sanitation* (pp. 1–25). Rugby: Practical Action Publishing Ltd.

Movik, S. & Mehta, L. (2009). *Going with the flow? Directions of innovation in the water and sanitation domain.* Brighton, UK: STEPS Centre.

Pacheco-Vega, R. (2005a). Applying the Institutional Analysis and Development framework to wastewater management policy in the Lerma-Chapala river basin. In UNU-INWEH (Ed.), UNU-INWEH/UNESCO-MAB-IHP International Workshop 'Water and Ecosystems: Water Resources Management in Diverse Ecosystems and Providing for Human Needs'. Hamilton, Ontario (Canada). Fort Collins, CO: US Geological Survey (USGS).

Pacheco-Vega, R. (2005b). Institutional analysis within the Lerma-Chapala Region: new challenges for watershed management. In UNU-INWEH (Ed.), UNU-INWEH/UNESCO-MAB-IHP International Workshop 'Water and Ecosystems: Water Resources Management in Diverse Ecosystems and Providing for Human Needs'. Hamilton, Ontario (Canada). Fort Collins, CO: US Geological Survey (USGS).

Pacheco-Vega, R. (2006). *Evaluación de políticas y programas dentro de la cuenca Lerma-Chapala* (p. 165). León, Guanajuato, Mexico: Centro de Investigación y Asesoría Tecnológica en Cuero y Calzado, A.C. (CIATEC, A.C).

Pacheco-Vega, R. (2009). Arreglos institucionales para el saneamiento de aguas residuales en México. Un caso de estudio en la cuenca Lerma-Chapala. In I. Sandré Osorio, R. L. do Carmo, S. Vargas-Velázquez, & N. B. Guzmán (Eds.), *Gestión del agua: una visión comparative entre México y Brasil.* Jiutepec, Morelos: Instituto Mexicano de Tecnología del Agua.

Pacheco-Vega, R. (2011). The promises and pitfalls of lesson drawing in water governance research: what can we learn about watershed management from river basin councils and organizations in Mexico. In Annual Meeting of the Canadian Association of Geographers. Calgary, Alberta, Canada: Canadian Association of Geographers.

Pacheco-Vega, R. (2012a). Arreglos institucionales dentro de la cuenca Lerma-Chapala: Una visión desde la política ambiental. In M. Sánchez Rodríguez, J. de J. Hernández López, J. M. Durán Juárez, & A. Torres Rodríguez (Eds.), *Los estudios del agua en la cuenca Lerma-Chapala-Santiago. Agricultura, industria y ciudad. Pasado y presente* (pp. 347–360). Guadalajara, Jalisco y Zamora, Michoacán.

Pacheco-Vega, R. (2012b). Governing wastewater: a cross-regional analysis within the Lerma-Chapala river basin in Mexico. In *Canadian Association of Latin American and Caribbean Studies (CALACS)* 2012. Waterloo, ON, Canada: Canadian Association of Latin American and Caribbean Studies.

Pacheco-Vega, R. (2013a). Geographies of wastewater: a comparative analysis of urban sanitation governance in the Mexican municipalities of Aguascalientes (Aguascalientes) and Leon (Guanajuato). In *2013 Meeting of the American Association of Geographers* (p. 25). Los Angeles, CA: American Association of Geographers.

Pacheco-Vega, R. (2013b). Los conflictos intratables por el agua en México: Aproximaciones teóricas y acercamientos metodológicos. In E. Kauffer (Ed.), *Taller temático 'Las dimensiones políticas de los recursos hídricos: miradas cruzadas para politizar el debate'* (pp. 1–30). San Cristóbal de las Casas, Chiapas, Mexico: Centro de Investigaciones y Estudios Superiores en Antropología Social.

Pacheco-Vega, R. & Vega, O. (2008). Retos y perspectivas en materia de política de tratamiento de agua y saneamiento en México. In R. Olivares & R. Sandoval (Eds.), *El agua potable en México: historia reciente, actores, procesos y propuestas* (pp. 173–186). México City: ANEAS (Asociación Nacional de Empresas de Agua y Saneamiento).

Solomon, L. D. (2011). *America's water and wastewater crisis.* New Brunswick, NJ and London: Transaction Publishers.

UNICEF/WHO/JMP. (2012). *Progress on drinking water and sanitation: 2012 update.* New York: UNICEF/WHO Joint Monitoring Programme for Water Supply and Sanitation.

7 Issues and challenges for water supply, storm water drainage and wastewater treatment in the Mexico City metropolitan area

Polioptro F. Martínez and Erick R. Bandala

Introduction

Supply of water in sufficient quantity and of appropriate quality is a growing concern for megacities. Factors such as population growth, changes in land use, deforestation, climate change and lack of sound planning combine in a complex, inter-related process to affect water supply and its management (Oswald, 2011). Several megacities in the developing world (e.g. São Paulo, Manila, Delhi and Cairo) face similar problems such as severe water scarcity, limitations in water supply and sewage systems, flooding and poor water management. The metropolitan area of the Valley of Mexico is an excellent case study of a megacity with such issues and challenges.

Mexico City is one of the largest cities in the world. Its hydro-geographical setting – in an endorheic basin (that is, having no natural outlets) – provides a specific context by which to understand water problems and their solutions. The city suffers from recurrent floods, droughts and other climate-related events, the handling of which is of interest in the international arena (Romero, 2010). The city has a long history of efforts made to address the challenges associated with high population and economic growth in this particular context.

In addition to the issues of water supply and storm water drainage, attention is given in this chapter to sewage and wastewater infrastructure, as well as to water reuse. Some of Mexico's largest water infrastructure projects are found in the nation's capital and the surrounding Valley of Mexico. The Atotonilco Wastewater Treatment Plant is a case in point: once in operation, the plant will be the largest in the country and one of the largest of its kind in the world. The chapter ends with a discussion on water governance and the need for a new institutional arrangement for water management in the Valley of Mexico. In light of the economic relevance of Mexico City to the nation, the proper consideration of these issues is not only a local or a regional concern, but rather is of national importance.

The historical setting of Mexico City's water issues

Mexico City is the most populous urban area in Latin America. It is situated within the Valley of Mexico, an endorheic basin. The basin was originally a lacustrine basin comprising three main lakes: Texcoco, Xochimilco and Chalco Xaltocan. Today, only small lagoons are found in Xochimilco and Zumpango. The Valley of Mexico, with an average elevation of 2,240m above sea level and an area of 9,560km^2, is surrounded by mountains of varying heights, several of which are more than 3,000m above sea level. Its rivers descend steep slopes that originally fed the basin's lakes, and produced severe floods. In the present day, with practically no lakes for storage and a large impervious area, combined with a huge population demanding water and discharging wastewater, it is easy to see that the water resources problems of Mexico City are great in scope.

In fact the city has struggled with water resources since its founding in 1325. In pre-Columbian times, the city faced serious water supply issues because water in the lake was soon contaminated by human waste from the city. The Aztecs resolved the supply problem by building a canal to convey spring water from Chapultepec, part of the urban area of modern Mexico City, to Tenochtitlan, now the city centre. The city's first water supply systems had to be constantly repaired, and others were built during the colonial era. Notable among the early water supply systems was the Santa Fe aqueduct, built in 1620, which brought water to the city from Santa Fe springs, located within the *Desierto de los Leones* National Park, which was then distributed by canoe to the population. In the pre-Columbian and the colonial periods, drainage systems and wastewater disposal systems generally did not exist.

Flooding has affected Mexico City since its establishment by the Aztecs, who built levees to protect their cities. The first levee in the city was built by King Ahuízotl after severe flooding in 1499; and the second, and largest, started to be built by King Netzahualcoyotl of Texcoco in the same year, and still bears his name. It divided Texcoco Lake, the origin of many floods that devastated Tenochtitlan (the Aztec city-state located on the site of modern Mexico City) throughout history. Building levees for flood control continued throughout most of the colonial period. The most important levee was the St Lazarus, which connected the Tepeyac and Tacuba causeways into the island of Tenochtitlan to the shore of the lake, thereby forming a large dyke protecting the west part of the city from flooding. This embankment was complemented by smaller dykes to protect the east of the city.

However, during colonial times Mexico City was just a metre above the average level of Lake Texcoco; during rainy years, flooding of the city was so frequent that colonial authorities considered relocating it altogether. As the city continued to grow, building more levees was considered ineffective. Therefore, in 1607–1608 the colonial government decided to drain the lakes, which were also perceived to be causing recurrent diseases that ravaged the population. After several costly and failed attempts, draining finally took place almost two centuries later, in 1788, along with the opening of an artificial outlet for the waters of the basin formed by channels that crossed the mountains in the area known as Nochistongo. This led to water surpluses in the neighbouring Tula River basin (Tortolero, 2000).

From the late eighteenth century, the flood protection strategy for the Valley of Mexico has focused on building additional outlets to transport surplus water from the endorheic basin to neighbouring basins via a combination of open channels and tunnels. However, drying out the lakes in the Valley of Mexico has deprived the city of virtually all capacity to store surface water. This has led to a serious over-exploitation of the aquifers.

Water supply

The Mexico City metropolitan area is managed by the Comisión Nacional del Agua (National Water Commission – CONAGUA) as part of the *Región Hidrológica del Valle de México* (Hydrological administrative region of Valley of Mexico – RHVM) (Figure 7.1). The RHVM is responsible for the Valley of

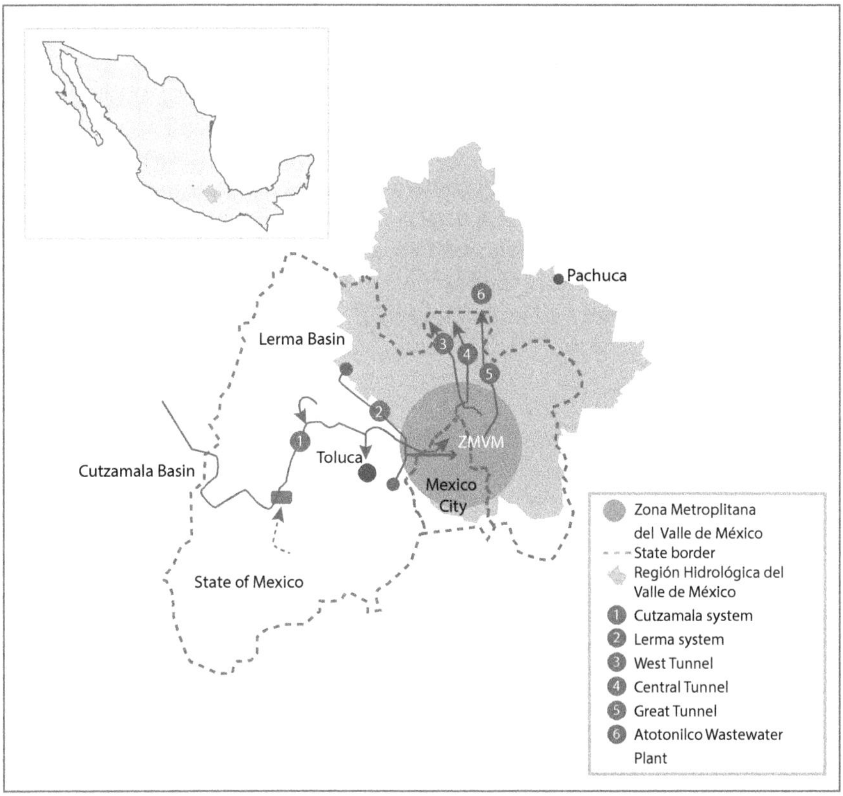

Figure 7.1 Geographical location of RHVM and schematic overview of the water supply sources and drainage outlets in the ZMVM

Source: Authors

Mexico basin and the neighbouring Tula River basin. These two independent basins are linked as the latter receives wastewater flows transferred from Mexico City. This hydrological administrative region now covers one of the largest metropolitan areas in the world – Mexico City, with its Federal District and conurbations of the states of Mexico and Hidalgo. The Zona Metropolitana del Valle de México (Metropolitan Zone of the Valley of Mexico – ZMVM) comprises 16 delegations of the Federal District and 18 municipalities of the State of Mexico. The urban area is 3,540km^2 – or 37 per cent of the total area of the RHVM.

The population of Mexico in 2010 was 113 million, while in the same year the population of the RHVM was about 20 million – or 18 per cent of the nation's total inhabitants (UNFPA, 2011). The high density of people in the RHVM, a highly populated and continuously growing urban area, has caused a huge imbalance between water availability and demand, as shown in Figure 7.2. The demand for water in the valley amounts to 83m^3/s, while the available flow, on average, is 37m^3/s (Aguirre-Díaz, 2012). To cover some of this deficit, water from surrounding basins is imported.

Most of the water deficit, however, is reduced by the over-exploitation of the aquifers in the valley. The average withdrawal is 59m^3/s, while the natural recharge is only 32m^3/s (Aguirre-Díaz, 2012). In this area where large lakes once existed, the water table is now located between 30 and 220m below ground level, except in a few discrete areas, the groundwater from the aquifers is still of high quality. The potential areas of contamination have had no significant effect on the aquifers as a whole (Ramos-Leal *et al.*, 2010). This may be due partly to the fact that wastewater from Mexico City, mixed with storm water during the rainy season, is exported from the basin.

The main contributor of water imported to the Valley of Mexico is the Cutzamala system (Figure 7.1). This complex hydraulic infrastructure system

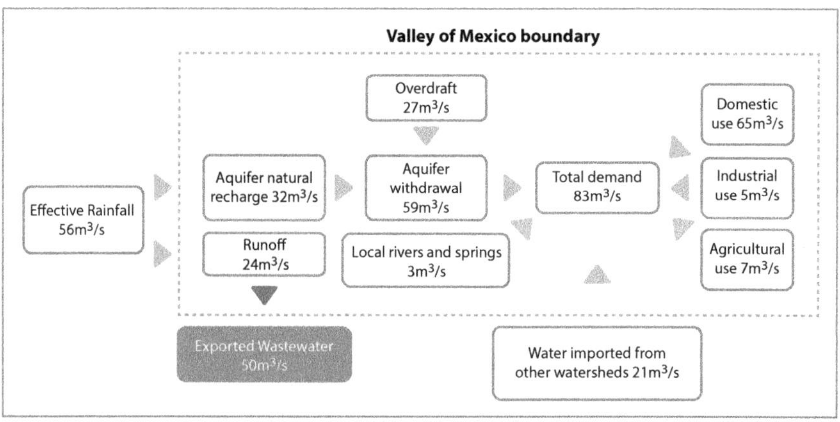

Figure 7.2 Water balance in the metropolitan area of the Valley of Mexico

Source: Authors, based on data from Ardavín (2013)

comprises four dams as well as pumping stations and pipes to elevate water from the Colorines Dam, at 1,629m above sea level, to the Santa Isabel holding tank, at 2,681m above sea level. The capacity of this system was designed for $19m^3/s$ but it usually operates at $17m^3/s$. Drinking-quality water from the Cutzamala system is transported and distributed to urban areas in the State of Mexico and in the Federal District by large transmission mains and an extensive distribution network (Luege, 2012).

However, the future poses great challenges for the RHVM. By 2030, the population is estimated to reach about 26 million – 6 million more than today. Current water demand in the region is approximately $149m^3/s$, while the sustainable supply yield is projected to be $114m^3/s$. By 2030, demand is estimated to reach $168m^3/s$, increasing the supply and demand gap to $54m^3/s$ (CONAGUA, 2012a). To a certain extent, the gap can be filled with additional exploitation of groundwater resources in the valley.

Given this scenario, government authorities responsible for water management in the RHVM have proposed a combination of policy measures: (i) increasing water efficiency, aiming at reducing water demand by a better understanding of the value of water; and (ii) developing new water sources outside the Valley of Mexico basin. The best water supply source for the basin is most likely the Tula River basin, which still has aquifers with surplus water. However, developing this option will be difficult because the neighbouring State of Hidalgo will need to agree to such an inter-basin water transfer project, and the local population of that state is currently in strong opposition to it.

Groundwater development in the Valley of Mexico has multiple consequences, including compaction of sand and clay layers forming the valley's subsoil, with ensuing substantial soil subsidence in some areas. For instance, the historic centre of Mexico City is sinking at a rate of 10cm/year, while in the Chalco Valley, which once held a lake of the same name, the subsidence rate is now up to 60cm/year. Accumulated settlement of the 1862 to 2011 period amounts to 13m at Mexico City's airport. During the same period in a large area of metropolitan Mexico City, subsidence is 6–9m (Arreguín, 2012). As will be discussed below, this phenomenon produces serious adverse effects on water infrastructure, particularly in the city's storm water and wastewater drainage system.

Storm water and wastewater drainage systems

Storm water and wastewater drainage systems in the Valley of Mexico are linked to the *Gran Canal*. The Great Channel, a gravity system in operation since 1910, has a capacity to evacuate $80m^3/s$. The other major components of the valley's drainage system are the Central Tunnel, with a design capacity of $170m^3/s$, and the West Tunnel, with a capacity of $30m^3/s$. As a result of soil subsidence in the central area of Mexico City, some sections of the Great Channel have decreased in slope, and some have even acquired a negative slope and come to be inverted channels. As a consequence, in 2007 the capacity of the Great Channel was reduced to only $15m^3/s$. In the same year, as a result of years of operation without major maintenance, the capacity of the Central Tunnel was reduced to 120m^3/s. Thus, the total capacity of the Valley of Mexico's drainage system (storm water and wastewater) has been reduced from 280m^3/s in 1975 to 165m^3/s in 2007 and it is expected to continue decreasing (Harding and Willis, 2010). This situation has significantly increased flooding risks and environmental health problems associated with the incapacity to effectively remove wastewater from urban areas.

In recent years, several measures have been taken to increase the valley's drainage capabilities and its resilience to flooding. A pumping plant was constructed for the Great Channel, which increased its capacity from 15 to 45m^3/s – an improvement, but still far from its 80m^3/s designed capacity. Because of continuous subsidence, the pumping plant is only a temporary solution, but it has enabled major maintenance work on the Central Tunnel. There is a technical consensus that the water evacuation capacity from the basin of the Valley of Mexico should be 315m^3/s, while current operational capacity is only 195m^3/s. Therefore, there is a current deficit of 120m^3/s (Table 7.1).

To increase the capacity of the valley's drainage system, the Eastern Discharge Tunnel project was launched in 2010. The tunnel, under construction, will be 62km long and 7m in diameter, and will transport 120m^3/s under normal conditions, operating as a gravity channel, and up to 150m^3/s in emergency conditions, operating under pressure (Harding and Willis, 2010). Once the Tunnel is in operation, the valley will be able to cope with storm events with a recurrent period of 50 years.The Eastern Discharge Tunnel comprises 24 vertical inlets with depths between 34.5 and 119m. The main tunnel will conduct flow out of the basin to the Atotonilco Wastewater Treatment Plant, which is its final stage of construction. Because of the geology and soil characteristics (including soft materials) in the construction area as well as hydraulic pressures of up to 4.2bars, the Tunnel is an extremely complex construction work. Without doubt it is one of the most interesting drainage tunnels currently under construction in the world.

The flood control system in the Valley of Mexico also includes a group of 29 small dams, with a combined capacity of about 31 million m^3, that control peak flows by temporarily retaining run-off from several rivers in the mountains surrounding the valley. It is expected the system will redirect the excess run-

Table 7.1 Drainage capacity of the Valley of Mexico system, 1975–2010 (m^3/s)

Component of the system	*Year*		
	1975	*2007*	*2010*
Great Channel	80	15	45
West Tunnel	30	30	30
Central Tunnel	170	120	120
Total	280	165	195

Source: Based on data from Ardavín (2013)

off to the deep drainage system currently under construction. However, due to informal urban settlements close to rivers and ravines, these flood control dams cannot be filled to their optimal levels. In addition, the loss of storage capacity due to sedimentation has reduced the capacity of such dams by more than 30 per cent (Cisneros-Iturbe and Domínguez-Mora, 2007). To improve the operational capacity of these dams, sediment is regularly removed by dredging, but this is an expensive and unsustainable solution, exacerbated by difficulties in finding suitable places to dispose of the dredged material.

Sewage and wastewater infrastructure

The Mexico City metropolitan area has a combined sewage and storm water network with a flow discharge of between 45 and 300m^3/s depending on the dry or rainy season (Jiménez-Cisneros and Chávez-Mejía, 1997). Sewage generated in the city is carried almost exclusively by a 68km-long main pipeline, the Emisor Central, built in 1975. The Emisor Central pipeline remains one of Mexico's largest infrastructure projects, but the designed hydraulic capacity of the pipeline has decreased significantly due to subsidence of about 10cm/year, which has changed the alignment and slope of the conduit. The overall system capacity has decreased from 280 to 165m^3/s (40 per cent) between 1975 and 2008 (Harding and Willis, 2010).

The average wastewater production of Mexico City's metropolitan area has been estimated to be as high as 106m^3/s, including municipal and industrial wastewater and storm water run-off (Downs *et al.*, 1999; Chávez *et al.*, 2011). In the late 1990s there were 16 secondary wastewater treatment plants in the urban area with a capacity to treat about 6 per cent of produced wastewater. The treated wastewater was used for the irrigation of city parks, gardens and leisure facilities (Jiménez-Cisneros and Chávez-Mejía, 1997). By 2011 there were 85 small wastewater treatment plants in the urban area, with a combined capacity to treat about 6.35m^3/s, or 23 per cent of the daily average wastewater generated (CONAGUA, 2011).

Despite the increase in wastewater treatment plant construction, more than 70 per cent of wastewater produced by the Mexico City metropolitan area is still discharged into the environment without any treatment, and this water is used downstream for agricultural irrigation in the Tula River basin (Siemens *et al.*, 2008) (Figure 7.1). At the same time, leakages from the sewerage system have become the most important source of groundwater contamination in the Valley of Mexico (Félix-Cañedo *et al.*, 2013).

Wastewater reuse for agriculture

Wastewater generated in Mexico City's metropolitan area flows by gravity to the Tula Valley, located in the State of Hidalgo north of the city (Figure 7.1). In the Tula Valley, raw wastewater is distributed through channels and dams for agricultural irrigation. The reuse of Mexico City's wastewater for irrigation started in the early 1900s and there are currently thousands of farmers irrigating an area of 84,500ha (Chávez *et al.*, 2011). Sewage reuse in the Mezquital Valley, a smaller valley within the Tula Valley, started in 1912 – this is the oldest and largest example of agricultural irrigation using municipal wastewater in the world (Jiménez and Chávez, 2004; Siemens *et al.*, 2008) (Box 7.1).

Reuse of properly treated wastewater has been identified as a very advantageous water management (see Box 7.1), whereas the use of raw wastewater for irrigation can seriously and adversely affect surface and subsurface water sources. High concentrations of microorganisms such as faecal coliform bacteria, somatic bacteriophages, *Giardia* spp. and helminth eggs have

Box 7.1 Wastewater reuse in the Mezquital Valley

The Mezquital Valley, 80km north of Mexico City, has approximately 450,000 inhabitants in 294 rural communities. Mean annual precipitation is 450mm, from May to October, while potential evaporation is 2,100mm (Lucho-Constantino *et al.*, 2005; DFID, 1998). The main activity in the valley is agriculture, in three irrigation districts. The main crops are corn and alfalfa.

Due to the valley's soil characteristics, irrigation water is needed in very high volumes – between 15,000 and 22,000m^3/ha/year – provided through the use of raw wastewater from Mexico City. The extended use of the wastewater for irrigation purposes in the valley has led to a large increase in recharge of the aquifer, estimated at 25m^3/s, or 13.3 times greater than its natural recharge rate (Jiménez *et al.*, 2000). As a consequence, the water table has risen steadily and spring water is outcropping with flows between 0.1 to 0.6m^3/s. Similarly, the average flow of the main river in the Tula Valley has increased from 1.6 to 12.7m^3/s in the past 50 years. These additional volumes of water have favoured the economic development of other regions downstream as the main source of water for irrigation (Jiménez and Chávez, 2004).

Several advantages of wastewater irrigation have been identified: higher crop yields, year-round production and ability to grow a wider variety of crops. Moreover, agricultural wastewater recycles organic matter and other nutrients to soils, and reduces the need for fertilization.The wastewater for irrigation strategy is a low-cost wastewater disposal method, avoids discharge of pollutants to surface water bodies, and increases the economic efficiency of investments in wastewater disposal and irrigation (Lucho-Constantino *et al.*, 2005; Jiménez, 2006). However, severe disadvantages may emerge when low-quality water is used without control, as is the case when raw wastewater from Mexico City is used for the irrigation of the Mezquital Valley (Cifuentes *et al.*, 1994; Downs *et al.*, 1999; Friedel *et al.*, 2000; Siebe, 1994; Justin-Cajuste *et al.*, 2001).

been widely found in surface and underground waters of the Mezquital Valley (Chávez *et al.*, 2011). Similarly, organic micro contaminants, mainly personal care products, pharmaceutics and other emerging organic compounds, as well as heavy metals, have been reported in high concentrations in the water bodies of the Mezquital Valley (Siemens *et al.*, 2008; Félix-Cañedo *et al.*, 2013; Calderón-Preciado *et al.*, 2011; Gallegos *et al.*, 1999).

Reuse of raw wastewater from Mexico City has changed soil properties in the Mezquital Valley, by the release of inorganic substances that negatively affect long-term soil productivity (Bahri, 1999; Lucho-Constantino *et al.*, 2005). Undesirable inorganic pollutants, including heavy metals and boron, have been documented in soil and groundwater within the Mezquital Valley during the past two decades.[1] Industrial pollutants generated in the Tula-Tepeji-Vito corridor have also been found on diverse plants and crops (Zambrano-García *et al.*, 2009; Downs *et al.*, 1999; Lucho-Constantino *et al.*, 2008; Prieto-García *et al.*, 2007).

The presence of chemical and microbiological contaminants in water, soil and crops from raw wastewater irrigation can have serious effects on human and animal health. *Escherichia coli* phenotypes have been detected in crops of salad vegetables, which are eaten uncooked, cultivated in the zone (Castro-Rosas *et al.*, 2012). The groups of people most affected by the wide variety of pollutants carried by the wastewater used for irrigation are agricultural workers, product consumers and inhabitants of the surrounding areas (with children and the elderly being most vulnerable) (Cifuentes *et al.*, 2000; Jiménez, 2006). There is strong analytical evidence for the occurrence of enteric infections – especially from helminths (e.g. *Ascaris lumbricoides* and *Trichuris trichiura*), and giardiasis – in agricultural workers exposed to untreated wastewater for irrigation (Cifuentes *et al.*, 1994, 2000), and a risk of cholera and typhoid fever in consumers of uncooked vegetables (Castro-Rosas *et al.*, 2012). Organic contaminants are known to be recalcitrant in water and show a wide variety of biological effects, from acute toxicity to endocrine system-disrupting capabilities, and many of them cannot be removed using conventional wastewater treatment processes (Bandala *et al.*, 2013a, b).

Government strategies to improve water management

CONAGUA's 2030 water plan for Mexico City proposes guidance to design and implement integrated and sustainable management of basins and aquifers; to strengthen efficiency in the use of water resources; and to improve water quality in basins and aquifers (CONAGUA, 2012a). There is a strong consensus on the critical importance of wastewater reuse as part of a sustainable water management strategy in Mexico City. Fortunately, there is also a general accord about the need for implementing innovative management practices for optimal and sanitary safe utilization of the large sewage flows generated in the city. To achieve the goals of the water plan, the following strategic areas have been identified.

Improving solid waste management

Improving solid waste management is closely related to sustainable water management. Solid waste generation in Mexico City is estimated to be as high as 1.45kg per person per day, of which a high percentage goes directly into the city's wastewater and drainage system. This not only contaminates the system but also contributes to flooding during the rainy season (Durán-Moreno *et al.*, 2013).

Controlling non-point contamination sources

Pollutants such as heavy metals and toxic organic matter in industrial effluents are difficult to remove from wastewater, but their contribution to the degradation of ecosystems is relatively easy to assess. However, non-point sources of contaminants are difficult to quantify. The amount of contaminants (organic, inorganic or microbiological) released into Mexico City's wastewater and later arriving in other areas (e.g. the Mezquital Valley) and pollutants from atmospheric deposition, run-off, agricultural drainage and soil infiltration have only been partially studied. Nevertheless, these contaminants and pollutants can contribute to generate economic, environmental and social problems. Estimating the contribution to the ecosystem degradation of non-point sources of contamination, and putting in place management practices for their control, will be critically important to the 2030 water plan.

Improving municipal and industrial wastewater treatment

The reduction of contaminants in both municipal and industrial wastewater effluent is probably the most important task ahead in the short to medium term. Because wastewater is used for irrigation, organic matter and nutrients should not be targeted for complete removal during wastewater treatment. The most important parameters to be considered regarding contamination control are toxic organic and inorganic matter and pathogens (Jiménez and Garduño, 2001).

Mexico's federal government has taken bold steps to improve the wastewater situation in Mexico City. In 2009, a US$1 billion investment contract for the Atotonilco Wastewater Treatment Plant project was awarded (CONAGUA, 2012b). The plant (158ha) will be the largest wastewater treatment plant in Mexico and one of the largest of its kind in the world. The Atotonilco facility will be capable of treating up to 35m^3/s wastewater and managing solids and bio-solids for energy production. It is expected to treat the sewage produced by a population of 16 million. The project is also designed to reduce recurrent flooding problems during the rainy season by accepting storm water and to improve the quality of life and the environment for current and future inhabitants of both urban and rural areas of the RHVM.

The engineering design of the Atotonilco plant includes state-of-art odour control systems and large-scale disposal of dewatered sludge. The plant will treat (through a secondary treatment process) approximately 2hm^3 year round, and an extra 1.64hm3 in the rainy season (CONAGUA, 2012b). The sludge produced by the plant will be stabilized by anaerobic digestion and the gases produced will be used for power co-generation. The installed capacity is 32.4MW, which will provide 60 per cent of the plant's power requirements. The stabilized sludge will be disposed of on-site in a dedicated land disposal operation.

Water governance

Like other megacities, Mexico City has serious difficulties in managing its water services (Tortajada, 2008). There are important gaps in water governance that should be considered in any proposed solutions to the problems of water supply, storm water drainage and wastewater treatment in the metropolitan area. Table 7.2 shows the results of applying the gap analysis of the Organization for Economic Co-operation and Development (OECD) to multilevel water governance in Mexico City (OECD, 2013).

The analysis shows that one of the main water policy issues in the ZMVM is the wide dispersion of responsibilities in water management. Groundwater withdrawals, for example, while authorized and supervised by the federal government, are usually overseen by the Federal District and the 18 municipalities within the Mexico City metropolitan area. In urban areas water supply systems

Table 7.2 Water governance gaps in Mexico

Gap type	*Description*
Administrative	Mismatch between administrative and functional units (water bodies, municipalities, metropolitan areas, regions, states) and hydrological boundaries and imperatives.
Information	Asymmetry of information across stakeholders limits standardization. Incomplete water rights public register and metering systems. Harmonization of data being a key concern.
Policy	Misaligned policies and silo approaches to water, energy, agriculture, territorial development and financing policies.
Capacity	High turnover of water professionals. Limited training programmes for technical, administrative and management staff.
Funding	Very limited own-source revenues at sub-national level. Large reliance on federal programmes and CONAGUA resources.
Objective	Lack of continuity of public policy at local level because of limited political mandates (three-year mayoral term).
Accountability	Limited stakeholder engagement in integrated water resources management (farmers and indigenous communities) and water and sanitation services (users and consumers). Limited official mechanisms to channel demands.

Source: Adapted by authors based on OCED (2013)

are under the responsibility of urban water utilities, which are usually part of the municipal government. Water transfers from other basins are operated by the federal government, as in the case of the Cutzamala aqueduct, or by the state authorities, as in the case of the Lerma aquifer wells that supply parts of Mexico City. Management of storm water drainage systems presents similar issues, and treatment plants are also in a comparable complex organizational scheme: smaller plants are administered by the Federal District or municipalities, while the new Atotonilco plant will be operated by the federal government. The federal government also liaises with the farmer organizations using wastewater.

Jurisdictional issues regarding water governance have caused conflicts between users, and even between the federal, state and municipal governments in Mexico. For instance, in 2003, the State of Mexico raised a constitutional injunction before the Supreme Court of Mexico, claiming that the federal government should pay several million dollars in compensation for environmental damage caused by the operation of the Lerma basin wells, which are administered and operated by the Federal District. The Supreme Court dismissed the case in 2009 (Perló and González, 2009).

As expected, these institutional arrangements are inefficient and lead to serious problems of management and coordination. The main challenges for better water governance include a lack of incentives for horizontal cooperation and the absence of common information systems (Akhmouch, 2012). Joint decision making by multiple authorities from the State of Mexico, the Federal District and the federal government thus becomes slow and difficult, for example in the case of new infrastructure development. Delays and complications also arise due to differences in priorities and financial capabilities among those authorities. Such was the case, for instance, of the East Tunnel already mentioned, the need of which was known at least since 2000 but was only initiated eight years later. Coordination and information problems are especially acute in emergency situations, particularly in response to flooding in the valley in rainy periods. In order to improve emergency response capabilities, protocols have been established, which are activated under predetermined meteorological conditions and establish rules of operation for each of the parties.

Because of its size and complexity, and in large part because of the difficulties in transporting water from different sources in an optimal way, Mexico City's large and diffuse water system produces problems of inequity. While many areas are well supplied, others, such as Iztapalapa in the eastern part of Mexico City, suffer from chronically poor water services. This causes social unrest and frequent conflicts. There is an acute need in the valley for new (and better connected) networks of water supply mains. Given the scarcity of water the region faces, a combination of new sources of supply, increases in water use efficiency as well as a better distribution infrastructure will be required in order to alleviate current water supply problems.

The fiscal gap – that is, the difference between the resources required to manage the water system and those available for this purpose through the collection of fees and government subsidies – is a serious problem. Among other effects, the hydraulic infrastructure of the city lacks maintenance, manifested in the frequent leaks that consume about 30 per cent of the volume of water passing through the system (Aguirre-Díaz, 2012). To reduce the fiscal gap, water rates that reflect full costs are necessary. Currently, the difference between the price of water to consumers and the full cost of the service is such that it produces the aforementioned imbalances. However, especially in the Federal District, the issue of water rates is politically very sensitive, and has not been adequately addressed by local authorities, who have traditionally applied a policy of water rates subsidies. In this context, information and awareness campaigns would be required in order to educate the public about the need to pay more for their water.

A recent estimate of water investment needs in only the Federal District area of Mexico City's water and wastewater system is US$10 billion over the next ten years; this is a capital expenditure that exceeds the long-term budget in current plans (Aguirre-Díaz, 2012). Water and wastewater services in Mexico (as elsewhere) are highly political, and it is unclear how Mexico City's water supply and wastewater investments will be financed. Tariffs, taxes and transfers all need to be examined seriously and honestly, and set apart from political party agendas as much as possible. In fact, Cox and Börkey (Chapter 3 in this book) point out the complexities behind the use of these economic instruments.

Many specialists in water governance are of the opinion that a new institutional arrangement for water management in the Valley of Mexico is sorely needed. Greater transparency and probably a new regulatory institution would also contribute towards a solution to the valley's water woes. The creation of a new agency that would be responsible for operating the entire water system throughout the metropolitan area has been proposed. While the proposal is under discussion, it faces serious legal and political challenges. Much may depend on the proposal for Mexico City in regard to what is arguably its most precious resource: water.

Conclusions

The current status of water resources in the Mexico City metropolitan area is to some extent the result of their historical management, but is more so the outcome of current national policies based on a geopolitical approach. The decision to erect the capital of New Spain over the ruins of the Aztec capital, taken by the conquistadors, sealed the fate of the city. Over the centuries since then, the concentration of power in a strongly centralized state has made Mexico City the axis of the economic, social, cultural and political life of a country with more than 2 million km^2 of territory.

The shaping force for the management of water resources in the city has been population growth, especially during the second half of the twentieth century. This has meant that water managers, in searching for new sources of supply and sanitation processes in a closed basin with recurring threats of flooding, have always been, to some extent, only catching up with the problems. Hydraulic infrastructure and sanitation processes have become increasingly expensive, and the difficulty in providing them has been exacerbated by the phenomenon of subsidence resulting in the over-exploitation of aquifers.

Decisions made throughout history by water managers in Mexico City have been limited, or even dominated, by political criteria and not by technical or scientific information. Today, the main pathway to solving the current water problems is a change in strategy, such that sustainability and water security are privileged in the decision-making process. The infrastructure and operations currently under planning and in execution will give the city a new lifespan. Nevertheless, unless conditions of sustainability are achieved (e.g. cessation of aquifer over-exploitation) and the existing gaps in water governance are closed, the sustainable development of the metropolitan area will continue to be compromised. Of all the gaps, the most important are those related to policy, administration and public participation.

The potential approaches to solving the gaps in water governance are improvement of coordination among the institutions involved, and the creation of an autonomous water management institution for the metropolitan area. The first approach has proven to be insufficient, while the second seems to have a better chance of success. Under any scenario, the future water policy for the Valley of Mexico should be focused on achieving water security and sustainability of the basin in the long term, moving away from a crisis-based policy design and response. The valley and its 20 million inhabitants have much to lose from the lack of sustainable approaches to managing its water issues.

Note

1 This issue has been well documented. Among the several authors that have addressed it, see the following: Ramírez-Fuentes *et al.*, 2002; Flores *et al.*, 1997; Mendoza *et al.*, 1996; Siebe, 1995; Siebe and Cifuentes, 1995; Flores-Delgadillo *et al.*, 1992; Cajuste *et al.*, 1991; Kabala *et al.*, 2011.

References

Aguirre-Díaz, R. (2012) 'Principales retos del sistema de aguas de la Ciudad de México', in V. Calvo (ed.) *El Gran reto del agua en la Ciudad de México,* Sistema de Aguas de la Ciudad de México, Mexico City.

Akhmouch, A. (2012) *Water governance in Latin America and the Caribbean: A multi-level approach*, OECD Publishing, Paris.

Ardavín, J. R. (2013) 'Programa de sustentabilidad hídrica en el Valle de México. El futuro hídrico en la cuenca de México', in *El agua y el Valle de México*, Comisión Nacional del Agua, Mexico City.

Arreguín, C. F. (2012) 'Uso estratégico del acuífero', in V. Calvo (ed.) *El gran reto del agua en la Ciudad de México,* Sistema de Aguas de la Ciudad de México, Mexico City.

Bahri, A. (1999) 'Agricultural reuse of wastewater and global water management', *Water Science and Technology*, vol. 40, no. 4–5, pp. 339–346.

Bandala, E. R., Cossio, H., Sánchez-Lopez, A. D., Córdova, F., Peralta-Hernández, J. M. and Torres, L. G. (2013a) 'Scaling-up parameters for site restoration process using surfactant-enhanced soil washing coupled with wastewater treatment by Fenton and Fenton-like processes', *Environmental Technology*, vol. 34, no. 3, pp. 363–371.

Bandala, E. R., Tiro, J. B., Lujan, M., Camargo, F. J., Sanchez-Salas, J. L., Reyna, S., Moeller, G. and Torres, L. G. (2013b) 'Petrochemical effluent treatment using natural coagulants and an aerobic biofilter', *Advances in Environmental Research*, vol. 2, no. 3, pp. 229–243.

Cajuste, L. J., Carrillo, G. R., Cota, G. E. and Laird, R. J. (1991) 'The distribution of metals from wastewater in the Mexican Valley of Mezquital', *Water, Air and Soil Pollution*, vol. 57/58, no. 1, pp. 763–771.

Calderón-Preciado, D., Matamoros, V. and Bayona, J. M. (2011) 'Occurrence and potential crop uptake of emerging contaminants and related compounds in an agricultural irrigation network', *Science of the Total Environment*, vol. 412–413, pp. 14–19.

Castro-Rosas, J., Cerna-Cortés, J. F., Méndez-Reyes, E., Lopez-Hernandez, D., Gómez-Aldapa, C. A. and Estrada-Garcia, T. (2012) 'Presence of faecal coliforms, *Escherichia coli* and diarrheagenic *E. coli* pathotypes in ready-to-eat salads, from an area where crops are irrigated with untreated sewage water', *International Journal of Food Microbiology*, vol. 156, no. 2, pp. 176–180.

Chávez, A., Maya, C., Gibson, R. and Jiménez, B. (2011) 'The removal of microorganisms and organic micropollutants from wastewater during infiltration to aquifers after irrigation of farmlands in the Tula Valley, Mexico', *Environmental Pollution*, vol. 159, no. 5, pp. 1354–1362.

Cifuentes, E., Blumenthal, U., Ruiz-Palacios, G., Bennett, S. and Peasey, A. (1994) 'Epidemiological panorama for the agricultural use of wastewater: The Mezquital Valley, Mexico', *Salud Pública de México*, vol. 36, no. 1, pp. 3–9.

Cifuentes, E., Gómez, M., Blumenthal, U., Tellez-Rojo, M. M., Romieu, I., Ruiz-Palacios, G. and Ruiz-Velazco, S. (2000) 'Risk factors for *Giardia intestinalis* infection in agricultural villages practicing wastewater irrigation in Mexico', *American Journal of Tropical Medicine and Hygiene*, vol. 62, no. 3, pp. 388–392.

Cisneros-Iturbe, H. L. and Domínguez-Mora, R. (2007) 'The detention dams of Mexico City: New challenges', paper presented at *NOVATECH*, Lyon, 25–28 June 2007.

CONAGUA (2011) *Estadísticas del Agua en México*, SEMARNAT, Mexico City.

CONAGUA (Comisión Nacional del Agua) (2012a) *Programa hídrico regional visión 2030. Región hidrológico-administrativa XIII Aguas del Valle de México*, SEMARNAT, Mexico City.

CONAGUA (2012b) *Memoria documental. Planta de tratamiento de aguas residuales de Atotonilco*, SEMARNAT, Mexico City.

DFID (Department For International Development) (1998) *Impact of wastewater reuse on groundwater in the Mezquital Valley, Hidalgo State, Mexico*, CNA, BGS, LSHTM and UB. Final Report – November 1998. BGS Technical Report WC/98/42.

Downs, T., Cifuentes-García, E. and Suffet, I. M. (1999) 'Risk screening for exposure to groundwater pollution in a wastewater irrigation district of the Mexico City region', *Environmental Health Perspectives*, vol. 107, no. 7, pp. 533–561.

Durán-Moreno, A., Garcés-Rodríguez, M., Velasco, A., Marín-Enriquez, J. C., Gutiérrez-Lara, R., Moreno-Gutiérrez, A. and Delgadillo-Hernández, N. A. (2013) 'Mexico City's municipal solid waste characteristics and composition analysis', *Revista Internacional de Contaminación Ambiental*, vol. 29, no. 1, pp. 39–46.

Félix-Cañedo, T. E., Durán-Álvarez, J. C. and Jiménez-Cisneros, B. (2013) 'The occurrence and distribution of a group of organic micropollutants in Mexico City's water sources', *Science of the Total Environment*, vol. 454–454, pp. 109–118.

Flores, L., Blas, G., Hernández, G. and Alcalá, R. (1997) 'Distribution and sequential extraction or some heavy metals from soils irrigated with wastewater from Mexico City', *Water, Air and Soil Pollution*, vol. 98, pp. 105–117.

Flores-Delgadillo, L., Hernández-Silva, G., Alcalá-Martinez, R. and Maples-Vermeersch, M. (1992) 'Total contents of cadmium, copper, manganese and zinc in agricultural soils irrigated with wastewater from Hidalgo, Mexico', *Revista Internacional de Contaminación Ambiental*, vol. 8, no. 1, pp. 37–46.

Friedel, J. K., Langer, T., Siebe, C. and Stahr, K. (2000). 'Effects of long term waste water irrigation on soil organic matter, soil microbial biomass and its activities in central Mexico', *Biological Fertilization of Soils*, vol. 31, no. 5, pp. 441–421.

Gallegos, E., Warren, A., Robles, E., Campoy, E., Calderon, A., Sainz, M. G., Bonilla, P. and Escolero, O. (1999) 'The effects of wastewater irrigation on groundwater quality in Mexico', *Water Science and Technology*, vol. 40, no. 2, pp. 45–52.

Harding, D. and Willis, D. (2010) 'Soft ground tunnelling on a Mexico City wastewater project', in L. R. Eckert, M. Fowler, M. F. Smithson, R. Bradford and F. Townsend (eds), *North American Tunnelling 2010 Proceedings*, SME, Littleton, CO.

Jiménez, B. (2006) 'Irrigation in developing countries using wastewater', *International Review for Environmental Strategies*, vol. 6, no. 2, pp. 229–250.

Jiménez, B. and Chávez, A. (2004) 'Quality assessment of an aquifer recharged with wastewater for its potential use as drinking source: El Mezquital case', Proceedings of the IWA Conference on Wastewater Reclamation and Reuse, Mexico City.

Jiménez, B. and Garduño, H. (2001) 'Social, political and scientific dilemmas for massive wastewater reuse in the world', in C. Davis and R. McGin (eds), *Navigating rough waters: Ethical issues in the water industry*, American Water Works Association, London.

Jiménez, B., Chávez, A., Barrios, E. and Pérez, R. (2000) 'Impact and potential of reused water in the Mezquital Valley', *Water 21*, no. 2.3, pp. 34–36.

Jiménez-Cisneros, B. and Chávez-Mejía, A. (1997) 'Treatment of Mexico City wastewater for irrigation purposes', *Environmental Technology*, vol. 18, no. 7, pp. 721–729.

Justin-Cajuste, L., Vázquez-Alarcón, A., Siebe-Grabach, C., Alcántar-González, G. and De la Isla de Bauer, M. L. (2001) 'Cadmio, níquel y plomo en agua residual, suelo y cultivos en el Valle del Mezquital, Hidalgo, México', *Agrociencia*, vol. 35, no. 3, pp. 267–274.

Kabala, C., Karczewska, A., Szopka, K. and Wilk, J. (2011) 'Copper, zinc and lead fractions in soils long-term irrigated with municipal wastewater', *Communications in Soil Science and Plant Analysis*, vol. 42, no. 8, pp. 905–919.

Lucho-Constantino, C. A., Prieto-García, F., Del Razo, M. L., Rodríguez-Vázquez, R. and Poggi-Varaldo, H. M. (2005) 'Chemical fractionation of boron and heavy metals in soils irrigated with wastewater in central Mexico', *Agriculture, Ecosystems and Environment*, vol. 108, no. 1, pp. 57–71.

Lucho-Constantino, C. A., Poggi-Varaldo, H. M., Del Razo, L. M., Cebrián, I., Sastre-Conde, I., Beltrán-Hernández, R.I. and Prieto-García, F. (2008) 'Arsenic in soil, plants and food chain issues: Effect of wastewater irrigation on arsenic concentration in soils and selected crops of Irrigation District 03, Hidalgo State, Mexico', in J. Bundschuh, P. Birkle, M. A. Armienta, P. Bhattacharya, J. Matschullat and A. B. Mukherjee (eds), *Natural arsenic in groundwaters of Latin America*, Taylor and Francis Publishers, Abingdon, UK.

Luege, J. L. (2012) *El sistema Cutzamala. El cárcamo de Dolores, en el agua y el Valle de México*, Comisión Nacional del Agua, Mexico City.

Mendoza, C. A., Cortes, G. and Muñoz, D. (1996) 'Heavy metal pollution in soils and sediments of rural developing District 063, Mexico', *Environmental Toxicolology and Water Quality*, vol. 11, no. 4, pp. 327–333.

OECD (Organization for Economic Co-operation and Development) (2013) *Hacer posible la reforma de la gestión del agua en México*, OECD Studies on Water, Paris.

Oswald, U. (2011) 'Aquatic systems and water security in the Metropolitan Valley of Mexico City', *Current Opinion in Environmental Sustainability*, vol. 3, no. 6, pp. 497–505.

Perló, M. and González, A. E. (2009) ¿Guerra por el agua en el Valle de México?, Universidad Nacional Autónoma de México, Mexico City.

Prieto-García, F., Lucho-Constantino, C.A., Poggi-Valardo, H., Alvarez-Suárez, M. and Barrado-Esteban, E. (2007) 'Caracterización fisicoquímica y extracción secuencial de metales y elementos trazas en suelos de la región Actopan-Ixmiquilpan del distrito de riego 03, Valle de Mezquital, Hidalgo, México', *Ciencia Ergo Sum*, vol. 14, no. 1, pp. 69–80.

Ramírez-Fuentes, E., Lucho-Constantino, C., Escamilla-Silva, E. and Dendooven, L. (2002) 'Characteristics and carbon and nitrogen dynamics in soil irrigated with wastewater for different lengths of time', *Bioresource Technology*, vol. 85, pp. 179–187.

Ramos-Leal, J. A., Noyola-Medrano, C. and Tapia-Silva, F. O. (2010) 'Aquifer vulnerability and groundwater quality in megacities: Case of the Mexico Basin', *Environ Earth Sci.*, vol. 61, pp. 1309–1320.

Romero, P. (2010) 'Water in Mexico City: What will climate change bring to its history of water-related hazards and vulnerabilities?', *Environment and Urbanization*, vol. 22, no. 1, pp. 157–178.

Siebe, C. (1994) 'Acumulación y disponibilidad de metales pesados en suelos regados con aguas residuales en el distrito de riego 03, Tula, Hidalgo, México', *Revista Internacional de Contaminación Ambiental*, vol. 10, no. 1, pp. 15–21.

Siebe, C. (1995) 'Heavy metal availability to plants and soils irrigated with wastewater from Mexico City', *Water Science and Technology*, vol. 32, no. 12, pp. 29–34.

Siebe, C. and Cifuentes, E. (1995) 'Environmental impact of wastewater irrigation in central Mexico: An overview', *International Journal of Environmental Health Research*, vol. 5, no. 2, pp. 161–173.

Siemens, J., Huschek, G., Siebe, C. and Kaupenjohann, M. (2008) 'Concentration and mobility of human pharmaceuticals in the world's largest wastewater irrigation system, Mexico City-Mezquital Valley', *Water Research*, vol. 42, no. 8–9, pp. 2124–2134.

Tortajada, C. (2008) 'Challenges and realities of water management of megacities: The case of the Mexico City Metropolitan area', *Journal of International Affairs*, vol. 61, no. 2, pp. 147–166.

Tortolero, V. A. (2000) *El agua y su historia. México y sus desafíos hacia el siglo XXI*, Siglo XXI Editores, Mexico City.

UNFPA (United Nations Population Fund) (2011) *Estado de la Población Mundial 2011*, Fondo de Población de la Naciones Unidas, New York.

Zambrano-García, A., Medina-Coyotzin, C., Rojas-Amaro, A., López-Veneroni, D., Chang-Martínez, L. and Sosa-Iglesias, G. (2009) 'Distribution and sources of bioaccumulative air pollutants at Mezquital Valley, Mexico, as reflected by the atmospheric plant *Tillandsia recurvata* L.', *Atmospheric Chemistry and Physics*, vol. 9, pp. 6479–6494.

8 Urban groundwater supply in Latin American cities

Panorama and the cases of Mexico City and São Paulo

Jürgen Mahlknecht, Ricardo Hirata and Rogelio Ledesma-Ruiz

Introduction

There is 50 times more groundwater within the planet than on its surface in rivers, lakes and other surface water bodies (NWC, 2012). Regarding water supply, groundwater is usually favoured over surface water when both are an option. This is because groundwater is often close to the area it supplies (usually directly underneath it), is well protected from surface contamination and thus needs minimum treatment to achieve high quality, and is less susceptible than surface water to drought and other climate-induced variability. It is also because of the nature of investments in groundwater development: if planned properly, groundwater infrastructure investments can be made more gradually than those for surface water; this allows for growing private, municipal and industrial demand to occur with lower investment and operation costs (Howard and Gelo, 2003). Groundwater is also more evenly distributed in aquifers over large regions. Therefore, it is not surprising that many of the most populous cities in the world, including some in Latin America, attribute their early origins and development to good quality groundwater obtained from shallow wells (Zektser and Everett, 2004).

Groundwater contributes a quarter of the total global water withdrawals, and nearly half of all drinking water in the world – a consequence of the rising cost of, competition for, and decreasing quality of surface water almost everywhere (UNESCO-UN/WWAP, 2009). More than half of the world's megacities depend on groundwater for at least a quarter of their water supply (e.g. Bangkok, Beijing, Buenos Aires, Cairo, Dhaka, Jakarta, Calcutta, Lagos, London, Manila, Mexico City, New Delhi, Shanghai and Tehran) (Morris *et al.*, 2003; Sahuquillo *et al.*, 2012). In Latin America, about 18 important and prominent cities rely significantly on aquifers (Margat, 2008) – Mexico City, Puebla, Chihuahua, Mérida (Mexico), San José (Costa Rica), Lima (Peru), São Paulo (Brazil), Santiago (Chile), Salta and Córdoba (Argentina) are just some of them. Given the prevalence of private, unregulated wells in both rural and urban areas in Latin America, the number of Latin Americans getting their water from groundwater sources might be considerably higher than the official estimates.

This chapter gives an overview of groundwater issues in urban Latin America. Groundwater has been and still is a critical, strategic resource for the sustained growth and prosperity of Latin American cities. Its use and issues regarding its sustainability require considerably more recognition, attention and protection than currently received. The chapter first addresses essential aspects of groundwater use in Latin America, and it then touches on more specific issues of urban groundwater management. Two case studies are considered to look at groundwater provision and management, and the associated challenges with each, in two of the largest cities in Latin America: Mexico City and São Paulo. In addition to currently being two of the world's most dynamic megalopolises, Mexico City and São Paulo have economies that are comparable in magnitude to medium-sized countries. What is critical to know is that the use and management of local groundwater resources is strategic to their continued sustainability and urbanization. Though this chapter stands on its own, there are links with Chapters 4 and 7, which focus on water affairs in São Paulo and Mexico City, respectively.

Aspects of urban groundwater

Urbanization and the attendant increase in demands and stresses that it brings to surface water supplies translates into groundwater being a key resource for humans in urban settings. But groundwater is a strategic resource in many settings and situations, such as the failure of other water sources due to drought, breakdowns in surface water infrastructure, and where surface water is polluted. It is important for irrigation, farming and energy production, and is a reliable resource for industry.

Groundwater, however, is still poorly understood (Llamas and Custodio, 2003; Vrba and Verhagen, 2011). In part this is because it is a hidden resource, one suffering from mischaracterizations and mistaken perceptions. Its appraisal and monitoring is more complex than that of surface water, so water managers usually do not pay enough attention to the management of groundwater. End-users, for example someone drinking from a mixed groundwater and surface water supply system, are often not aware they are using groundwater. The nature of groundwater fits with the old proverb 'out of sight, out of mind'.

From an environmental and ecosystem sustainability perspective, the groundwater resource is almost completely overlooked. Groundwater's important ecological functions are almost universally unrecognized and unheeded, although groundwater provides base flows to springs, streams, lakes, wetlands and areas of phreatophytes.[1] The average contribution of groundwater to surface water supplies amounts to approximately 50 per cent, while from 5 to 10 per cent of subsurface groundwater flow moves directly into the oceans (Vrba and Verhagen, 2011).

Estimates for 2010 indicate that total global groundwater abstraction was about 1,000km^3. Approximately 67 per cent was used for irrigation, 22 per cent for domestic purposes and 11 per cent for industry (van der Gun, 2012).

Global groundwater withdrawal is increasing at an annual rate of between 1 and 2 per cent. Two-thirds of groundwater abstraction is in Asia, 11 per cent in North America (not including Mexico), 8 per cent Europe and 7 per cent Latin America (with Mexico). In Latin America, about half of the groundwater withdrawn is used for irrigation, one-third is for domestic use, and the rest is for industrial use.

Assessments of groundwater resources in terms of recharge, storage, irrigation use, depletion and footprint have been documented on global and regional scales (Gleeson *et al.*, 2012). International organizations such as the United Nations, World Meteorological Organization and World Bank offer platforms and centres where documents and tools for groundwater can be consulted. However, data on urban groundwater are still limited and often unreliable because water utilities typically do not account for and monitor groundwater separately from their surface water resources. Moreover, data from private wells (most of them being for self-supply purposes and used without permission) represent a large missing component from overall groundwater data assessments (van der Gun, 2012). Comprehensive data on current global urban groundwater use and all of the complex issues associated with it are simply not available.

Urban groundwater development and depletion

Over the past century, technological advances, population growth and economic progress have fostered the successful development of groundwater resources where hydrogeological conditions are favourable. Llamas and Martínez-Santos (2005) called this development the 'silent revolution', mainly because of the decision millions of farmers made to pump groundwater with no control and planning by governmental authority. Groundwater resources development has contributed significantly to the welfare of many countries, but during the past few decades, concern has grown – including in Latin America – about the anthropogenic impact agricultural development and urbanization have had on the resource. Groundwater quality has gradually deteriorated as a result of lacking or leaking sewerage systems, improper operation of water treatment facilities, and inadequate solid and liquid waste disposal, especially in low- and middle-income countries. The deterioration of groundwater quality is a problem in practically all cities in the region (UNEP, 2002). Although the causes are numerous, the bulk of anthropogenic pollution comes from in situ sanitation such as septic tanks and latrines (Kuroda and Fukushi, 2008). Wells drilled in shallow, unconfined aquifers are especially prone to pollution, and they cause long-lasting impacts on public health.

In addition to the deterioration of groundwater quality in many Latin American urban areas produced by human activities, the unsustainable use of groundwater – from a yield perspective – has caused equally serious problems (UNEP, 2002; WWF, 2011; van der Gun, 2012). Uncontrolled heavy groundwater withdrawals have continuously lowered the water table in many Latin American cities, from

Mexico City to Santiago de Chile, increasing the cost of extraction due to the need to drill to ever-growing depths. There are other negative effects such as drying up streams and springs, and irreversible land subsidence. The former affects local ecosystems, and the latter impacts urban infrastructure.

Often the quality and quantity issues associated with groundwater withdrawals are linked, such as when water quality deteriorates due to groundwater extraction from deeper geological formations enriched with arsenic, fluoride or both (e.g. Mexico City and Buenos Aires). In coastal areas of rapid urban growth, saltwater intrusion has occurred as a result of heavy extractions and unsustainable groundwater resources development (e.g. Lima and Buenos Aires). Salinity encroachment on groundwater is a major issue for the sustainability of a number of Latin American cities, as desalinization or dilution of saline waters is an expensive and energy-intensive proposition.

The most sobering scenarios of groundwater deterioration and depletion have a common basis: uncontrolled growth. One of the most destructive of all behaviours affecting groundwater – and this is certainly the case throughout Latin America – is the construction of private, unregulated wells by individuals or groups acting independently (and according to self-interest) in response to a deficient or expensive public water supply service (Hirata *et al.*, 2006). Unregulated, illegal and private 'off the radar' well proliferation is a perfect if unfortunate example of the 'tragedy of commons' (Hardin, 1968) – unfortunate in that it almost always depletes the common groundwater resource, and thus interferes deeply with the sustainability of cities.Human activities alter profoundly the water cycle in regard to quantity and quality over time and space, the groundwater cycle is no exception. Human activities may reduce water availability yet also create new sources of water. The construction of buildings and paving of streets reduces soil water infiltration, increases rain interception and reduces natural aquifer recharge. The overall water balance, however, may show a groundwater recharge increase, resulting from leakage from water supply and sewage networks and storm water systems. The combination of various factors could lead eventually to rising groundwater levels ('groundwater rebound'), as has happened, for example, in Buenos Aires (Foster and Garduño, 2002).

Groundwater and climate change

It is generally recognized that climate change has a relevant impact on water resources. Most climatologists expect precipitation to intensify and groundwater recharge to increase with a warming atmosphere (Grönwall *et al.*, 2010). There are important knowledge gaps, however, concerning the water-related impacts of climate change, especially in regard to groundwater. The uncertainty is due to not only the limited knowledge of projected precipitation and its geographical distribution, but also the fact that knowledge of groundwater recharge is poor. There are many factors contributing to the uncertainty about the effect of climate change on groundwater. One of them is the accuracy of predictions of future greenhouse gas emissions and how these predictions translate into global climate change and groundwater modelling.

Nevertheless, there is general agreement about how climate change may alter long-term (through the 2050s) average recharge and thus renewable groundwater resources. It is expected that groundwater recharge will increase in the northern latitudes, but will decrease strongly, by 30 to 70 per cent or even more, in some current semi-arid zones (Döll, 2009; van der Gun, 2012; Hirata and Conicelli, 2012). The latter areas include central and northern Mexico, north-east Brazil and Argentina; but of course, uncertainties exist in these predictions. While acknowledging the difficulties associated with long-term predictions from studies and modelling of climate change effects, when these are coupled with projections for regional and global population growth, it is clear that both climate change and population growth are likely to have significant impacts on the world's water resources (Vörösmarty *et al.*, 2000; Loaiciga, 2009).

Conjunctive management that takes into account both surface water and groundwater may play an important role in overcoming the problems expected from global climatic changes (Dillon *et al.*, 2009; van der Gun, 2012). Groundwater and surface water are clearly complementary in many hydrogeological aspects, but there are important differences in management approaches to each of them (Table 8.1). Understanding, and then taking advantage of, surface water and groundwater complementarity is crucial to increasing water availability for societal and economic activities, and to protecting and sustaining the environment. The problem of water supply in Latin American cities, and the maintenance of crops, is in many cases much more related to seasonal rain variation than to adequacy or management of the local water infrastructure. Here is where groundwater and surface water complementarity come into play: aquifers can store large amounts of water, but their use occurs through wells, which generally have low to moderate yields. In contrast, rivers and other surface water bodies have relatively low storage capacities but can provide large yield. This characteristic of complementarity is a key for dealing with the problems arising from climate change.

Groundwater management in Latin America

During the past three decades, many Latin American countries have obtained enormous social and economic benefits from groundwater development (Foster and Garduño, 2009). However, updated data on urban groundwater consumption and use are difficult to obtain. Available data, generally presented at the country level, show that about half of the total drinking water originates underground. This varies by country (Figure 8.1): in Mexico, Chile, Peru, Venezuela, Nicaragua, Costa Rica, El Salvador, Surinam, Guyana and French Guyana, the figure is 50–100 per cent; in Brazil, Colombia, Argentina, Bolivia, Guatemala, Belize and Honduras, it is 25–50 per cent; while Paraguay, Uruguay, Ecuador and Panama are less dependent on groundwater (Morris *et al.*, 2003).

The magnitude of groundwater demands imposed by urban expansion in the most arid areas of Latin America (northern and central Mexico, north-eastern

Table 8.1 Comparison of groundwater and surface water according to hydrological characteristics and socio-economic factors

Feature	*Hydrological characteristics*	
	Groundwater	*Surface water*
Storage volumes	Very large	Small to moderate
Resource areas	Relatively unrestricted	Restricted to water bodies
Flow velocities	Very low	Moderate to high
Residence times	Generally decades/ centuries	Mainly weeks/months
Drought propensity	Generally low	Generally high
Evaporation losses	Low and localized	High for reservoirs
Resource evaluation	High cost and significant uncertainty	Low cost and often less uncertainty
Abstraction impacts	Delayed and dispersed	Immediate
Natural quality	Generally (but not always) high	Variable (but generally needs treatment)
Pollution vulnerability	Variable natural protection	Largely unprotected
Pollution persistence	Often extreme	Mainly transitory
	Socio-economic factors	
	Groundwater	*Surface water*
Public perception	Mythical, unpredictable	Aesthetic, predictable
Development cost	Generally modest	Often high
Development risk	Less than often perceived	More than often assumed
Style of development	Mixed public and private	Largely public

Source: Adapted from Cap-Net *et al.* (2010, p. 14)

Brazil, the coasts of Peru and Chile and the pre-Andean region of Argentina) has generated dispute among experts about the sustainability of groundwater use in these areas (Mahlknecht, 2013). Concern about the withdrawal and over-exploitation of groundwater in these regions is exacerbated by land use patterns which do not follow the regional planning that is in place. The situation generates multiple pollution hazards for groundwater resources. Adequate groundwater management – including robust and enforced protection measures – is one of the most complex challenges currently facing Latin American society (Foster and Garduño, 2009).

As its population grows and urbanizes, and increases its demand for both basic and 'middle class' services, Latin America arguably is moving towards worsening the over-exploitation and degradation of its water resources. The

Figure 8.1 Estimated extent of groundwater use for drinking water in Latin America

Source: Morris *et al.* (2003, p. 2)

content of certain chemical constituents in groundwater may increase from the excessive groundwater abstraction that comes with accelerated population growth and urbanization. Often, this pressure for more water means tapping this resource from deeper geological formations. Arsenic, for example, is gradually mobilized through water supply systems from the weathering effect on young volcanic rocks, exposing 14 million people in 14 out of 20 countries in Latin America (Bundschuh *et al.*, 2012). Fluoride is often mobilized in concert with arsenic.

Unfortunately, few groundwater protection and management programmes are currently in place in Latin America. Those that do exist often do not take into account aquifer vulnerability to anthropogenic pollution, or the need to define wellhead protection areas – the latter concept being directly aligned with watershed land protection techniques. Groundwater protection and management programmes in Latin America do not often incorporate groundwater abstraction with the aim of avoiding aquifer over-exploitation, even when excessive groundwater pumping reduces the base flow to rivers and wetlands, causes subsidence, or introduces saltwater intrusion. Most of the problems caused by uncontrolled exploitation of groundwater could be avoided or mitigated by stronger and more appropriate governmental action from those agencies functioning in the evaluation and control of groundwater use (Llamas and Martínez-Cortina, 2007).

As mentioned earlier, one of the central and critical issues confronting groundwater management throughout Latin America is the existence of a large number of unregistered wells, many operating illegally. In Brazil, for example, although there are laws and regulations addressing the drilling and use of private groundwater wells, more than 70 per cent of wells are unregistered, and have been dug in response to poor water service or as a strategy to reduce water supply costs (Hirata, 2012). This situation is not unique to Brazil, though. Drilling for groundwater often ends up being a simpler, cheaper way of producing a local water supply for a neighbourhood. One of the ironies of the wide-ranging existence and use of unregistered wells in Latin America is that while hydrologists and decision-makers in the region discuss and hope, from a strategic perspective, to preserve and protect the region's groundwater resources for the future, without measuring, monitoring or having firm data on unregulated well use, it is impossible to do so with any degree of assurance.

It is worth pointing out that the key issue is not a lack of laws, which are indeed in place throughout Latin American countries and cities, but rather a lack of political will (Zektser and Everett, 2004; GWP, 2012; World Bank, 2013). First, groundwater problems are usually slow to materialize, so the political response from the population is muted and correspondingly slow, in comparison with surface water issues. But there is also a general absence of groundwater management mechanisms (such as instruments for monitoring and implementation) and local resources (institutional incapacity to monitor and advise on groundwater use). There are even conflicts between water and land management authorities that complicate satisfactory groundwater management implementation. At the least, integrated water resource management planning that includes instruments and resources for the sustainable use of groundwater should exist. This would give Latin American aquifer and basin councils more authority over this precious resource.

Climate change and urban population growth cry out for sustainable and efficient long-term groundwater management throughout Latin America. Traditionally, such management has been based on meeting growing demand through augmented supply (the supply-side approach). But now that there is uncertainty about climate change and population growth effects on Latin

American groundwater resources, new and better management aimed at achieving efficiency and sustainability will be required (Hansen, 2012). The importance of groundwater in overall water planning and policy will only increase in the years to come; its serious consideration by high-level decision-makers and full integration into planning and policy that includes surface water concerns must occur (Bates *et al.*, 2008).

Case studies

Mexico City, Mexico

The metropolitan area of Mexico City is located in a relatively small, endorheic basin (Valley of Mexico) with modest rainfall.[2] Over the past century, the city's exponential growth has virtually depleted its groundwater resources. Heavy groundwater extraction started in the 1940s with the first massive drilling of deep wells, and the overdrafting of the safe yield of the basin's groundwater became common in the 1960s. Additional surface water sources from neighbouring basins have been developed and incorporated into Mexico City's water supply system to cover the deficit (Breña-Puyol and Breña-Naranjo, 2009; CONAGUA, 2009). Figure 8.2, however, shows that the trend of aquifer over-exploitation has not been reversed. Today, the abstraction rate from the aquifer is nearly twice the recharge rate.[3]

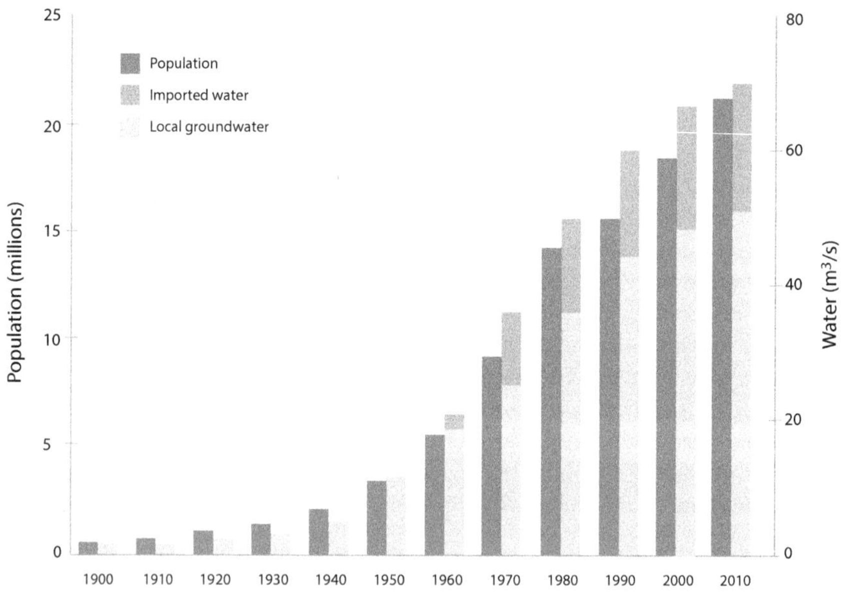

Figure 8.2 Population of and extraction and importation of water to Mexico City, 1900–2010

Sources: Adapted from Mazari-Hiriart and Mazari-Menzer (2008, Figure 1), updated with data from Rosales (2014)

The overuse of groundwater resources in the Valley of Mexico has two causes. The first has to do with hidden incentives that encourage both water operators and water users to waste water. According to the World Bank (2013), Mexico City's water services are underwritten up to 50 per cent by government subsidies. This situation reduces the water utility's autonomy and motivation to increase physical and financial efficiencies, or even measure and regulate water extractions; it also gives water users the false impression that water is less expensive than it is by underwriting its cost. In moving forward with the management of the Valley of Mexico's groundwater, it will be important to develop a transparent model, which considers full costs of not only assets, personnel and current operations, but also future investments.

The second cause relates to the significant increase in the number of private wells drilled in response to the intermittent service received from water authorities. About 66 per cent of current water consumption in the Valley of Mexico comes from local aquifers – through 1,581 deep wells – and the rest from surface systems (24 per cent from the Cutzamala system, 8 per cent from the Lerma system, and 2 per cent from a dam and springs system) (CONAGUA, 2010). Authorities estimate that 3,000 private wells are operating without permits in Mexico City (CEAA, 2013).

Groundwater extraction has produced multiple adverse effects. Among these are the following ones: decline of the groundwater table; cessation of spring flow; land subsidence, groundwater contamination; loss of phreatophytes and, in the end, a supply of lower quality water (Huizar-Alvarez *et al.*, 2009). At the same time, groundwater recharge has decreased, mainly due to the expansion of the urban area and the consequent loss of vegetation cover. A decree enacted by the Comisión Nacional del Agua (National Water Commission of Mexico – CONAGUA) in 2003, imposing severe limits on the use of groundwater, has not reversed the annual groundwater deficits in the Valley of Mexico, which reached 25 m^3/s by 2012 (World Bank, 2013) and 27 m^3/s by 2013 (Ardavín, 2013). Chapter 7 in this book also touches upon these issues.

Current concerns about Mexico City's aquifer system focus not only on continued land subsidence, but also on production wells that extract water at an ever-growing depth and yield a lower flow rate and/or inferior quality water. In the 1960s, water supply wells were drilled to between 100 and 150m; they are now drilled to between 300 and 450m (SACMEX, 2011). The replacement or rehabilitation of existing wells risks tapping into less productive geological aquifer formations at even deeper levels – up to 800–900m (SACMEX, 2012). With an average decline of groundwater levels at a rate of 1m/year and declining water quality, Mexico City's water authorities are now being forced to conceptualize and implement advanced and improved water purification systems.

Against this background, several recent sanitation and conservation programmes – financed by national and international sources (including private funding) have been implemented. Former President Felipe Calderón launched the 2007–2012 US$2.8 billion Water Sustainability Programme for the Valley of Mexico. This aimed to reduce groundwater over-exploitation (among other objectives) by importing additional water, exchanging wastewater for clean water at present used for irrigation, and rehabilitating existing infrastructure (Ardavín, 2013). In parallel with that programme, Mexico City's Federal District government launched the Green Plan, a 15-year plan whose objectives include the reduction of the Valley of Mexico aquifer's deficit through reforestation and the building of recharge structures such as infiltration trenches, gabion dams and absorption wells (SACMEX, 2012).

However, it is very unlikely that the above programmes and other laudable ones – such as the reuse of treated wastewater and the reduction of water operating inefficiencies – and even the incorporation of new sources, can stop the groundwater depletion that has taken place in Mexico City. Furthermore, according to water authorities, the incorporation of new water supply sources with the aim of reducing groundwater extractions and recovering the current deficit in the medium term is practically unavoidable (CONAGUA, 2012; World Bank, 2013). What can best be hoped for and achieved now is a stabilization of the groundwater table, which, when combined with new measures, will help lead to a more sustainable use of water within Mexico City over the long term.

Over the decades, the rising demand for water in Mexico City has put enormous pressure on local and neighbouring surface water and groundwater supply sources, causing both economic and environmental damage. The now fully developed practice of importing water to meet urban demand, coupled with water scarcity, has led to a series of social and political conflicts over the distribution and management of water resources in Mexico City (Cohen and Reynoso, 2005). But Mexico City's water supply problems are not just about finding new water sources (Carrera-Hernández and Gaskin, 2009). The overall water system is extremely complex and faces very strong limitations: ageing infrastructure; operational costs; and lack of investment in maintenance and rehabilitation. The deterioration of water quantity and quality in the Valley of Mexico have led Mexico City's supply system to a limited economic and physical operability (Tortajada and Castelán, 2003; WWF, 2011).

Most groundwater extraction in the Valley of Mexico is performed directly by numerous utilities operating under different municipal and state entities, and the Federal District. Information on extraction is not reliable and there is no evidence of systematic measurement and follow-up on extracted volumes and groundwater table data coordinated overall by CONAGUA, despite the critical level of groundwater over-exploitation (World Bank, 2013). There is no basin-level, autonomous entity that has the capability not only to consolidate information on water use, aquifer status and evolution, consumption patterns, and more, but also to design and implement policy programmes for the sustainable use of the Valley of Mexico's groundwater. A strong and independent utility should be established that interacts with both governmental entities and other stakeholders.

São Paulo, Brazil

Groundwater supply plays an important role in the Upper Tietê Basin region of Brazil (Figure 8.3), where about 20 million people live in the city of São Paulo and 35 other municipalities. The total water consumption is 2,545 million m^3/year (80.7 m^3/s). Of this amount, 2,195 million m^3/year (69.6 m^3/s) comes from surface water reservoirs (SABESP, 2014) and 350 million m^3/year (11.1 m^3/s) from about 10,000 private wells that are in operation (out of more than 13,000 drilled wells) (Servmar, 2013). It is estimated that 100 to 200 new wells come into operation each year in the Upper Tietê Basin. The São Paulo State Basic Sanitation Company (SABESP) operates a complex water supply system: the Cantareira reservoir (1,041 million m^3/year; 33m^3/s) supplies more than 50 per cent of São Paulo's needs, followed by the Guarapiranga (489 million m^3/s year; 14m^3/s), Alto Tietê (328 million m^3/year; 15m^3/s) and Rio Grande (183 million m^3/year; 5m^3/s) systems. Smaller additional amounts come from the Rio Claro and Upper and Lower Cotia systems (SABESP, 2014).

The state of São Paulo has a decentralized water resource management system consisting of 22 water basin management units; the Upper Tietê Basin is one of them. Each unit is governed by a basin committee, with the participation of state government institutions, municipalities and civil society. In the case of the Upper Tietê Basin, the basin committee plans and manages the surface water and groundwater in the Upper Tietê Basin, and the basin agency implements its policies (FUSP, 2009). Despite the creation of the basin committee system in 1991, its implementation is still very poor in practice. Actual water resource

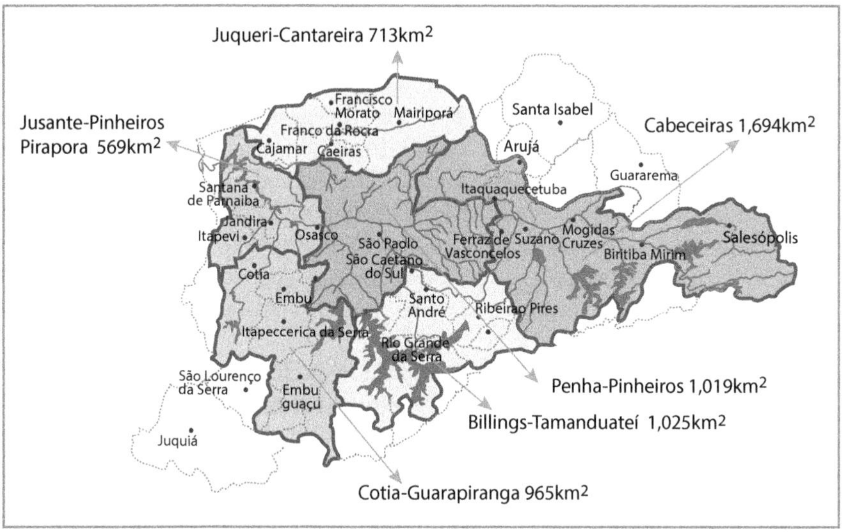

Figure 8.3 Location of the Upper Tietê Basin in the state of São Paulo, as well as its municipalities and sub-basin units

Source: Adapted from UBC (2010)

management is still centralized in state government institutions. Within the state government, the State Secretary for Water Supply, Sanitation and Energy is in charge of water supply and sanitation policy, and the State Secretary for the Environment is responsible for environmental policy, including environmental aspects of water resources management (Hirata *et al.*, 2014).

Although groundwater resources represent only 14 per cent of the total water produced in the Upper Tietê Basin, this share maintains the balance between demand and supply. The loss of the Upper Tietê Basin's groundwater resource, either by contamination or by over-exploitation, would create significant stress on the Upper Tietê Basin's water supply system. SABESP cannot produce additional water without huge investments in a system that transports water from a distant basin (the São Lourenço system) and that will take years to come on line. This new supply will capture 4.3m^3/s of water from the Cachoeira do France dam (Vargem Grande do Sul), which is located 83km from the city of São Paulo (SABESP, 2014).

From the legal status perspective, of existing wells, the Upper Tietê Basin presents a troublesome situation. It is estimated that more than 60 per cent of the wells are illegal (with no extraction controls) (Servmar, 2013), and many exist because of the lack of enforcement of existing groundwater use regulations. It is also important to emphasize that both private unregulated wells and the aquifer itself face the risk of contamination. Many wells are constructed in substandard ways, causing quality problems. More than 50 per cent of the known groundwater contamination cases within the state of São Paulo are within the Upper Tietê Basin – a result of the São Paulo region's strong industrialization and dense, rapid and unplanned urban settlement.

Rainfall within the Upper Tietê Basin's agricultural and non-urban areas (43 per cent of the total watershed area; 150–300 mm/year) recharges its aquifers. In the vast São Paulo urban area, natural recharge is reduced by soil impermeability and buildings. The total recharge, however, increases up to 350 mm/year due to leakages from water mains and sewers. In those urban areas without sewage transport and treatment systems, the sum of natural and anthropogenic recharge can reach up to 450 mm/year (FABHAT-Servmar, 2013). Groundwater flows naturally towards surface water bodies, and in the Upper Tietê Basin; the Tietê River is the main flow point. However, intense groundwater extraction has created some hydraulic head depressions that have changed the natural water flow.

Until the decade 1970–1980, groundwater in the Upper Tietê Basin was predominantly exploited from the upper sedimentary aquifer system underlying the region. Economic expansion increased the use of and subsequently diminished the upper sedimentary aquifer system, and so extraction from the deeper crystalline aquifer system followed. Industry is still the largest user of groundwater in the Upper Tietê Basin, although today most wells are drilled for condominium and other residential properties so as to avoid the cost of connecting to and paying for the maintenance of the public water supply system (Hirata *et al.*, 2014).

The potential vulnerability of São Paulo's groundwater system is evidenced in the following statistics. There are 2,320 contaminated sites in the Upper Tietê Basin's urban areas (CETESB, 2012). Petrol stations are responsible for 73 per cent of these, while 27 per cent are related to industrial areas, of which 12 per cent are contaminated by chlorinated solvents. Contaminated sites are located in the sub-basins Penha-Pinheiros (58 per cent) and Billings-Tamanduateí (23 per cent) (FABHAT, 2014). About 37 per cent of the contaminated sites are in areas of high aquifer vulnerability, as classified by the pollution index GOD[4] (Foster and Hirata, 1988), 41 per cent in areas of medium vulnerability, and 22 per cent in areas of medium to low vulnerability (FABHAT-Servmar, 2013). The São Paulo Environmental Agency potential contaminant load inventory (SIPOL), presents 82,682 activities with the potential to generate contaminant load to aquifers (including those using chlorinated solvents and heavy metals). Based on the POSH[4] method (Foster and Hirata, 1988), 53,456, 11,228 and 9,560 activities were classified as high, moderate and low potential, respectively, of contaminating groundwater.

These statistics show the risk of losing São Paulo's groundwater to contamination. Currently, groundwater is a complementary resource for the overall water supply system. Any problems that restrict its use will cause an increase of public system water demand. Because São Paulo's many supply distribution systems are not integrated, water within the Upper Tietê Basin cannot be transferred easily from one system to another. Risks of shortages of surface water in the Upper Tietê Basin exist too, thereby calling into question the presumed capability of the Upper Tietê Basin to increase surface water production to compensate for the loss of groundwater resources. High risks of shortages are expected in the Cantareira, Guarapiranga, Lower Rio Grande and Cotia production systems, which provide water for 78 per cent of the Upper Tietê Basin's population (Hirata *et al.*, 2014).

The São Paulo State Government has implemented a policy of identifying those areas with a higher risk of aquifer over-exploitation and contamination by anthropogenic activities (using the POSH method). Based on the distribution of production wells and contaminant loads, regional studies have been undertaken to identify areas that should be studied in more detail. If necessary, these areas will have restrictions placed on the use of groundwater.

Although there are regulations controlling the drilling for and exploitation of groundwater in the Upper Tietê Basin, there are serious problems with the enforcement of these regulations. State government agencies claim that a lack of technical staff and administrative infrastructure is to blame. However, the problem is much more complex and relates to a number of factors. First, there is not much societal and governmental awareness of the true value (and cost) of groundwater. Second, groundwater regulation is flawed because it only penalizes illegal users (thousands) and not the companies that drill the illegal wells (a much smaller number to enforce laws on). Third, groundwater users see no benefit in having wells legalized, because of the bureaucratic hurdles and fees that must be paid to do so. Finally, the groundwater (over-exploitation and contamination) conflicts are not apparent, or are difficult to detect; therefore, users who are affected by such problems do not recognize them nor pressure the state for control measures (Hirata *et al.*, 2014).

It is thus imperative to design and implement an effective programme leading to the regulation of illegal wells. A key issue that needs to be taken into account in its design is the drilling and exploitation limit permits and their enforcement. It is also necessary to monitor groundwater use in order to update information from registered users and, most importantly, to identify illegal users. A coordinated, cooperative and integrated effort by all Upper Tietê Basin water agencies is needed to make this a reality.

These top-down initiatives should be complemented with bottom-up initiatives, including incentives and penalties – sanctions that go side-by-side with communication and education programmes. Mechanisms for groundwater stakeholder participation are usually much less defined in urban areas such as São Paulo than in rural areas. But the representation and engagement of major stakeholder groups is an essential component of any action plan for the Upper Tietê Basin's groundwater resource management and protection; thus there is a need to build social consensus.

Comparison of the case studies

The case studies present two Latin American megacities with their own biophysical particularities and complexities in providing water. Mexico City, located in the endorheic Valley of Mexico, a high plateau with a subtropical highland climate, is built mostly on dewatered lakes and has underlying sedimentary and volcanic rocks. Local groundwater was for a long time practically the only source of Mexico City's water supply, until significant sources of imported water were added in the 1950s. Today, local and imported groundwater makes up about 75 per cent of Mexico City's public water supply, and is provided by a large number of utilities. São Paulo, in contrast, has a humid subtropical climate and sits on both sedimentary and crystalline aquifer systems. This city gets only 14 per cent of its water supply from local groundwater sources, although this percentage allows São Paulo to maintain the balance between current demand and the need for new water production.

These cases not only show how Latin America's two most populated cities rely on groundwater in different ways and to different degrees, but also show groundwater development at different stages and with both different and similar challenges. São Paulo continues to add hundreds of wells each year, although parts of its aquifers are now over-exploited and contaminated. Mexico City has managed to maintain well capacity during the past decade, even though all of its aquifers are now over-exploited. In order to meet its current water supply demands, Mexico City imports substantial amounts of water from neighbouring water basins. Unless both surface water and groundwater supplies are managed thoughtfully and sustainably, São Paulo may need to make large investments to import water from adjacent water basins. In both Mexico City and São Paulo, the greatest problems behind groundwater management are the unregulated or illegal wells that tap into the resource. The proportion of unregulated or illegal wells ranges from 60 per cent (São Paulo) to 65 per cent (Mexico City) when compared to the total number of wells within each of the two systems. These wells are often not constructed using rigorous or even acceptable standards, causing important additional abstraction and water quality problems; no records are kept of these two situations. Current concerns about Mexico City's aquifer system focus on continued land subsidence and increasing extraction depth with lower productivity, while São Paulo's interests focus more on groundwater issues. General groundwater awareness in the latter city is much lower than in Mexico City.

Coordination and integration of water management is an issue in both Mexico City and São Paulo. In the Valley of Mexico, groundwater is managed and extracted for public use by a large number of uncoordinated water utilities. Mexico's national water authority, CONAGUA, is responsible for defining and implementing overall water resources management. The Basin Council of the Valley of Mexico, represented by governmental institutions and civil society, has only limited competencies when compared to CONAGUA and other government agencies.

In the Upper Tietê Basin, groundwater extraction for São Paulo is undertaken by the water utility SABESP. But water resources management is planned and managed by a basin committee made up of government institutions and civil society, and this basin committee's planning and policies are implemented by the basin's water agency. Water resources management and regional planning are not harmonized. This indicates a lack of consensus among the federal, state and local authorities as well as society. These situations hamper the sustainable development of the two cities, especially concerning the management of their groundwater resources. It is believed that Mexico City's water supply system needs a major institutional redesign to overcome its water crisis, while São Paulo's priority must be a programme to register all illegal wells.

Conclusions

Historically, groundwater has been a key factor and critical resource underlying the social and economic development of many cities, especially when used in conjunction with surface water resources. In many cases, and certainly for both Mexico City and São Paulo, the growth of megacities is unimaginable without groundwater being considered an accessible, flexible and seemingly inexhaustible resource. Groundwater is far more significant from a water supply perspective than is commonly appreciated – it is one of those 'invisible links' between various facets of the urban infrastructure. Unfortunately, though, government agencies at all levels that are responsible for urban water supply and natural resource management often have a poor understanding of groundwater and do not collect comprehensive data on groundwater in comparison to surface water. The lack of understanding about, and poor monitoring and management of, groundwater needs to be corrected.

There is an increasing concern about anthropogenic impacts on groundwater resources in urban Latin America. Leakage from in situ sanitation such as septic tanks and latrines is the most common source of groundwater contamination. Moreover, land use plans that are often aimed in part at protecting groundwater resources are not followed, and generate multiple pollution hazards for these resources. The most important driver for groundwater deterioration throughout the region is the uncontrolled proliferation of shallow private wells. Often the number of the unregistered wells in a city is greater than the number of registered wells. In terms of water quantity, uncontrolled groundwater withdrawals from rapid urban growth have continuously lowered the water table in many cities, affecting adversely extraction costs, local ecosystems, soil firmness and even water quality.

Municipal water utilities in Latin America typically have been primarily focused on a supply-side approach to groundwater – the construction of water wells and their operations – and have not shown much interest in understanding and managing groundwater as a basic resource. This is certainly the case in Mexico City and São Paulo. Few groundwater protection and management programmes are currently in place in Latin America, and the programmes that do exist often do not take into account aquifer vulnerability to anthropogenic pollution, or undertake activities such as defining wellhead protection areas. Both climate change and population growth will affect the region's groundwater resources, so it is crucial that decision-makers consider seriously the role of groundwater and integrate it in overall water planning and policy.

The cases of Mexico City and São Paulo provide useful lessons for water policy. Mexico City's water crisis started early in the 1960s when local groundwater resources were overdrafted. The overuse is and always has been fostered by perverse incentives for water users and water utilities to waste water. The resulting damages to ecosystems and infrastructure from the lowering water table are difficult to quantify but thought to be severe; in addition, of course, to social and political conflicts.

São Paulo's water supply system is not less complex. Although the share of groundwater in its overall water use is much less than in Mexico City, it is an important strategic resource to safeguard the balance between the city's demand and production of public supply. Similarly to Mexico City, the number of private, unregistered wells outweighs the number of public wells, as private wells offer a cheaper and more secure alternative to public water supply. This points to uneven policies, difficulty in enforcing regulations, and institutional incapacity. As for Mexico City, there is a general unawareness of the role groundwater plays in the urban water cycle. It is imperative for São Paulo to design a programme to register illegal wells. This programme should be based on improving the control capability of the authorities and the involvement of decision-makers and stakeholders.

Integrated monitoring and planning for the use of groundwater is essential, and must include a broad range of stakeholder participation if it is to help with the wide range of issues confronting groundwater resources, including rapid urbanization and climate change. More than ever, it is necessary to integrate the entire public network, including self-suppliers of water resources. A new generation of tools and management approaches is needed for Latin America's groundwater resources. But before any 'new generation' work can start on Latin American groundwater issues, consensus is needed between the authorities involved – not only from water and sanitation management but also from land use management – and society so as to create the institutional conditions and incentives for the sustainable use of groundwater. Mexico City and São Paulo are good places to start.

Notes

1 Phreatophytes are long-rooted plants that absorb their water from the water table or the soil above them.
2 For a general description of the geographic and historical setting, see Chapter 7.
3 For the water balance, water supply and drainage scheme of the Valley of Mexico, see Chapter 7.
4 GOD and POSH are groundwater vulnerability assessment methods. Groundwater vulnerability is a measure of how easy or how hard it is for pollution on the surface to reach the water table. These methods assemble information on important factors affecting vulnerability (e.g. groundwater confinement, overlaying strata, depth to groundwater) to develop an index. The outcome – a map of distribution of vulnerability classes – serves for identifying and protecting groundwater resources.

References

Ardavín J. R. (2013) 'Programa de sustentabilidad hídrica en el Valle de México. El futuro hídrico en la cuenca de México', in Comisión Nacional del Agua (eds), *El agua y el Valle de México*, Comisión Nacional del Agua, Mexico.

Bates, B. C., Kundzewicz, Z. W., Wu, S. and Palutikof, J. P. (2008) *Climate Change and Water*, Technical Paper of the Intergovernmental Panel on Climate Change, IPCC Secretariat, Geneva.

Breña-Puyol, A. and Breña-Naranjo, J. (2009) 'Problemática del recurso agua en grandes ciudades: zona metropolitana del Valle de México', *Contacto S*, vol. 74, pp. 10–18.

Bundschuh, J., Litter, M. I., Parvez, F., Román-Ross, G., Nicolli, H. B., Jean, J. S., Liu, C. W., López, D., Armienta, M. A., Gómez-Cuevas, A., Cornejo, L., Cumbal, L., Guilherme, L. R. G. and Toujaguez, R. (2012) 'One century of arsenic exposure in Latin America: a review of history and occurrence from 14 countries', *Science of the Total Environment*, vol. 429, pp. 2–3.

Cap-Net, African Groundwater Network (AGW-Net) and Ground Water Management Advisory Team (GW-MATE) (2010) 'Groundwater management in IWRM. A training manual developed by the collaborative efforts of Cap-Net, AGW-Net and GW-MATE', www.gwp.org/Global/ToolBox/References/Groundwater%20Management%20in%20IWRM.%20Training%20Manual%20(CAP-Net,AGW-Net,GW-Mate,%202010).pdf, accessed 21 May 2014.

Carrera-Hernández, J. J. and Gaskin S. J. (2009) 'Water management in the Basin of Mexico: current state and alternative scenarios', *Hydrogeology Journal*, vol. 17, pp. 1483–1494.

CEAA (Comisión Estatal de Agua y Alcantarillado) (2013) 'Acuífero Cuautitlán', http://cuencavalledemexico.com/wp-content/uploads/2013/02/GEO-83-Modelaci%C3%B3n-subterr%C3%A1nea-C-Pachuca-2.pdf, accessed 8 May 2014.

CETESB (Companhia Ambiental do Estado de São Paulo) (2012) 'Contaminated areas in the state of São Paulo in 2012 (last inventory)', www.cetesb.sp.gov.br/areas-contaminadas/rela%C3%A7%C3%B5es-de-%C3%A1reas-contaminadas/4-Rela%C3%A7%C3%B5es-de-%C3%81reas-Contaminadas, accessed 22 February 2014.

Cohen Perló, M. and González Reynoso, A. E. (2005) *¿Guerra por el agua en el Valle de México? Estudio sobre las relaciones hidráulicas entre el Distrito Federal y el Estado de México*, Coordinación de Humanidades, PUEC, UNAM and Fundación Friedrich Ebert', www.scielo.org.mx/pdf/igeo/n62/n62a11.pdf, accessed 31 May 2014.

CONAGUA (2009) 'Estadísticas del agua de la Región Hidrológica-Administrativa XIII, Aguas del Valle de México', www.conagua.gob.mx/conagua07/noticias/ocavm-1-eaocavm2009.pdf, accessed 31 May 2014.

CONAGUA (2010) 'Sistema Cutzamala', *Organismo de Cuenca Aguas del Valle de México*, Mexico City.

CONAGUA (Comisión Nacional del Agua) (2012) 'Acciones de infraestructura de drenaje y abastecimiento de agua en el Valle de México 2007–2012', Coordinación General de Proyectos Especiales de Abastecimiento de Agua Potable y Saneamiento del Valle de México, www.conagua.gob.mx/conagua07/Noticias/InformeDifusionIngenieriaPSHCVM.pdf, accessed 18 May 2014.

Dillon, P., Pavelic, P., Page, D., Beringen, H. and Ward, J. (2009) *Managed Aquifer Recharge: An Introduction*, Waterlines Report Series No. 13, National Water Commission Australian Government, Australia.

Döll, P. (2009) 'Vulnerability to the impact of climate change on renewable groundwater resources: a global-scale assessment', *Environmental Research Letters*, vol. 4, no. 3, p. 5.

FABHAT (Servmar Fundação Agência da Bacia Hidrográfica do Alto Tietê) (2013) *Mapeamento de áreas com potenciais riscos de contaminação das águas subterrâneas da UGRHI-06 e suas regiões de recarga*, Servmar Ambiental, Final Report, vol. 2.

FABHAT (Fundação Agência da Bacia Hidrográfica do Alto Tietê) (2014) 'Plano da bacia do alto tieté-Relatorio Final', http://www.fabhat.org.br/site/images/docs/24.%20p%E1g.%20 299%20a%20300.pdf, accessed 31 May 2014.

Foster, S. and Garduno, H. (2002) 'Mitigation of groundwater drainage problems in the Buenos Aires conurbation: technical and institutional way forward. Sustainable groundwater management, lesson from practice, case profile collection', http://documents.worldbank.org/curated/en/2002/07/7441025/argentina-mitigation-groundwater-drainage-problems-buenos-aires-conurbation-technical-institutional-way-forward, accessed 31 May 2014.

Foster, S. and Garduño, H. (2009) 'Gestión apropiada. El recurso hídrico subterráneo en América Latina. Lecciones de experiencias internacionales', *Aqua-LAC*, vol. 1, no. 1, p. 5.

Foster, S. and Hirata, R. (1988) *Groundwater Pollution Risk Assessment: A Methodology Using Available Data*, Pan American Sanitary Engineering and Environmental Science Center (CEPIS/PAHO/WHO*)*, Lima, Peru.

FUSP (Fundação da Universidade de São Paulo) (2009) *Plano da bacia hidrográfica do Alto Tietê,* Final Report, Fundação da Universidade de São Paulo, Brazil, vol. 4.

Gleeson, T., Wada, Y., Bierkens, M. F. and van Beek, L. P. (2012) 'Water balance of global aquifers revealed by groundwater footprint ', *Nature*, vol. 488, no. 7410, pp. 197–200.

Grönwall, J. T., Mulenga, M. and McGranahan, G. (2010) *Groundwater and Poor Urban Dwellers – a Review with Case Studies of Bangalore and Lusaka*, International Institute for Environment and Development – International Institute of Environment and Development, London.

GWP (Global Water Partnership) (2012) 'Situación de los recursos hídricos en Centroamérica: hacia una gestión integrada', www.gwp.org/Global/GWP-CAm_Files/SituaciondelosRecursosHidricos.pdf, accessed 26 May 2014.

Hansen, J. K. (2012) 'The economics of optimal urban groundwater management in southwestern USA', *Hydrogeology Journal*, vol. 20, pp. 865–877.

Hardin, G. (1968) 'The tragedy of the commons', *Science,* vol. 162, no. 3859, pp. 1243–1248.

Hirata, R. (2012) 'PPLegal – Projeto Poço Legal – ornando a gestão dos recursos hídricos subterrâneos mais participativos', www.sigrh.sp.gov.br/sigrh/ARQS/NOTICIA/CRH/CTAS/648/pplegal%20ctas%20esp%20310112%20f.pdf, accessed 18 May 2014.

Hirata, R. and Conicelli, B. (2012) 'Groundwater resources in Brazil: a review of possible impacts caused by climate change', *Anais da Academia Brasileira de Ciências*, vol. 84, no. 2, pp. 297–312.

Hirata, R., Foster, S. and Oliveira, F. (2014) 'Águas subterrâneas urbanas no Brasil: uma avaliação crítica para o desenvolvimento de políticas sustentáveis de gestão', *IGc-USP Editora* (in press).

Hirata, R., Stimson, J. and Varnier, C. (2006) 'Urban hydrogeology in developing countries: a foreseeable crisis', International Symposium on Groundwater Sustainability (ISGWAS), Instituto Geológico y Minero de España, Alicante, Spain.

Howard, K. W. F. and Gelo, K. K. (2003) 'Intensive use in urban areas: the case of megacities', in M. Ramon Llamas and E. Custodio (eds), *Intensive Use of Groundwater. Challenges and Opportunities*, CRC Press, Boca Raton, FL.

Huizar-Alvarez, R., Hernández-García, G. and Carrillo-Rivera, J. (2009) 'Simulation of groundwater flow and the hydrological balance of the Tecocomulco Lagoon, Central México', *Open Environmental Sciences*, vol. 3, pp. 1–13.

Kuroda, K. and Fukushi, T. (2008) 'Groundwater contamination in urban areas', in S. Takizawa (ed) *Groundwater Management in Asian Cities*, Springer, Tokyo.

Llamas, M.R. and Custodio, E. (2003) *Intensive Use of Groundwater: Challenges and Opportunities*, Balkema, The Netherlands.

Llamas, M. R. and Martínez-Santos P. (2005) 'Intensive groundwater use: silent revolution and potential source of social conflicts', *Journal of Water Resources Planning and Management*, vol. 131, pp. 337–341.

Llamas, R. and Martínez-Cortina, L. (2007) 'Specific aspects of groundwater use in water ethics', in R. Llamas, L. Martínez-Cortina and A. Mukherji (eds), *Water Ethics,* CRC Press, Taylor & Francis Group, Abingdon, UK.

Loaiciga, H. A. (2009) 'Long-term climatic change and sustainable ground water resources management', *Environmental Research Letter*, vol. 4, no. 3, p. 11.

Mahlknecht, J. (2013) 'Los recursos hídricos y la gestión del agua en América Latina y el Caribe: un panorama', in J. Mahlknecht and E. Pastén (eds), *Diagnóstico de los recursos hídricos en América Latina,* Pearson Education, Mexico City.

Margat, J. (2008) *Exploitations et utilisations des eaux souterraines dans le monde*, United Nations Educational, Scientific and Cultural Organization and Bureau de Recherches Géologiques et Minières, Paris.

Mazari-Hiriart, M. and Mazari-Menzer, M. (2008) 'Efectos ambientales relacionados con la extracción de agua en la megaciudad de México', *Agua Latinoamerica*, vol. 8, no. 2.

Morris, B. L., Lawrence, A. R., Chilton, P. J., Adams, B., Calow, R. and Klinck, B. A. (2003) *Groundwater and Its Susceptibility to Degradation: A Global Assessment of the Problems and Options for Management*, Early Warning and Assessment Report Series, RS, 03–3, United Nations Environment Programme, Nairobi, Kenya.

NWC (National Water Commission) (2012) 'Groundwater essentials', http://www.nwc.gov.au/__data/assets/pdf_file/0020/21827/Groundwater_essentials.pdf, accessed 18 May 2014.

Rosales, A. (2014) 'State investment in municipal water systems: the case of Mexico City', *Putting Public in Public Services: Research, Action and Equity in the Global South International Conference*, April 13–16, 2014, Cape Town, South Africa.

SABESP (Saneamento Basico do Estado de São Paulo) (2014) 'Mananciais da Bacia do Alto Tietê', http://site.sabesp.com.br /site/interna/Default.aspx?secaoId=36, accessed 22 February 2014.

SACMEX (Sistema de Aguas de la Ciudad de México) (2011) 'Estudio de las capas profundas de la cuenca del Valle de México para la búsqueda de un acuífero profundo como una alternativa de abastecimiento de la Ciudad de México', Simposio: Las ciencias de la tierra en el estudio del agua subterránea, Mexico City.

SACMEX (2012) *El gran reto del agua en la Ciudad de México. Pasado, presente y prospectivas de solución para una de las ciudades más complejas del mundo*, Sistemas de Agua de la Ciudad de México, Mexico City.

Sahuquillo, H. A., Custodio, G. E. and Llamas, M. R. (2012) 'La gestión de las aguas subterráneas', in Fundación Nueva Cultura del Agua (ed.), *Panel Científico-Técnico de Seguimiento de La Política de Aguas*, Universidad de Sevilla-Ministerio de Medio Ambiente, Spain.

Servmar (Servmar Ambiental e Engenharia) (2013) *Mapeamento de áreas com potenciais riscos de contaminação das águas subterrâneas da UGRHI 6*, Fundação da Agência da Bacia do Alto Tietê (FABHAT), São Paulo, Brazil.

Tortajada, C. and Castelán, E. (2003) 'Water management for megacity: Mexico City', *Ambio*, vol. 32, no. 2, pp. 124–129.

UBC (University of British Colombia) (2010) 'São Paulo metropolitan region: the Upper Tietê Watershed Management Committee', www.chs.ubc.ca/consortia/outputs3/CaseStudy-São_Paulo_Metropolitan_Region-Upper_Tiete.pdf, accessed 26 May 2014.

UNEP (United Nations Environmental Program) (2002) 'State of the environment and policy retrospective', www.grida.no/geo/geo3/english/pdfs/chapter2-5_freshwater.pdf, accessed 18 May 2014.

UNESCO-UN/WWAP (United Nations Educational, Scientific and Cultural Organization-United Nations/World Water Assessment Programme) (2009) *The United Nations World Water Development Report 3: Water in a Changing World. World Water Assessment Programme*, United Nations Educational, Scientific and Cultural Organization, Paris and Earthscan, London.

van der Gun, J. (2012) *Groundwater and Global Change: Trends, Opportunities and Challenges*, United Nations Educational, Scientific and Cultural Organization, Paris.

Vörösmarty, C. J., Green, P. J., Salisbury, J. and Lammers, R. B. (2000) 'Global water resources: vulnerability from climate change and population growth', *Science*, vol. 289, no. 5477, pp. 284–288.

Vrba, J. and Verhagen, B.T. (2011) *Groundwater Emergency Situations. A Methodological Guide*, United Nations Educational, Scientific and Cultural Organization, Paris.

World Bank (2013) 'Agua urbana en el Valle de México: ¿un camino verde para mañana?', http://www.agua.unam.mx/sacmex/assets/docs/AguaUrbana_ValleMexico.pdf, accessed 28 April 2014.

WWF (World Wildlife Fund) (2011) 'Big cities. Big water. Big challenges: Water in an urbanising world', www.wwf.se/source.php/1390895/Big%20Cities_Big%20Water_Big%20Challenges_2011.pdf, accessed 25 May 2014.

Zektser, I. S. and Everett, L. G. (2004) *Groundwater Resources of the World and their Use*, United Nations Educational, Scientific and Cultural Organization, Paris.

Part III
Climate change, risk management and planning

9 Flash floods in the Monterrey metropolitan area, Mexico

Lessons from Hurricane Alex[1]

Nicholas P. Sisto and Aldo I. Ramírez Orozco

Introduction

The Monterrey metropolitan area (MMA), with approximately 4 million inhabitants, is the third most populous urban area in Mexico. It constitutes an industrial and economic hub of great national importance, particularly for the manufacturing sector, and it is also a significant pole of attraction for foreign direct investment. The MMA embraces the capital of the state of Nuevo León and is home to most of the state government institutions as well as various regional offices of the federal government. Its educational, medical and cultural facilities extend their influence across the entire north-eastern portion of the country, and even beyond.

The geographic location of the MMA and the topography of its setting make it vulnerable to a particular type of natural disaster: flash floods. The MMA lies at the foot of the Sierra Madre Oriental mountain range on a semi-arid plain that is criss-crossed with rivers, valleys and ravines which carry away the run-off from the nearby mountains. Those mountains, which reach heights of 2,000m, capture and distil the moisture carried by tropical storms as they move in from the Atlantic Ocean. Major storms (hurricanes in particular) can cause sudden and heavy precipitation along the Sierra, sparking extraordinary freshets that tumble with great destructive force into the generally dry riverbeds and gullies of the MMA.

Since its founding more than four centuries ago, Monterrey has suffered a long history of flash floods (Cavazos, 1961). These infrequent but recurring events have caused great damage. The worst occurred in 1909, killing as many as 5,000 people (Sánchez and Zaragoza, 1989; Salazar, 2010). In recent times, two catastrophic events have hit the MMA: in 1988, as a result of Hurricane Gilbert, and in 2010, in the wake of Hurricane Alex. This chapter focuses on the more recent of these events, which did great damage to the urban infrastructure of the MMA.

The chapter is organized as follows. The first section describes the Hurricane Alex event in meteorological terms and quantifies its immediate impact concerning flooding in the MMA. The second section describes and analyses the damage caused by Alex, including property losses, both public and private, and disruptions to economic activity. The third section addresses the underlying causes of the disaster, with specific reference to the relationship between the hydrological reality of the MMA and its pattern of urban development. The fourth section proposes lessons that can be drawn from Hurricane Alex and that might contribute to reducing the risk of flash flooding in the MMA. The chapter concludes with a discussion of the challenges and opportunities that the MMA will face in the future when an event such as Hurricane Alex strikes again.

Flooding from Hurricane Alex

Alex was the first storm of the 2010 hurricane season in the Atlantic Ocean. It made landfall in Belize, crossed the Yucatán Peninsula and the Gulf of Mexico, and flung itself at the north-east coast of Mexico (Figure 9.1) as a Category 2 hurricane on the Saffir–Simpson scale. Because of the size and the path of Hurricane Alex, its remnants generated intense rainfall over the north-eastern region of Mexico, and in particular on the Gulf of Mexico side slopes of the Sierra Madre Oriental, the site of the MMA. Figure 9.2 shows the isohyets of cumulative precipitation recorded during the 48 hours between 29 June and 1 July 2010 in the state of Nuevo León. That accumulation was a record for the MMA, reaching more than 600mm in the mountainous area that rises to the south-east, precisely where the Santa Catarina River, the main watercourse of the MMA, has its source.

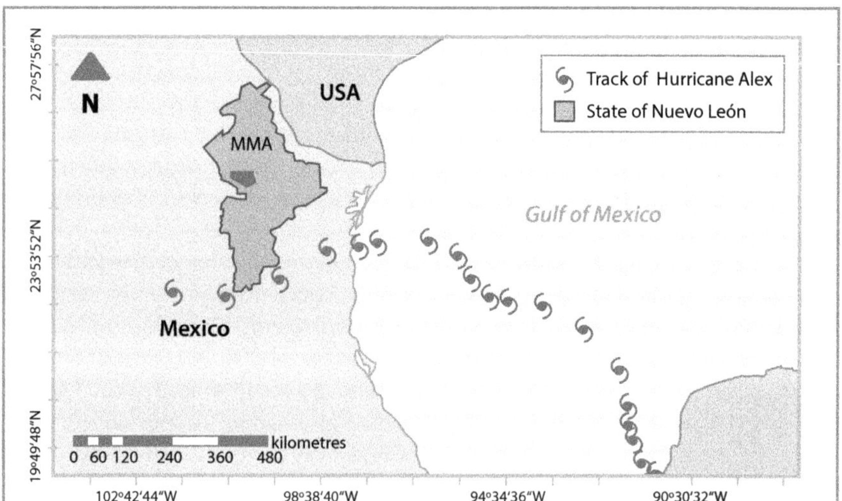

Figure 9.1 Track of Hurricane Alex and location of the Monterrey metropolitan area (MMA)

Source: Authors

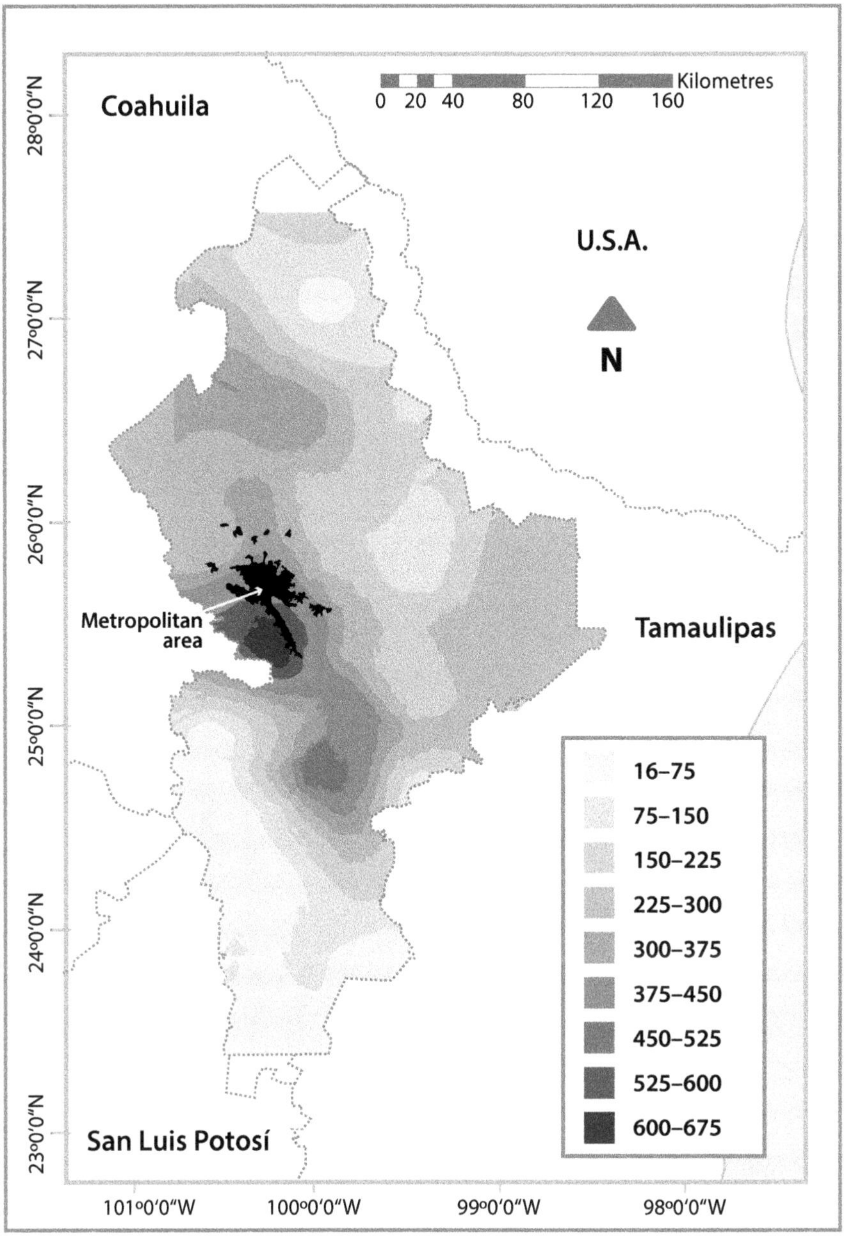

Figure 9.2 Isohyetal lines of cumulative precipitation (mm) in the 48 hours between 29 June and 1 July 2010, Nuevo León

Source: Authors, with data from CONAGUA (2010)

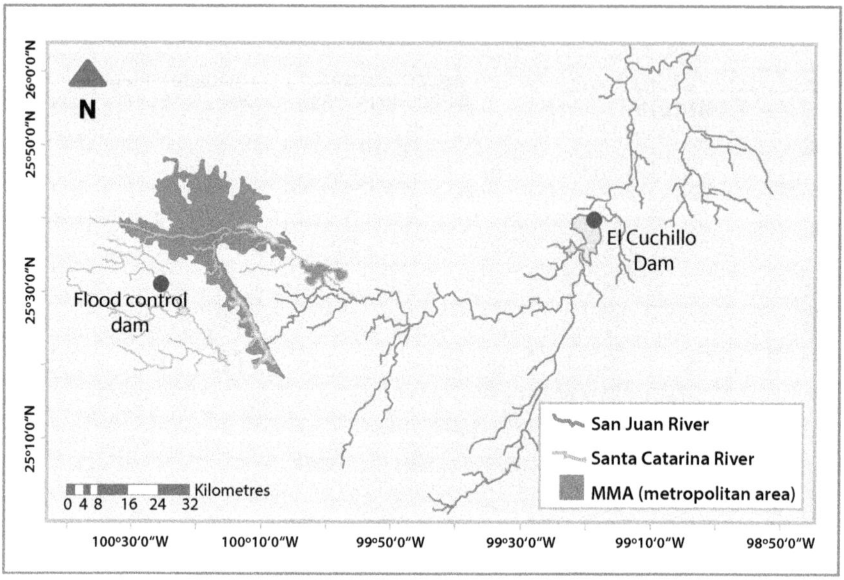

Figure 9.3 Hydrological and hydraulic features in the Monterrey metropolitan area (MMA) and surrounds

Source: Authors

Figure 9.3 illustrates the principal hydrological and hydraulic features of the MMA and its adjacent areas. Upstream from the MMA, in the upper valley of the Río Santa Catarina, is a flood control dam known as the *Corral de Palmas*, designed to reduce the peak levels of freshets sparked by heavy rains. The dam was completed in 2004 at a cost of US$46 million (Ferriño-Fierro *et al.*, 2010) and it regulates a portion of the upper watershed of the Río Santa Catarina. Further downstream, the Río Santa Catarina traverses the MMA from west to east, serving as the area's main element for rainwater drainage, and finally empties into the Río San Juan. More than 100km downstream, to the east of the MMA, is a large-scale storage reservoir, El Cuchillo, which is the principal source of drinking water for the MMA.

The rains that came with Hurricane Alex sparked exceptional run-off in the rivers and gullies of the MMA. Figure 9.4 shows the maximum flow for each year from 1960 to 2010 in the Río Santa Catarina. The peak flow associated with Alex in 2010 reached 2700m^3/s, the second highest recorded over that period of time, exceeded only by the 4400m^3/s generated by Hurricane Gilbert in 1988.

The hydrological regime of the Río Santa Catarina is characterized by long periods during which flows are extremely low. Table 9.1 shows the frequency of peak flows for the period 1960–2010. In general (more than 80 per cent of the time), the river was virtually dry during that period, with peak flows of less than 250m^3/s. In fact most of the river channel normally remains free of water throughout the year as normal flows run through a narrow natural canal. Yet

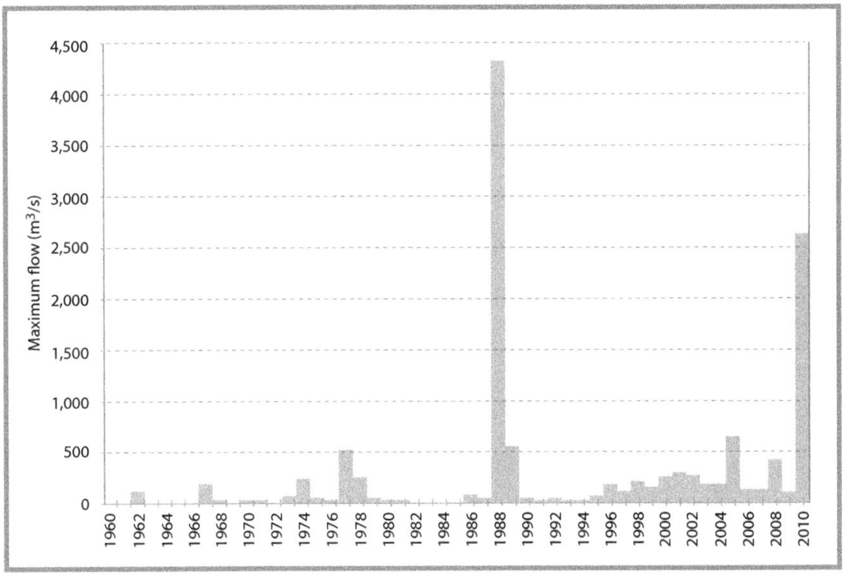

Figure 9.4 Maximum flow recorded during years between 1960 and 2010, Río Santa Catarina, Monterrey metropolitan area

Source: Authors, with data from CONAGUA (2010)

Table 9.1 Frequencies of maximum flow recorded during years between 1960 and 2010, Río Santa Catarina, Monterrey metropolitan area

Range (m3/s)	*Number of events*	*Frequency (%)*
<250	41	80.4
250–1,000	8	15.7
1,000–2,000	0	0.0
>2,000	2	3.9
Total	51	100

Source: Authors, with data from CONAGUA (2010)

on two occasions there were disastrous freshets that exceeded 2,000m^3/s and abruptly filled the whole river channel. The annual probability of such events stands at less than 4 per cent.

Damage caused by Hurricane Alex[2]

Damage to public infrastructure

The flooding caused by Hurricane Alex did major damage to public infrastructure in the state of Nuevo León, and particularly to road infrastructure in the MMA. In just a few hours, the MMA's most important arterial route – formed by two

Figure 9.5 Flash flood in the Río Santa Catarina, Monterrey metropolitan area, following Hurricane Alex

Source: Courtesy of Centro de Calidad Ambiental and Centro del Agua para América Latina y el Caribe, Tecnológico de Monterrey, Monterrey (2010)

steady-flow avenues that run along either side of the Río Santa Catarina – lost 70 per cent of its traffic handling capacity as entire lanes collapsed into the swollen river. Official estimates for the state as a whole place road damage at 7.8 million m^2 of paving destroyed, 2.7 million m^3 of mud and rocks dumped on roadways, 100 vehicle bridges and 154 culverts destroyed or severely damaged. Figure 9.5, taken from a point on the Río Santa Catarina in the centre of the MMA, shows the floodwaters pounding the Santa Bárbara culvert crossing, which was obliterated.

The storm also had a considerable impact on drinking water, sewage and sanitation services. According to the municipal water utility Servicios de Agua y Drenaje de Monterrey (SADM; 2010), Hurricane Alex destroyed more than 54,000m of water pipes and 45,000m of sanitary sewer collectors and sub-collectors. Seven pumping stations were damaged or destroyed, as were some 10,000 home water and sewage connections, and 11 sewage treatment plants were disrupted.

Table 9.2 summarizes the local and federal funding contributions that were approved to address the damage caused by Hurricane Alex. Of the total of 16,580 million Mexican pesos (approximately US$1.3 billion[3]), 97 per cent was allocated to the rebuilding of highways and of urban and hydraulic infrastructure that had been damaged or swept away. The total amount was the equivalent of 1.83 per cent of the gross domestic product (GDP) of the state of Nuevo León in 2010 which, according to the Instituto Nacional de Estadística, Geografía e Informática (National Statistics and Geography Institute – INEGI), was 907,733 million Mexican pesos (US$69,825 million[3]) (INEGI, 2014).

Table 9.2 Reconstruction budget approved for Nuevo León following Hurricane Alex

Item	*Mexican pesos (millions)*	*US dollars (millions)*[a]	*Share of total (%)*
Roadworks	6,132.43	471.73	37.0
Urban infrastructure	4,457.12	342.86	27.9
Hydraulic infrastructure	5,417.93	416.76	32.7
Other items	572.56	44.04	3.5
Total	16,580.04	1,275.39	100

Note: a = at an exchange rate of 13 Mexican pesos per US dollar
Source: Authors, with data from CERNL (2014)

Damage to private property

Although considerable, losses of private property as a result of Hurricane Alex were relatively minor in comparison with the impact on public facilities. A total of 15,811 houses were listed as reporting flood damage to their furnishings and appliances. Of those homes, 4,964 (or 31 per cent) were 'irregular'; that is, squatter dwellings without legal land title. Generally speaking, they were built in risk-prone areas not suitable for housing, such as along the banks of rivers and creeks. In terms of structural damage, 652 homes were reported to be totally destroyed, and another 3,835 to be minimally to partially destroyed. Insurance companies paid out some 2,500 million pesos (roughly US$192 million) to policyholders for losses to homes, businesses and vehicles (Aguilar-Barajas *et al.*, 2013).

Damage to private property, however, represented only a fraction of the impact on individuals. In many aspects, Hurricane Alex was a major disaster for the population: 15 lives were lost; 12,361 people had to be rescued or evacuated; and 68,314 families needed emergency supplies of food and water, 148,000 were left without electricity and 50,000 without water service for days and even weeks. In addition to such disruptions, there was the impact of the chaotic driving conditions which prevailed for months after the event – hundreds of thousands of people faced much longer and roundabout daily commutes. The millions of hours thus wasted represented a sizeable economic loss.

Impact on economic activity

In comparison with 2009, the year 2010 saw strong economic growth in Nuevo León despite Hurricane Alex, with state GDP rising by 7.21 per cent. Growth recorded in the two most important sectors of the state economy, manufacturing and commerce, was even more pronounced (14.8 per cent and 17.5 per cent, respectively). Alex, then, cannot be said to have sparked an economic crisis in the state, but a detailed analysis of monthly economic statistics shows that it did

Table 9.3 Manufacturing Production Index (MPI, Nuevo León) and Retail Sales Index (RSI, Monterrey metropolitan area), April to September 2010 (real terms, with base in 2003 of 100)

Month	*MPI*		*RSI*	
	Index value	*Monthly variation (%)*	*Index value*	*Monthly variation (%)*
April	122.43	–3.0	123.74	–2.4
May	127.81	4.4	136.47	10.3
June	128.08	0.2	125.39	–8.1
July	120.79	–5.7	127.98	2.1
August	129.73	7.4	132.04	3.2
September	131.35	1.2	124.05	–6.1

Source: Authors, with official data from INEGI (2014)

indeed have some short-term effects. Table 9.3 summarizes the manufacturing production index and the retail sales index for the months from April to September 2010. In the month following Alex (July 2010), manufacturing showed a significant drop of nearly 8 per cent, although it recovered in the month of August.

Figure 9.6 places this drop within a wider timeframe of 36 months. It is apparent that the magnitude of the hurricane's impact was comparable to normal downturns in the business cycle during that time. In the case of retail sales, both July and August of 2010 were growth months according to the data in Table 9.3. In fact, this indicator's performance in the months following Alex was similar to that in 2009 and in 2011.

Generally speaking, the impact of Hurricane Alex on economic activity in Nuevo León and the MMA made itself felt primarily in output, and it lasted less than three months. The event does not seem to have significantly dampened consumption or employment.[4] It may be noted that the MMA's resilience in terms of its economic activity, and in particular of its industrial production, derives from the simple fact that its manufacturing facilities and industrial parks are located away from the danger zones. Such effects as were recorded were indirect and were caused by the transportation and communication disruptions that persisted in the MMA in the months following the hurricane.

Diagnosing the causes of the disaster triggered by Hurricane Alex

The immediate cause of the extensive damage from Hurricane Alex was the lack of hydraulic capacity in the riverbeds and gullies of the MMA for handling the extraordinary run-off that occurred. This lack of capacity can be laid, in turn, on various characteristics of the urban structure of the MMA that make it

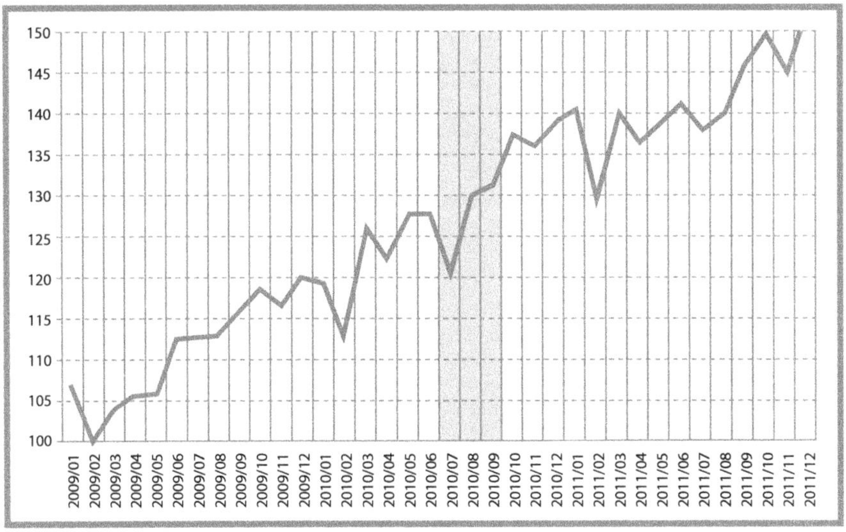

Figure 9.6 Monthly Industrial Production Index (real terms, with base in 2003 of 100), Nuevo León, January 2009 to December 2011
Note: The shaded area is the three months following Hurricane Alex

Source: Authors, with data from INEGI (2014)

vulnerable to such events. Of these characteristics, the two most important ones are the narrowing of the waterways and the inappropriate use of their banks, in both cases reflecting urban development patterns that have turned out to be fragile and vulnerable.

For example, the cross-sectional profiles of the Río Santa Catarina have been severely reduced at several points along its course by the horizontal layout and alignment of the city's principal arterial roads: the Constitución (north bank) and Morones Prieto (south bank) avenues. Figure 9.7 (an aerial photograph taken before Hurricane Alex struck) shows the drastic narrowing of the riverbed resulting from this road alignment. Bottlenecks of this kind, together with the flimsiness of the roadways themselves (e.g. the weakness or lack of reinforcement of the riverbanks), meant that serious damage from Alex was inevitable all along the river's channel.

Housing development patterns have also contributed to the lack of hydraulic capacity in the MMA. For example, Figure 9.8 shows how in one section of the Río Santa Catarina the channel width has been dramatically reduced (from 260m to 140m) by the presence of a residential zone. In this case, the culprit is a series of squatter dwellings whose inhabitants have no property title because their dwellings are located in a federally designated off-limits area where, for obvious reasons, this type of land use is prohibited. Note that squatter settlements in the MMA come in different types, from assortments of temporary refuges to neighbourhoods of proper houses with access to piped water and electricity. The residential zone in Figure 9.8 belongs to the latter type of settlement.

Figure 9.7 Narrowing of river channel by road infrastructure, Monterrey metropolitan area, 2007

Source: Ramírez (2011b)

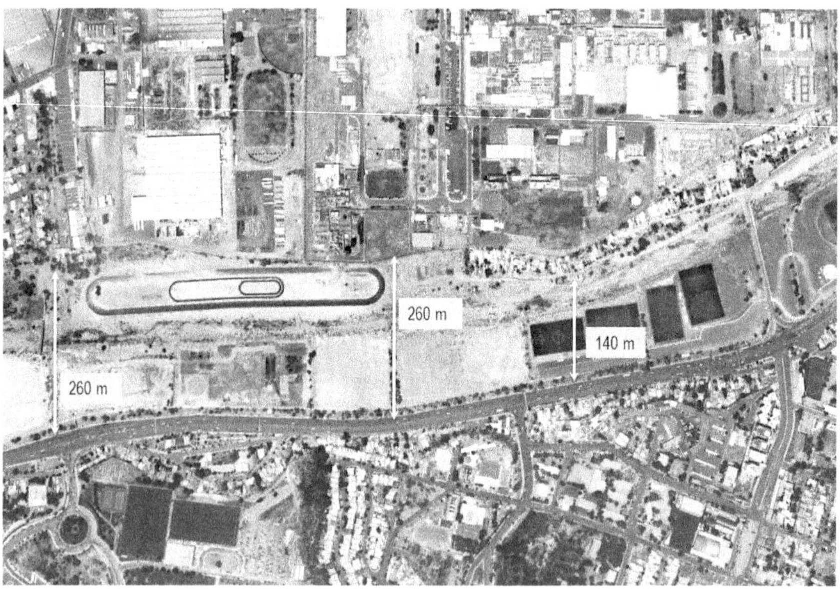

Figure 9.8 Narrowing of river channel by residential infrastructure, Monterrey metropolitan area, 2007

Source: Ramírez (2011b)

Figure 9.9 Impact of Hurricane Alex on river channel, Monterrey metropolitan area, 2010

Source: WorldView-2, obtained from Digital Globe by Centro del Agua para América Latina y el Caribe, Tecnológico de Monterrey, Monterrey (2010)

Nevertheless, it must be recognized that inappropriate location and inadequate protection against flash floods are not restricted to squatter settlements in the MMA. On the contrary, more than two-thirds of the dwellings affected by Hurricane Alex were located on lots where there was no legal restriction on their use as building sites. All of these considerations point to serious failures – in both the enforcement of federal standards and the design of local urban development plans – to take account of hydrological realities in the MMA.

Figure 9.8 illustrates another urban phenomenon which (if to a lesser extent) also affected the MMA's capacity to cope with the extraordinary run-off from Hurricane Alex: the invasion of the Río Santa Catarina riverbed by a plethora of recreational and commercial structures. When Alex hit, numerous areas of the riverbed were occupied by sporting and recreational installations, including soccer fields, a miniature golf range, a full-scale golf driving range, batting practice cages, a go-kart track and a cycling and jogging path. At various points on the riverbed there were also makeshift markets, food stalls, parking lots and other public facilities.

Figure 9.9 illustrates the impact of the flash flood from Hurricane Alex on the section of the Río Santa Catarina represented in Figure 9.8. Structures and facilities that previously occupied the riverbed, including a culvert crossing, were completely swept away by the flooding river and transformed into rubble that was later deposited downstream in the riverbed. Interestingly, the only thing left standing was the aforementioned residential zone, which was saved by its elevation relative to the rest of the riverbed. A slightly more powerful peak flow however would have washed it away.

Lessons from Hurricane Alex

Hurricane Alex offers a number of useful lessons for urban development planning in the MMA, in light of the latent risk of flash floods. We focus here on three fundamental lessons that point to three lines of defence for the MMA against such events: flood prevention works upstream from the MMA, urban design within the MMA and emergency response capabilities. In each case, the discussion stresses both the progress achieved and the tasks still outstanding.

Preventive engineering

Hurricane Alex demonstrated the effectiveness of the Corral de Palmas flood control dam (Figure 9.3). During the event, the maximum flow entering the dam's reservoir is estimated to have reached 1440m^3/s, while the maximum outflow was around 655m^3/s. This significant cut of more than 50 per cent in the flow of the Río Santa Catarina in its upper portion lowered its crest in the centre of the MMA by between 50 and 60cm (Ramírez, 2011a). While it is difficult to quantify the damage thus avoided, it is clear that without the dam the disaster would have been even greater, and there could well have been major flooding in densely populated areas.

The Corral de Palmas dam became operational in 2004, a decade and a half after Hurricane Gilbert struck in 1988, and it remains the only structure protecting the MMA from the great freshets that are occasionally produced upstream. The Alex experience suggests the need for more works of this type, in particular to control flows along the entire upper reaches of the Río Santa Catarina. It would also be wise to identify the best prevention options in other watersheds and valleys upstream from the MMA, including detention and retention basins, small check dams and gabion barriers, depending on the particular features of each site.

A full prevention system based on flood control works would certainly reduce the risk of sudden flooding in the MMA. Such a solution does not pose any significant technical challenge, but it will be costly to implement, and is unlikely to produce much in the way of political dividends in the short term. By their very nature, works of this kind tend to escape the notice of voters, because they are remote and they prove their worth only on rare occasions. The political and administrative initiatives that ultimately led to construction of the Corral de Palmas dam demonstrate nevertheless that it is possible to consider multiple engineering works for protection of the MMA.

Urban design

Hurricane Alex laid bare numerous shortcomings in the urban design of the MMA. Those shortcomings could be readily appreciated in the many works, structures and buildings that, because of their inappropriate specifications or location, were severely damaged or swept away. The event demonstrated the need to integrate hydrological risk into the urban design of the MMA. Such integration requires three basic elements: a sufficient hydraulic capacity to accommodate the freshets generated by an event such as Hurricane Alex in the MMA; a robust design for urban facilities that cannot be moved out of harm's way; and the relocation of urban facilities that are exposed to risk but that cannot be adequately reinforced.

The reconstruction effort in the wake of the hurricane provided significant progress on these three fronts. For example, some 7.35 million m^3 of materials and sediment were removed along 27km of the Santa Catarina riverbed, 4km of erosion protection walls were constructed, 4.2km of slopes were rehabilitated, and 13 homes were relocated (CONAGUA, 2013). The opportunity was also seized to replace damaged or destroyed road infrastructure with much more robust and functional structures; for example, the new bridges at Agua Fría–Zuazua, Cosmópolis–Metroplex and El Realito (CERNL, 2014). In all three cases, elevated, four-lane bridges replaced lower, two-lane culverts crossings that were old and dilapidated and did not have sufficient capacity to carry the flow of vehicular traffic. In addition, four years after the hurricane, the Santa Catarina channel is still free of recreational and commercial encroachments.

Notwithstanding these achievements, the basic problems with the urban design of the MMA have not been resolved fully. For example, there has been little change in the layout of the *Morones Prieto-Constitución* east–west arterial route along the banks of the Río Santa Catarina, and the river channel is still severely narrowed in places, especially in the central area of the MMA. Moreover, new constrictions have appeared, particularly where the *Gonzalitos* north–south arterial crosses the river. Obviously, it is very difficult under any circumstances to redesign an established urban fabric on the basis of strict hydrological criteria. Fortunately, the MMA has an alternative, provided it recognizes this fundamental point: the more it invests in preventive works upstream, as described in the previous section, the less problematic its urban design will be, and the less it will have to spend on reinforcing it.

Response capacity

The MMA demonstrated great response capacity in the face of the many challenges that were presented by Hurricane Alex and its aftermath. Three days after the storm had passed, 85 per cent of the population had water service, and within 15 days, service had been restored virtually throughout the MMA (SADM, 2010). Multiple programmes, involving not only the public authorities but also civic and business associations, were promptly put into action to help the victims. Within a short time an emergency traffic plan was activated for the MMA, the main feature of which was to restrict circulation to one direction on both roadways of the east–west arterial, thereby allowing more efficient use of what had survived the battering by Alex.

Three weeks after the event, the Consejo Estatal de Reconstrucción de Nuevo León (CERNL – State Council for the Reconstruction of Nuevo León) had already been constituted, bringing together representatives of the three levels of government (federal, state and municipal), the private sector and local higher education institutions. The story of that nascent Council's work and the many challenges it had to face would by itself merit an in-depth case study. The final outcomes were certainly easy enough to see: three years after Hurricane Alex, it was difficult for an outside or uninitiated observer to detect any trace of the disaster in the MMA.

The impressive response capacity of the MMA reflects its relative economic prosperity, the quality of its human capital, and its long-standing experience with such issues. Nevertheless, relying on response as a strategy for adapting to the risk of flash floods could turn out to be unsustainable, given the random and unpredictable nature of the phenomenon. Statistically speaking, there is nothing to prevent two or more events of this kind from following each other in fairly quick succession. It is alarming, then, to consider the improbable but possible scenario in which a series of disasters could overwhelm the MMA's response capacity, with consequences that are difficult to imagine.

Conclusions

Hurricane Alex represented a repeat of the havoc wreaked more than 20 years previously, in 1988, by the even more powerful storm Gilbert. The apparent failure to learn the lessons of that earlier event betrays the particular model of risk adaptation that has taken hold in the MMA: a reactive model, focused on restoring lost or damaged infrastructure, and one that seizes the opportunity to make some technical improvements but falls short of the radical changes that would integrate hydrological risk into the MMA's urban design.

Although this model has so far allowed the MMA to survive a long history of disastrous flash floods, it is problematic. On one hand, beyond the obviously burdensome reconstruction effort that the model entails, it generates sizable hidden costs. The clearest and most recent example of those costs is the millions of hours that hundreds of thousands of citizens lost in their daily commute because of the traffic chaos that prevailed for months after Hurricane Alex passed by, with all the foregone economic value this implies, not to mention the disruption and the stress suffered by these people. On the other hand, the relatively successful history of the model offers the MMA no guarantee for the future, given the very nature of the threat. Flash floods are random and unpredictable, and there is always the possibility of a succession of storms such as Alex that would completely overwhelm the MMA's response capacity.

To boost the MMA's capacity to withstand flash flooding, the technical solutions are clear and simple enough: a package of upstream flood control works, and another package of works to improve urban design within the MMA, based on the criterion of minimizing hydrological risk. The purpose, of course, is to give the MMA the capacity to manage its rivers and creeks as an integrated system for evacuating extraordinary run-off.

Making decisions concerning these solutions is unfortunately not a simple matter under any circumstances. The works required carry an immediate cost, but their benefits are as random and unpredictable as the threat they are intended to ward off. Institutionally speaking, even in the best of cases – that is, with only one metropolitan authority that would have the necessary and sufficient powers and responsibilities to tackle what is essentially a metropolitan problem – the decisions would still be complex. To begin with, there are the dilemmas posed by the question of what is the best balance to strike between preventive works upstream from the MMA versus works within the MMA, which would affect a great many people directly and indirectly.

The problem is compounded by the fact that the institutional reality in the MMA is far from the ideal case just mentioned. On the contrary, it is characterized by an architecture that involves a broad range of stakeholders: federal authorities (in particular, the Ministry of Communications and Transportation and the National Water Commission), multiple state authorities and a dozen municipal governments. While this complicated framework – in which intersect so many interests, powers, institutions, programmes and laws at the different levels of government – has been able to respond to the challenges posed by Hurricane Alex, it nevertheless appears by its very design incapable of producing in-depth and long-term solutions for the MMA.

It is perhaps ironic that the passage of Hurricane Alex, which came just as the rainy season was beginning, served to fill the reservoirs that deliver water to the MMA, thus guaranteeing more than two years' supply. This made it possible to alleviate the terrible consequences that might have arisen from the climatic conditions of the following year, 2011, which was the thirteenth driest year since 1941 and the least rainy of the previous 14 years (CONAGUA, 2012). Had it not been for the beneficial effects of Alex, the MMA – the nationally important productive hub of Mexico's north-east – would have suffered a severe water shortage.

The MMA has historically demonstrated its adaptability to the extreme climatic conditions of its semi-desert setting. So far it has been able to resolve its water supply problems, thanks largely to the downstream reservoirs and intake works which in their time reflected leading-edge engineering and a long-term vision. But on the other side of the climatic coin, flash floods represent what is objectively a much greater challenge and one that the MMA has yet to resolve. Surely the most pressing issue that must be addressed is to design and establish an institutional structure that will allow decision-making commensurate with the risk and implementation of the works needed for effective management of that risk. Otherwise, the MMA will be doomed to relive the past, or worse.

Notes

1 This chapter is based in part on the research project 'Strengthening Water Governance in a Context of Climate Change: The Case of the Metropolitan Zone of Monterrey', which was sponsored by the United Nations Human Settlements Programme (UN-HABITAT) and carried out during 2012–2013.
2 In this section, data on damage often refer to the whole state of Nuevo León, not just the MMA, and unless otherwise indicated, are taken from CERNL (2014). State-wide data represent an acceptable approximation for the MMA, as the latter accounts for more than 86 per cent of the state's population and an even greater proportion of its infrastructure and economy. Moreover, the MMA and surrounding areas received the highest amounts of rain during the Hurricane Alex event (Figure 9.2) and thus suffered most of the damage recorded in the state.
3 At an exchange rate of 13 Mexican pesos per US dollar.
4 For an exhaustive analysis of the economic impacts of Hurricane Alex using Auto-Regressive Integrated Moving Average (ARIMA) statistical models, see Aguilar-Barajas *et al.* (2013).

References

Aguilar-Barajas, I., Magaña-Rueda, V., Ramírez Orozco, A. and Sisto, N. P. (2013) 'Fortalecimiento de la gobernanza hídrica en contextos de cambio climático: el caso de la zona metropolitana de Monterrey', unpublished technical report for UN-HABITAT, Monterrey.

Cavazos, I. (1961) *Historia de Nuevo León*, Biblioteca de Nuevo León, UANL, Monterrey,.

CERNL (Consejo Estatal para la Reconstrucción de Nuevo León) (2014) *Juntos reconstruyendo Nuevo León. Memorias CERNL*, Gobierno de Nuevo León, Monterrey, www.conl.mx/, accessed 1 May 2014.

CONAGUA (2010) 'Informe interno sobre los impactos del huracán Alex en Monterrey', Subdirección General Técnica y Organismo de Cuenca Río Bravo, Mexico City.

CONAGUA (2012) *Reporte del clima en México. Reporte Anual 2011*, CONAGUA, Mexico City, http://smn.cna.gob.mx/climatologia/analisis/reporte/Anual2011.pdf, accessed 1 May 2014.

CONAGUA (Comisión Nacional del Agua) (2013) 8ª Reunión de Planeación del Consejo Estatal para la Reconstrucción de Nuevo León, 18 de julio de 2013, unpublished technical report, Palacio de Gobierno Estatal, Monterrey.

Ferriño-Fierro, A., Guerra-Cobián, V., Bruster-Flores, J., De Lira-Reyes, G. and Cavazos-González, R. (2010) *Control de Inundaciones en Zonas Urbanas, Caso de Estudio: Presa Rompe Picos Corral De Palmas Ubicada en Santa Catarina, Nuevo León, México, Ciencia – FIC*, no. 1, pp. 5–9.

INEGI (Instituto Nacional de Estadística, Geografía e Informática) (2014) *Banco de Información Económica (BIE)*, www.inegi.org.mx/sistemas/bie/default.aspx, accessed 1 May 2014.

Ramírez, A. (2011a) 'Evaluación del funcionamiento de la presa rompepicos', technical report commissioned to integrate *Reconstrucción del río Santa Catarina después del huracán Alex*, CONAGUA, Monterrey.

Ramírez, A. (2011b) 'Hidrología e hidráulica en el río Santa Catarina', technical report commissioned to integrate *Reconstrucción del río Santa Catarina después del huracán Alex*, CONAGUA, Monterrey.

SADM (Servicios de Agua y Drenaje de Monterrey) (2010) *Tormenta tropical Alex – el agua: prioridad inmediata*, SADM, Monterrey.

Salazar, A. (2010) 'Pérdidas del patrimonio arquitectónico de Monterrey. Desastres naturales por lluvias', *Ciencia UANL*, vol. 13, no. 1, pp. 11–15.

Sánchez, O. and Zaragoza, A. (1989) 'El río fiera, bramaba', *Cuadernos del Archivo General del Estado de Nuevo León*, no. 35, pp. 1–125.

10 The Río Bogotá environmental recuperation and flood control project

Carlos R. Costa Posada, Carlos Felipe Urazán Bonells and Edder A. Velandia Durán

Introduction

The Río Bogotá is an iconic river of Colombia. It waters the country's most fertile plain and then flows through the city of Bogotá, only to hurl itself over the Tequendama Falls before disgorging into the Río Magdalena. Yet the quality of water in the Río Bogotá leaves much to be desired. When it reaches its namesake city, with an average flow of $10m^3/s$, it is suitable only for irrigation; in passing through the city, it receives some $20m^3/s$ of virtually untreated sewage. At that point it is deemed a 'dead' river.

The basin of the Río Bogotá is the scene of varied economic activities – farming, industry and the service sector – which together with the prosperity of Bogotá, Colombia's capital, have attracted the migration of many people into the region.[1] The estimated population of the basin stands at 9.3 million in 2012 (SDP, 2012). The city of Bogotá, which lies in the middle part of the basin, not only contains the majority of its population (around 7.5 million inhabitants or approximately 20 per cent of the country's total population) but also has exhibited the highest annual population growth. This growth is exerting great pressure on the environment, a situation that is aggravated by the phenomenon of migration into the region, and the subsequent spread of irregular settlements.

Although Bogotá grew with its back to the river and has always discharged its sewage into the watercourse, today the city has colonized the entire river bank, exposing the riparian population to health problems and frequent flooding, both within the city and downstream. These concerns led the Government of Distrito Capital (DC – the Capital District), Empresa de Acueducto y Alcantarillado de Bogotá (EAAB – Water Supply and Sewerage Company de Bogotá), Corporación Autónoma Regional de Cundinamarca (CAR – Autonomous Regional Corporation of Cundinamarca) and the national government to formulate the Hydraulic Upgrade and Environmental Recuperation Project for the Río Bogotá. The project represents a comprehensive plan to improve the quality of water in the river; to protect the urban area from 100-year flood events; and to restore areas along the banks of the river that serve as floodplains, ecological habitats and public recreation areas. This is an example of how a project for mitigating flood damage and cleaning up a watercourse can become an urban development project that will improve the quality of life for city residents.

The Río Bogotá watershed

The Río Bogotá headwaters are on the Guacheneque Plateau at an altitude of 3,300m above mean sea level in the municipality of Villapinzón (Department of Cundinamarca), and flows for about 336km in a south-easterly direction to join the Río Magdalena at an altitude of 280m, in the municipality of Girardot (Department of Cundinamarca). Along its course it receives water from 15 tributaries. The Río Bogotá supplies water to Bogotá and another ten municipalities. The biggest user is the EEAB. Rivers of importance in the urban context of Bogotá are the Tunjuelo, the Fucha and the Salitre, which run through the city from east to west, receiving water from a number of smaller streams and creeks before flowing into the Río Bogotá (SDP, 2011). Overall, these rivers constitute the main mechanism for urban drainage and the city's principal ecological feature. The Río Bogotá watershed has been conceptually divided into three sections known as the 'Cuenca Alta', 'Cuenca Media' and 'Cuenca Baja' – the upper, middle and lower basins – as shown in Figure 10.1. This classification reflects considerations relating to topography, water quality, hydraulic control structures and territorial jurisdiction.

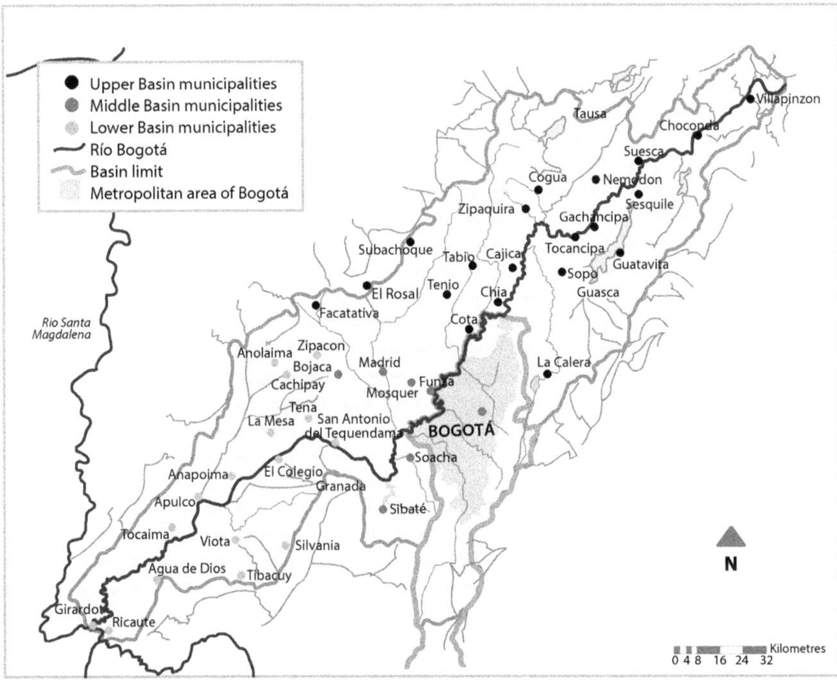

Figure 10.1 The Río Bogotá watershed and the location of municipalities in its upper (Cuenca Alta), middle (Cuenca Media) and lower (Cuenca Baja) basins

Source: Adapted from CAR (2014)

The upper basin

The upper basin is located to the north of Bogotá and extends from the municipality of Villapinzón to the La Virgen station. The stretch of the river in this part of the basin is 145km long, and it is divided among 16 municipalities, with an estimated combined population of 538,000 in 2012 (DANE, 2012a). Because of the physical characteristics of its upper basin, which give the river a steeply inclined course and a consequently swift current, the Río Bogotá is considered a mountain river. Water quality in the river is compromised by discharges from population centres and productive activities. The basin has a relatively low flow, regulated by three reservoirs: Sisga, Tominé and Neusa (Velandia, 2006).

The middle basin

In the 68km stretch between Puente La Virgen station and the Alicachín sluice gates near the Muña Reservoir, the river runs along the western city limit of Bogotá. The population of this part of the basin was approximately 8,473,000 in 2012 (DANE, 2012a). It is in this middle basin that the waters from the Salitre, Fucha and Tunjuelo rivers meld with the Río Bogotá. The middle basin has the following tributaries: Río Juan Amarillo, Río Fucha, Río Tunjuelo, Río Balsillas, Río Soacha, the discharge from the El Salitre sewage treatment plant and the storm water and sewage pumping plants Fontibón, Navarra, Saucedal II and Gibraltar as well as the sewage system collectors in the southern zone of Bogotá and the municipality of Soacha.

The lower basin

The lower basin corresponds to the last 123km of the watercourse, from Alicachín on the Muña Reservoir to the junction with the Río Magdalena. In this part of the basin, the Río Bogotá passes through the municipalities of Agua de Dios, Anapoima, Anolaima, Apulo, Cachipay, El Colegio, Girardot, La Mesa, Quipilé, Ricaute, San Antonio, Tena, Tocaima, Viotá and Zipacón, which taken together had a population of 284,000 in 2012 (DANE, 2012a).

Río Bogotá water quality

As the river descends along its course, the quality of water deteriorates steadily. This is the combined result of several aspects of the basin: human and productive activities, the low coverage of sewers and wastewater treatment, the gradual deterioration of soils from over-exploitation, and the dumping of garbage and agricultural wastes. According to the National Bureau of Statistics of Colombia, 37 per cent of the population living in the Río Bogotá basin is not served by sewers (DANE, 2012b). Outside the city of Bogotá itself, where sewer coverage is 100 per cent, this figure rises to 45 per cent. The greatest deficit is to be found in the lower basin.[2]

At the lower end of the middle basin the quality of water has deteriorated to the point that the river is now essentially an open sewer, with a high toxicity that poses a risk to human health. Water quality suffers in particular from the discharge of sewage from Bogotá. Industrial and domestic wastes from a number of other municipalities are also discharged into this section of the river. According to water quality monitoring measurements reported in the Consejo Nacional de Política Económica y Social (CONPES – National Economic and Social Policy Council) (DNP, 2004) and the study conducted in 2008 (CAR, 2009), water quality in the Río Bogotá has been deteriorating steadily.

Discharges in the middle basin, starting at a point approximately 178km downstream from the river's source, have a significantly higher level of total suspended solids (TSS), thus boosting levels in the lower basin, than in the upper basin. The situation is similar for concentrations of coliform bacteria, chromium, lead and nickel. The segment of the Río Bogotá along the urban perimeter of Bogotá is most affected by poor water quality. During 2009, people in the municipalities along the banks of the Río Bogotá contributed about 428,000 tonnes of TSS and organic load, with the city of Bogotá accounting for roughly 85 per cent of this load (CAR, 2009).

By contrast, water quality in the lower basin is partially restored by the process of sedimentation in the riverbed in the middle basin, and in the Muña Reservoir, by the increased reaction time of organic material, the aeration of waters as they tumble down steep slopes and over high waterfalls,[3] and the fact that there is less dumping of discharges by the riparian population. Since 2004, the first phase of the Salitre wastewater treatment plant (WWTP) has been in place in Bogotá. The plant has a capacity of $4m^3/s$, and secondary treatment removes up to 40 per cent of the organic load and 60 per cent of TSS from the water treated. Wastewaters from the Juan Amarillo drainage sub-basin to the north of Bogotá are conveyed to this plant, located in the north-west of the city (DNP, 2004).

The poor quality of water in the Río Bogotá, especially where it passes alongside the city limits, generates various adverse impacts. Among the most important are the effects on economic activities, degradation of the riverbed and the landscape, production of noxious odours through decomposition of the organic load, and flooding of lands, which has limited the growth of Bogotá along the western bank of the river and in the neighbouring municipalities. The value of land in urban areas has diminished and large portions of the environmental buffer zones along the river have been rendered useless for recreation.

The fact that the water is unfit for domestic and agricultural use has had impacts on the quality of life of the riparian population, who must compete for water even if it is polluted. The environmental degradation of ecosystems associated with the Río Bogotá poses serious health problems not only for people living near the river, who generally belong to the lowest socio-economic strata, but also for people who consume food crops irrigated with its waters, as is the case with the La Ramada irrigation district. The micro-organisms found in the middle basin are the source of waterborne diseases with high percentages of morbidity and mortality, especially among the most vulnerable members of the population – children, the elderly, and women trapped in pockets of poverty.

The poor quality of the water has adverse environmental impacts all along the Río Bogotá, especially where it passes by the city. Among these impacts are the following: the loss of aquatic ecosystem organisms; displacement of wildlife; proliferation of insects and rodents; and growth of aquatic plants on the river's surface and its banks, with detrimental effects on hydraulic dynamics, the landscape and water quality itself.

Flooding of the Río Bogotá

Flooding in the Río Bogotá watershed exhibits varying behaviour and has different impacts in each of the three basins into which it is subdivided. In the upper basin, flooding is rare because flows are regulated by reservoirs, and the impacts of any flooding that does occur are moderate because there are relatively few human settlements near the river. However, during the La Niña meteorological phenomenon the river spills over its banks, even in its upper reaches, as a result of excessive restrictions imposed on its channel. To the north of the city of Bogotá, flooding is associated with: (i) the breaching of dykes[4] or the hydraulic capacity of the riverbed being exceeded as a result of shortcomings in the design, construction or maintenance of the dykes; (ii) a reduction in the river's cross-sectional area through works constructed in its main channel or along its floodplain; or (iii) the loss of hydraulic profile through sedimentation.

Flooding in the middle basin occurs as a result of direct overflows of the Río Bogotá and the Fucha and Tunjuelo rivers (which cross the city from west to east). In this portion of the river basin, flooding is more frequent because of the low gradient of the riverbed and the river's meandering course. Urban development has changed the basin's response to precipitation, and increased flash flooding occurs as a result of impermeable surfaces, removal of vegetation cover, lack of seasonal water storage, shrinking wetlands, channelling of storm water, and a reduction in the land's natural roughness through levelling. In the lower basin, the problems associated with flooding have to do with the instability of the riverbed and its banks.

According to the Sistema de Información para el Manejo de Riesgos y la Atención de Emergencias de Bogotá (SIRE – Risk Management and Emergency Response Information System of Bogotá), one of the causes of flooding in the city is inappropriate urban development on the *rondas*[5] along the rivers. This urbanization generates sewage discharges that contribute to increasing the flow beyond the capacity of the buffer zones. The occupation of wetland ecosystems leads to the loss of environmental services such as a reduction in buffering capacity to mitigate flooding. Some of the areas that suffer most from flooding are illegal or squatter settlements that were constructed below the 10-year flood line in the buffer zones. These settlements have been legalized by the

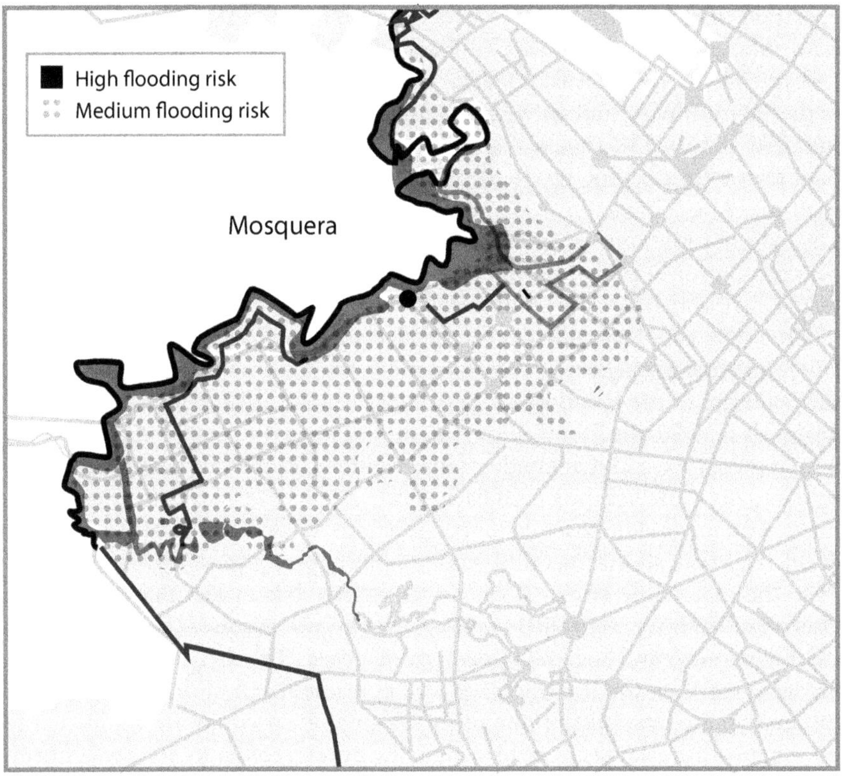

Figure 10.2 Areas at risk of flooding in one part of Bogotá

Source: Adapted from SDP (2013)

authorities since they were built, but there have been no works to mitigate the vulnerability of dwellings in the high-risk zones. Figure 10.2 shows the limits of the flood-prone area in western-most Bogotá, which overlaps with human settlement areas legalized by district authorities.

Pollution and flood control strategies for the Río Bogotá

Chronology and milestones, 1974–2014

There is a long history of efforts made to deal with pollution and flooding of the Río Bogotá (CAR, n.d.). In 1974 a study was conducted on ways to improve the river. In 1989 it was proposed the construction of three wastewater treatment plants, at Salitre, Fucha and Tunjuelo. In 1994 a build-operate-transfer (BOT) contract was signed with Degremont and Lyonnaise des Eaux for the first phase of the Salitre WWTP. A land-use plan (*Plan de Ordenamiento Territorial*, POT) for the city was adopted in the year 2000. In 2004 CONPES designed a strategy for the environmental clean-up of the Río Bogotá and EAAB took over direct operation of the Salitre WWTP. Two years later, the CAR adopted the River Basin Management and Land-Use Plan (*Plan de Manejo y Ordenamiento de una Cuenca*, POMCA) for the Río Bogotá, which called for expansion and upgrading of the Salitre WWTP and set water quality objectives for the basin. Also in 2006, the CAR and the DC signed '*Convenio* 171' in which the two parties agreed to pool efforts for cleaning up the Río Bogotá. In 2010 an agreement was signed with the World Bank for a loan to finance the first phase of the Hydraulic Upgrade and Environmental Recuperation Project for the Río Bogotá (see the next section for details of this project and an update of activities to restore the river's environment and to control flooding).

Over the past four decades, successive municipal governments and the environmental management authority, the CAR, have taken a keen interest in the pollution of the Río Bogotá. This interest has translated into a number of proposals that did not always enjoy consensus among the players responsible for deciding the type and magnitude of interventions. In seeking a comprehensive solution, the CAR and the EAAB have jointly led the river clean-up programme, focused on the middle basin, comprising actions for the ecological recuperation of the river with short-, medium- and long-term projections.

In the case of the city of Bogotá, actions have been taken that include the clearing of lands for controlled flooding, the removal of aquatic vegetation from the watercourse and the reinforcement of dykes. The CAR has undertaken dredging works in the Río Bogotá as well as maintenance of the dykes in order to improve the hydraulic profile of the channel. Inspections are conducted to prevent nearby residents from breaching the dykes in order to draw water from the river. The EAAB has taken steps to regulate run-off in the urban areas of Bogotá and to integrate operations with the power generation system.

Hydraulic Upgrade and Environmental Recuperation Project for the Río Bogotá

The CAR is responsible for executing this project, which falls within the strategy for environmental management of the Río Bogotá proposed by CONPES in its Document 3,320 and the actions agreed between the CAR and EAAB for recuperation of the river and the prevention of flooding in Bogotá. These actions will be applied in the middle basin of the Río Bogotá, in the 68km stretch between the Alicachín sluice gates in Soacha municipality and the Puente La Virgen station in Cota municipality (Figure 10.3).

Objectives

Within this frame of reference, on 14 December 2010 a loan was signed with the World Bank (World Bank, 2010). Environmental recuperation and flood control of the Río Bogotá were the chief objective. More particularly, environmental recuperation was seen as an asset for Bogotá and its surrounding region.

Figure 10.3 Location of the principal projects included in the first and second phases of the Hydraulic Upgrade and Environmental Recuperation Project for the Río Bogotá

Source: Adapted from World Bank (2010, p. 110)

Structure

The project is structured in four main components:

1 *Wastewater treatment.* This involves an investment of US$335 million to upgrade and expand the Salitre WWTP from a primary treatment plant with a capacity of $4m^3/s$ to a secondary treatment plant with a capacity of $8m^3/s$. The affluent will be discharged into the Río Bogotá and could potentially be reused in the La Ramada irrigation district.
2 *Environmental restoration.* This entails an investment of US$140 million for the construction of dykes, the rehabilitation of riparian habitats, the purchase of lands to clear illegally occupied areas, and the establishment of parks and multifunctional zones along the Río Bogotá, including the creation of links connecting the river and the wetlands.
3 *Priority environmental studies.* An amount of US$7 million has been earmarked to finance plans and analytical studies, such as an update of the comprehensive management plan for the Río Bogotá watershed and the master plan for management of biosolids in the metropolitan region of Bogotá.
4 *Project administration.* This has an allocation of US$5 million.

Wastewater treatment works

In 1990 the municipal government of Bogotá, the National Planning Ministry, EAAB, the Government of the Capital District and the CAR formed an interagency committee to define policies and actions for restoration of the Río Bogotá. Since that time there have been a number of proposals for cleaning up the river, including a failed attempt to improve conditions in the watershed through a project financed by the Inter-American Development Bank (IDB) calling for the establishment of 24 sewage treatment plants in the middle and upper basins.

Among different options, the Salitre WWTP was selected as part of a plan to restore the Juan Amarillo wetland and to control flooding in the Río Salitre valley. The Salitre WWTP is a facility that has reduced the discharge of pollution from Bogotá into the Río Bogotá. This plant was built between 1997 and 2000 by the firm Bogotana de Aguas y Saneamiento (BAS) Suez–Lyonnaise des Eaux–Ondeo Degremónt. That firm was responsible for operating the WWTP until June 2004, when it was taken over by EAAB with a view to cutting operating costs. The Salitre WWTP is currently a chemically assisted primary treatment plant with a capacity of $4m^3/s$. In this first phase of the project it will be upgraded and expanded into an activated sludge plant with a conventional capacity of $7.1m^3/s$.

Environmental restoration works

The works under this component are of two types:

1 Construction of dykes or *jarillones* (embankments), and dredging of the river channel to raise its hydraulic capacity from the current 80–100m^3/s to 100–200m^3/s. This increase considers a 100-year flood event. The carrying capacity of the channel will be increased through these works from 8 to 14 million m^3.
2 The retrieval of lands taken over by squatters in order to return them to their natural function as floodwater receptacles. Restoration of these areas will not only minimize losses of life and property during flooding but will also allow them to be used as green spaces and recreation areas as well as for other purposes.

Some of the channelling and flood control works could incorporate recreational infrastructure, as was done for example with the Río Salitre where, as shown in Figure 10.4, the buffer zone along the river offers added value through footpaths and cycling lanes. Recovery work on the Juan Amarillo wetland included the planting of trees along the perimeter, thereby enhancing the landscape, offering refuge to animal species, improving the quality of life for local residents, and returning lands to their natural state.

Figure 10.4 Recovery works on the Río Salitre, Juan Amarillo, north-western Bogotá

Source: Authors

Challenges and recommendations

Implementation of the environmental recuperation project for the Río Bogotá has been delayed by circumstances such as radical changes to some of the works, disagreements over operation of the works and their cost, and questions about the works that are in operation today. These issues are discussed below in greater detail.

Obstacles to interagency coordination

As various entities are responsible for the river's water quality, conflicting viewpoints inevitably make coordination difficult, despite the multidisciplinary and comprehensive approach of the working teams. Examples of such situations include:

- Disagreement in 2004 between EAAB and the CAR as to the efficiency of the Salitre WWTP and the type of modifications and extensions needed.
- Disagreement between the CAR and the national government on the one hand and the DC and EAAB on the other concerning the volume of water that should be treated at the Canoas WWTP.
- Disagreement between EAAB and the power generating utility – *Empresa de Energía de Bogotá* (EMGESA) – over financing operating costs for the Canoas pumping station.

Outside mediation has been a key element in resolving these situations. On some occasions the national government has served as a mediator between local institutions, while on others mediation has been through multilateral institutions that have provided technical studies or co-financing for the investments.

Lack of coordination between urban development policies and environmental policies

Desired outcomes as envisioned by environmental entities tend to be different from those pursued by municipal entities. The municipal authority frequently claims that it is denied the opportunity to plan for environmentally sustainable growth in areas that the environmental authority wants to protect. There are also differences over de-pollution targets and the investments needed, and operating costs incurred to achieve those targets.

In the context of this discussion, it is worth noting that, in Colombia, the environmental authorities are hierarchically superior to the urban authorities on these aspects, and are therefore legally entitled to impose their views. However, their decisions are not always carried out; either because the investments are delayed or because irregular settlements are present (citizens or communities have appropriated territory without proper permits). Fostering open public debate has been very useful for clarifying the positions of the parties. Another element that has contributed to agreement has been the co-financing of investments by the national government and the regional environmental authority of Cundinamarca.

Continuous amendments to projects

The lengthy periods needed to carry out the proposed solutions to the river's problems have witnessed changes of government, new technological developments, and changes in the ownership and occupancy of the lands needed for executing the works. All of these changes, together with political and private interests of various kinds, has generated significant modifications and amendments to the solutions previously proposed. An example is the change from the initial proposal that called for construction of three WWTPs (Salitre, Fucha, Tunjuelo) to the solution currently being pursued – constructing only two large plants (Salitre and Canoas). There have also been important changes in the sources of financing and in private sector participation, including the shift from private foreign financing with long-term BOT-type contracts to direct financing and loans from the multilateral banks.[6] Resolving these situations has required the intervention of entities with solid technical credibility, primarily the National Planning Department, and multilateral institutions such as the World Bank and the Inter-American Development Bank.

The impacts of industrial discharges

The legislation requiring treatment of industrial wastewaters before they are discharged into the public sewers is not being enforced, and this poses a significant threat to the effectiveness of the investments made in sewage collection and treatment systems, which must comply with regulations governing domestic wastewater discharge. However, the experience in Bogotá shows that a combination of awareness-raising campaigns to reduce industrial consumption and the imposition of penalties can have a salutary effect. Through such campaigns and measures, EAAB has managed to change water consumption habits significantly and has invested in water saving measures (with final consumers) that have reduced average per capita consumption of drinking water.

Sewage treatment costs

One of the main concerns with respect to the sustainability of investments in sewage collection and treatment systems is their operating costs and the need to pass on a greater portion of those costs to users. This may represent a significant increase in utility bills, with social and political implications that tend to delay decision-making. At the present time, alternatives are being examined to subsidize the costs of energy used by the WWTPs as a way of reducing their operating costs and generating energy through the recovery of methane gas.

There is also a proposal to revise regulations governing the use of residual sludge from the WWTPs, which can now be deposited only at high cost in sanitary landfills.

Conclusions

The Río Bogotá watershed is of national significance to Colombia. Overall, Bogotá is a growing and prosperous metropolis. This growth, however, is exerting great environmental pressure on the watershed. The stretch of the river that flows along the city's western perimeter is badly polluted, and this has an adverse impact on the economy, health of residents, growth of the city, and environment. Moreover, urbanization in the city has sparked a change in the basin's capacity to cope with run-off and this, together with the spread of settlements on the river's floodplains without any mitigation measures, has increased the frequency, depth and duration of flooding caused by the river and its tributaries.

In this context, the Hydraulic Upgrade and Environmental Recuperation Project for Río Bogotá have been established, with works for water processing, flood control, urban renewal, and to improve quality of life. Sewage treatment projects normally hold little attraction for public administrations because of their high cost of investment and subsequent operation, without any obvious impact on the quality of life for the people making the investment. The Río Bogotá case runs counter to experience observed elsewhere. The strategy has been designed to transform a sewage treatment and flood control project into an urban development project, with a direct impact on the quality of life of the people who pay for the works.The Río Bogotá project offers valuable lessons for the Latin American and international context. A relatively small budget increase can convert a storm water channelling facility into a landscaped recreation area where users will exert social control, preventing other citizens from settling in the zone or using it as a garbage dump. The recovery of wetlands as buffer zones against flooding can convert them into environmental conservation zones with recreational uses. Because they have multiple uses, these investments will benefit a greater number of interest groups which, in turn, will be motivated to defend and protect them. Sewage treatment and flood mitigation works, with little additional budgeting, can add value to the areas where they are located, promote urban renewal and improve the quality of life for residents.

These benefits, however, will not come about automatically. A successful approach to sewage treatment requires the cooperation of all levels of government (national, regional and local) to guarantee both the financing and the sustainability of the infrastructure and its operation. External agencies such as development banks can be crucial for achieving consensus on the most appropriate technical solution and on interagency coordination mechanisms for financing, construction and operation of such a solution. A social awareness-raising process is also essential to gain popular backing for the investments that have been or must be made. The modifications that projects undergo is a result of the long timeframe needed to execute them, during which there may be many and varied changes of circumstance.

It seems clear that the works under the Río Bogotá environmental recuperation and flood control project have great potential to improve the quality of life and reduce public health risks as well as the economic and human losses occasioned by flooding. The improvement in urban conditions for the beneficiary sectors is having a positive impact on Bogotá and its surrounding region by increasing its potential to attract investment. In this way, environmental clean-up projects are 'good businesses' for both the economy and society.

Notes

1 In Bogotá, some 50,000 households (or 2.3 per cent of the total) have no title to their homes, or occupy them de facto under collective ownership (SDP, 2011). The metropolitan area of Bogotá receives the bulk of the population displaced within the country, and it is expected to continue growing (ACNUR, 2012).

2 Sanitary sewer coverage in the three sections of the watershed is as follows: upper basin, just above 65 per cent; middle basin, 75 per cent; lower basin, just under 50 per cent.

3 There is a drop of some 2,000m to the confluence with the Río Magdalena.

4 *Jarillones* are longitudinal embankments built along the river courses for flood control.

5 *Rondas* are buffer zones established in the floodplain, extending in parallel fashion along both sides of the permanent channel, and extending generally 15–30m inland from the average annual crest level.

6 One alternative considered would have involved the construction of interceptors for all wastewater from Bogotá, which would be discharged into the Río Bogotá downstream from the city, where it would receive primary treatment to the required quality standard at the Tibitoc drinking water plant, with subsequent reliance on the terrifying potential of oxygenation as the river descends to the Magdalena.

References

ACNUR (Agencia Colombiana de las Naciones Unidas para los Refugiados) (2012) *Mapa interactivo*, http://www.acnur.org/t3/operaciones/situacion-colombia/mapa-interactivo/, accessed 10 August 2012.

CAR (2009) *Adecuación hidráulica y recuperación ambiental del río Bogotá*, CAR Gestión Ambiental, Colombia, http://www.car.gov.co/recursos_user/Proyectos%20Especiales/RIO%20BOGOTA/Evaluacion%20Ambiental%20Volumen%20I.pdf, accessed 1 June 2014.

CAR (Corporación Autónoma Regional de Cundinamarca) (2014) Gestión de la calidadhídrica en la cuenca río Bogotá, CAR, Colombia, http://www.minambiente.gov.co/documentos/3799_170709_taller_rh_car_180909.pdf, accessed 28 April 2014.

CAR (n.d.) *Proyectos Especiales – Río Bogotá*, www.car.gov.co/index.php?idcategoria=10063, accessed 16 January 2014.

DANE (Departamento Administrativo Nacional de Estadística de Colombia) (2012a) *Proyecciones de población*, DANE, Gobierno de Colombia, www.dane.gov.co/index.php/poblacion-y-demografia/proyecciones-de-poblacion, accessed 10 August 2012.

DANE (2012b) *Registro de Necesidades Básicas Insatisfechas* (RNBI – desagregados por cabecera, resto, municipio y nacional), DANE, Gobierno de Colombia, www.dane.gov.co/index.php/estadisticas-sociales/necesidades-basicas-insatisfechas-nbi, accessed 16 July 2012.

DNP (Departamento Nacional de Planeación República de Colombia) (2004) *Estrategia para el manejo ambiental del río Bogotá*, CONPES 3320 Ministerio de Ambiente, Vivienda y Desarrollo Territorial, Gobierno de Colombia, Colombia.

SDP (2011) 21 Monografías de las localidades, Distrito Capital, Bogotá.

SDP (2012) Información general de la ciudad de Bogotá, Distrito Capital, Bogotá, www.sdp.gov.co/portal/page/portal/PortalSDP/ciudad, accessed 28 April 2014.

SDP (Secretaría Distrital de Planeación de Bogotá) (2013) Plano de amenaza de inundación por desbordamiento, Alcaldía Mayor de Bogotá, Colombia, www.sdp.gov.co/portal/page/portal/PortalSDP/POT_2020/Documentos/04_Inundacion_V2_2013.pdf, accessed 28 April 2014.

SIRE (Sistema de Información para la Gestión de Riesgos y Atención de Emergencias de Bogotá) (2012) *Gestión de riesgos por inundación en Bogotá*, http://www.sire.gov.co/portal/page/portal/sire/gestionRiesgo/Inundaciones, accessed 25 August 2012.

Velandia, E. (2006) Marco metodológico para la modelación hidráulica e hidrológica de crecientes en ríos, Caso río Bogotá, El Espino – Alicachín. M.Sc. thesis, Universidad de Los Andes, Bogotá.

World Bank (2010) *Project Appraisal Document on a Proposed Loan in the Amount of US$250 Million to the Corporación Autónoma Regional de Cundinamarca with a Guarantee from the Republic of Colombia for a Río Bogotá Environmental Recuperation and Flood Control Project*, Sustainable Development Department, The World Bank, Washington DC.

11 Protection and environmental restoration of the slopes of Pichincha in Quito, Ecuador

Xavier Vidal, Lucía Burgos and Othón Zevallos

Introduction

Quito stands on the eastern flank of the Pichincha volcano in Ecuador. The city's strong population growth is leading to the uncontrolled spread of settlements on steep slopes surrounding the volcano, including environmentally vulnerable areas such as the rims of ravines and the banks of creeks. This pattern of urban encroachment has brought with it the removal of vegetation cover and the consequent exposure of slopes to erosion by wind and water, the obstruction of watercourses, and the loss of biodiversity and natural landscape.

The occupation of hillsides in Quito due to poor management of the urban environment increases the risk of natural disasters, and is indeed a case typically seen in urban areas of Latin America. The *Empresa Pública Metropolitana de Agua Potable y Saneamiento* (EPMAPS – the Metropolitan Water and Sewage Company) has played a significant role in efforts to mitigate the risks posed by this situation through engineering works, which are operated and maintained jointly with the municipal government. At the same time, the community has become aware of its responsibility in making proper use of this infrastructure.

This chapter describes the implementation of the integrated management plan for urban drainage and associated risks that was launched in Quito nearly 20 years ago, to address the environmental and urban problems associated with population growth, particularly on the city's hillsides. Besides being an example of a successful risk management plan, this case illustrates a new stage of public policy in which prevention efforts are in progress of being effectively institutionalized. The chapter starts by considering the geographical setting and the human settlement pattern of Pichincha's slopes in Quito. It then focuses on the protection and environmental restoration programmes for this area. Major lessons learnt and conclusions are presented in the end.

Geography of Quito and the Pichincha volcano

The city of Quito is the capital of the Metropolitan District of Quito and of the Republic of Ecuador. The metropolitan area has a population of 1.9 million (in 2013). It is home to the country's most important governmental, cultural and financial organizations and is a major production centre. Quito stands at the foot of the Andean Cordillera, on the eastern flank of the volcano Mount Pichincha. The city has 32 districts ('parishes') that make up the consolidated urban area, which covers approximately 19,000ha, stretching along a valley 42km long and 4.5km wide at an altitude that varies between 2,800 and 3,200m above sea level.

Average rainfall ranges from 1,200mm on the urban plain to 1,500mm in the upper reaches of the city. The average number of days with rain per year is 177. The rainy season runs from January to May, and the greatest amount of precipitation comes in April and accounts for nearly 20 per cent of the annual total. At its most intense, precipitation can be 40–45mm per hour. The average temperature is 14°C (Zevallos, 1996).

The slopes of the Pichincha-Atacazo massif comprise part of the natural drainage system of the area, with 85 ravines and stream beds that descend from the flanks of the volcano to the city, cross the urban area, and disgorge into the Machángara, San Pedro and Monjas rivers. The drop in elevation from the summit of 'Rucu Pichincha' (4,627m) to the lower part of the city (2,700m) occurs over distances ranging from 1 to 10km; the inclines are consequently very steep, between 30 and 60 per cent. Heavy rain falling on the highly erosion-prone slopes produce torrential flows that have hollowed out the ravines to depths of 10–30m and even more (Zevallos, 1996).

Land use is determined in part by the city's unique geography, with its hills and ravines. In recent years, the growing population has been pushing its settlements towards the hillsides surrounding the city (Figure 11.1). This

Figure 11.1 Urban encroachment on the hillsides of central Quito

Source: Courtesy of EPMAPS

Figure 11.2 Urban expansion of Quito, 1535–2009
Note: Image shows the urban expansion throughout the last centuries. This process includes slope zones converted into urban areas

Source: Adapted from EPMAPS (2011)

expansion can be appreciated in Figure 11.2, which shows the shifting pattern of land occupancy in the city from 1535 to 2009. The city is subject to threats arising from volcanic activity, the area's high seismicity, and the instability of the mountainsides.

People have always been drawn to settle on the mountainous slopes surrounding cities. This has been true of Ecuadorian cities such as Quito, Ambato, Portoviejo, Esmeraldas and Guayaquil, as well as many cities in South America, including Caracas, Medellín and La Paz (Zevallos, 2009). Yet hillside settlements are especially vulnerable to disasters associated with the climate, water and morphology. Peltre (1989) reviewed records in Quito up to 1989 and identified morphoclimatic incidents had occurred in 64 of the 85 ravines that descend the western slopes of Quito. The eastern flanks of Mount Pichincha are very steep. The vegetation covering these slopes shelters biologically diverse species that help to conserve conditions favourable to the water supply of the city, essential for keeping it habitable (Secretaría del Ambiente, 2012).

Inappropriate occupation of the hillsides has left the city subject to increasingly frequent slides of sediment and rubble that originate in the high valleys and cascade down the ravines and stream beds that dig into the steep hillsides of the volcano. Such events are becoming more serious with the encroachment of human settlements and mining, forestry and farming activities that are ill-suited to the environmental conditions on the slopes. The eastern slopes of the Pichincha-Atacazo massif above Quito are a typical example of increased risk of disaster through improper management of the urban environment and the degradation of 'common goods' such as soil, water and landscape (Zevallos, 1996).

Settlement on the Pichincha slopes

Historical context

Before the Spaniards arrived at Quito, the hills were cultural icons of veneration and respect (Del Pino and Marín de Terán, 2005). The Spaniards, however, had a different attitude to the hills and they came to be seen only as a source of material – rock, water and trees – from which to build the city. As the hillsides began to be regarded as obstacles to the expansion of the city, the ravines and stream beds were occupied. This approach has persisted until today, despite evidence that the enclosure of these areas and their inappropriate settlement results in frequent washouts during the rainy season that disrupt traffic in the city, affect businesses, and cause material damage and even the loss of human lives – to say nothing of the additional time needed by residents to perform daily activities under such conditions.

In the 1970s, occupation of the slopes intensified as a result of the oil boom. The boom exerted great pressure on urban land in Quito and Guayaquil, which became focal points of growth and development and attracted waves of migrants from the countryside, especially from the central Sierra (Unda, 1998). The expansion of northern Quito has been, and is still, characterized by a steady process of encroachment into the hillsides. The historic centre of Quito has seen the spread of slums, while the southern part of the city has undergone a significant change in land use, with broad swathes of farmland being sacrificed for urban development.

Urban, semi-urban and rural ways of life coexist on the slopes of Pichincha. Because of the varying origins and cultures of the population living there, working with these people requires an understanding of their different lifestyles and the recognition that they do not perceive the use and the distribution of space in the same way. Consequently, there can be no 'one-size-fits-all' management model or working strategy for these population groups. What is needed, rather, is to take into account all factors in order to propose a common future – a consensus-based agenda in which risk management focuses on the use, control and conservation of the slopes according to the nature and suitability of their soils.

Major threats and challenges

The hillsides are exposed to three types of threat: geomorphological, climatic and anthropic. *Geomorphological* threats are related to geological formation and structure, stratification planes, natural fractures and faults, weathering, slopes and reliefs, seismicity, volcanic and tectonic action, and fracturing (MDMQ, 2010). Threats of the climatic type have been associated with extreme rainfall events that contribute to and reinforce the action generated by geomorphological threats, as detailed in the study by Peltre (1989). Zevallos (2009) found that between 1900 and 1999 there were 465 hydro-geomorpho-climate disasters of varying magnitudes that affected Quito and caused great material and economic

Figure 11.3 House at risk of collapse in the lower part of the El Tejar ravine

Source: PSA (2006)

losses. Jiménez (2011) showed that of the 272 natural accidents and disasters that struck La Paz, Lima and Quito between 1970 and 2007, 79 per cent were caused by hydro-geomorphological phenomena. *Anthropic* (human-made) factors that constitute threats to the slopes include the spread of illegal squatter settlements, the exploitation of quarries, the improper disposal of solid wastes and rubble, deforestation, and reforestation with inappropriate species (ill-suited to the slopes) that result in changes to soil humidity conditions. Urban occupation of the hillsides results in blocking the existing drainage courses (Zevallos, 2009). An estimated 100ha of forest cover disappears every year through urban encroachment (Zevallos, 2009). Urbanization places many dwellings at risk (Figure 11.3).

Landslide events have increased very dramatically over recent years. From 1902 to 1998 Quito recorded 34 landslide events as the result of long periods of rain. These caused flooding in the streets and left homes and heritage buildings in danger of collapse because of the release of rocks and mud. There were heavy losses of both human life and property (Peltre, 1989). Over the 20-year period between 1990 and 2009, 426 incidents were recorded on the eastern slopes of the Pichincha volcano – an average of 21 per year.

The slopes and ravines of the Pichincha-Atacazo massif expose a disorderly pattern of urban development and changes in land use. Extensive areas of ecological importance have been converted for farming and livestock activities that generate pollution through the use of pesticides. The soil changes in permeability and is given less time to recover, and isolation and compacting phenomena are evident. These impacts are incurred through meeting the growing demands of the local agricultural market. On the other hand, illegal occupation of the Pichincha slopes has been promoted by land traffickers, who encourage invasion and squatting with the slogan 'land for the poor'. The subdivision of properties obtained through inheritance is another important factor that encourages the densification of housing and illegal squatting in natural areas due to the increased number of spaces for new settlers.

On the hillsides of southern Quito the principal mechanism behind informal urban sprawl is the cooperatives (or communes). These cooperatives erect dwellings disregarding the minimum technical specifications of construction (Zevallos, 2009) which spread into the protective buffer zones of the ravines. In order to increase the area available for building, they fill in watercourses without considering the hydrological impact of such behaviour. The picture becomes even worse when the municipality is lax in its control over land use and when its property register is in constant need of updating.

Disorderly urban sprawl and the topographic conditions of the hillsides complicate the delivery of water and sewage services by EPMAPS. The situation is just as complicated or perhaps even worse for street cleaning and garbage collection, undertaken by the municipal company *Empresa Pública Metropolitana de Aseo* (Municipal Garbage Collection Company, EPMASEO). It is very difficult for garbage trucks to make their way through the narrow streets, and they have no access at all to the many dwellings that can be reached only by stairs. Manual collection by teams equipped with pushcarts is also problematic, as haulage on the steep slopes requires greater physical effort with the attendant decline in labour productivity. Because garbage collection is so difficult and thus irregular, or non-existent, some hillside residents have no alternative but to dispose of their rubbish on the slopes and in the ravines.

People living in dwellings on the hillsides typically construct their own sewage disposal drains and run them into the streets or into the watercourses. This leads to the deterioration of surface water quality, destabilization of the slopes through changes in soil humidity, and damage to the environmental quality of adjacent green spaces. Traditionally, the hillsides were a tourist attraction because of the views they offered over the city and its surroundings, but pollution has seriously compromised their scenic appeal and has turned some areas into high crime zones.

Protection and environmental restoration of the Pichincha slopes

EPMAPS has had to cope with landslides on the slopes of Pichincha that have damaged and obstructed local water intake and conduction structures, thereby reducing the city's capacity to capture and transport water. To this situation must be added the rise in flow rates during times of heavy rainfall, leading to widespread flooding and its consequences, including the blockage of sewers and the inundation of dwellings, streets, pavements and urban infrastructure in general. This is the context for the *Programa de Protección de Laderas del Pichincha y Programa de Saneamiento Ambiental* (Slope Protection for Pichincha Programme and Environmental Sanitation Programme, PSA).

In 1996, the *Municipalidad del Distrito Metropolitano de Quito* (MDMQ – the metropolitan government) instructed EPMAPS to coordinate the actions needed to mitigate the consequences of the challenges described above. In response, EPMAPS designed plans and priority projects to guarantee the availability of drinking water in Quito, clean up the rivers of Quito, reduce water losses and consumption and to adapt to climate change, integrate management of hillsides of the Pichincha-Atacazo massif, and manage and control floods and optimize urban drainage.

EPMAPS, then, has been responsible since 1996 for the protection and restoration of the slopes and watercourses on the lower flanks of the Pichincha and Atacazo volcanoes, with a view to reducing the risk of flooding and washouts; reducing operating and maintenance costs for the collector systems; and instilling a culture consistent with the well-being of the community (Aguinagua and Paucar, 2011). The interventions undertaken by EPMAPS have the following objectives:

- To protect natural drainage systems through appropriate urban planning and the introduction of watercourse conservation and protection strategies.
- To install storm run-off systems that will be sustainable over the long term.
- To minimize risks and damage to public and private property caused by flooding, landslides, and discharges of mud and rubble.
- To maintain pleasant natural and urban settings so that people can enjoy the mountain landscape.
- To strengthen the social fabric on the slopes through training and management support for the local administrative authorities (the District Administrations).

The flood control programme incorporates those projects that will optimize storm drainage capacity. The approach taken is both comprehensive and socially inclusive. EPMAPS activities are not confined to structural interventions (water control works) but also include complementary measures such as:

- Construction of works to restore public spaces that will help address the shortage of recreational areas for children and adults alike.
- Environmental and civic education campaigns.
- Construction of infrastructure to improve pedestrian and vehicular movement across ravines and watercourses.
- Physical delimitation of the urban boundary in order to avoid sprawl into 'no build' zones of high vulnerability.
- Institutional strengthening through creation of the *Comisaría de Laderas* (hillside control commission) to prevent inappropriate land use on the slopes.
- Creation of natural reserves along ravines and stream banks as a site mitigation measure to encourage passive use of the soil in fragile zones.
- Resettlement of families at risk in cases where there is no alternative but to remove buildings.
- Preparation of a comprehensive slope management plan for the Pichincha-Atacazo massif, with designation of the zone as a natural heritage area for the city.
- Municipal capacity-building through training for the local administrations of the Metropolitan District of Quito.

With respect to social inclusion, EPMAPS has developed some participatory methodologies. It has successfully enlisted participation by the principal stakeholders and institutions in most of its interventions on the slopes and in the ravines over the entire cycle of its projects. Recognizing that all municipal institutions must be involved in managing the hillsides, EPMAPS has conducted training activities to share its methodologies, in particular those relating to risk management, and those methodologies are now being applied by risk management bodies in every one of the city's District Administrations.

Programme financing

The funds allocated to the works and activities for protection and environmental restoration of the Pichincha slopes mainly come from loans extended to the MDMQ by the Inter-American Development Bank (IDB), with supplementary funding by the local government and by EPMAPS itself. The three IDB-financed programmes are summarized in Table 11.1.

The Pichincha slopes programmes

Slope Protection for Mount Pichincha Programme

The first stage of remedial work on the slopes of the Pichincha-Atacazo massif, in the western part of Quito, took place during 1996 to 2002. The total investment was US$30 million, of which US$25 million came from a loan provided by the IDB, and the remaining US$5 million was a local contribution. The beneficiary

Table 11.1 Investments in the Pichincha slopes, 1996–2014

Programme	*Intervention zones*	*Area (ha)*	*Project*	*Period*	*Total programme (US$ millions)*[a]	*Environmental conditioning projects investment (US$ millions)*
Slope Protection for Mount Pichincha	North-western slopes	7,000	EC0143 935/ OC-EC	1996–2002	30.0 (25.0)	30.0
Environmental Sanitation Phase I	Central-western slopes	3,600	EC0200 1424/ OC-EC	2002–2006	53.3 (40.0)	9.4
Environmental Sanitation Phase II	Western and eastern slopes	6,500	EC-L1022 1802/ OC-EC	2007–2014	112.8 (67.1)	12.0

Note: a = Investments through IDB loans are in brackets
Source: From Loan Contracts signed between EPMAPS and IDB in the mentioned years

population numbered approximately 100,000. The focus of attention was the slopes lying between Miraflores Creek in the north-central part of the city and El Rancho Creek to the north-west of the city, comprising a natural area of 5,000ha and an urban area of 2,000ha.The programme included both engineering works and institutional strengthening, with the following activities:

- Civil works for the protection of watersheds.
- Creation of the *Comisaría de Laderas* to control use and occupation of the slopes.
- Work with the community through the project 'Strengthening community self-management in the squatter settlements of Ecuador'.

Environmental Sanitation Programme, Phase 1

The objectives of this programme were to reduce the incidences of flooding, mudslides and landslides; to expand water and sewer services in priority areas; and to strengthen institutional capacity for the efficient management of services. In designing the slope management and flood control components of the programme, due attention was paid to the lessons learnt from the Slope Protection for Mount Pichincha Programme, which concluded in 2002.

Phase I of the programme involved the construction of water regulation works supplemented by environmental stabilization works in the ravines and the recovery of open spaces. These works would allow the local population to

Figure 11.4 Bridging and environmental conditioning works, Quebrada El Tejar

Source: Courtesy of EPMAPS

move about more freely, to enjoy green spaces and recreational areas, and to have greater personal and property security (Figure 11.4). On the institutional front, further steps were taken for the coordination and strengthening of municipal institutions, boosting their capacity to manage conservation and soil use on the slopes (with training for the technical staff of these institutions), and clarifying the role of each body in ensuring proper land management. It was determined to be necessary to intensify communication with the beneficiaries of the works to ensure that the community was fully aware of its rights and duties.

Phase I took an approach that was at once comprehensive and socially inclusive, considering the different visions throughout the cycle of each project, especially those geared to the restoration of public spaces. The diagnostic assessments and project designs were discussed with the community, with explanation of how the alternatives might meet the requirements put forth by the local population. This consultative process continued through the final design, pre-contract and actual construction stages. During the construction stage, the local population often organized volunteer work squads (*mingas*[1]) to clean up the creek beds. As part of the social outreach activities, public meetings were held during construction to report on progress, to demonstrate accountability, to resolve disputes concerning construction, and to establish understandings for seeing the work through to completion.

Total investment in Phase 1 was US$53.3 million, of which US$40 million was financed by the IDB and US$13 million was the local contribution. The project zone covered a natural area of 2,400ha and an urban area of 1,200ha. Investment in works and projects under the slope management component amounted to US$9.4 million. Phase I of the programme was carried out in several ravines, all of them adjacent to areas of the city's historic and cultural heritage. This phase included water regulation works such as dykes, expansion and renovation of collectors, stabilization of slopes and other environmental stabilization works, diffusion wells (for energy dissipation), diversion channels, and design and implementation of the family resettlement plan. Phase 1 reached approximately 450 beneficiaries, of whom 100 were hillside dwellers.

The owners and occupants of properties adjacent to the affected ravines were invited to an information session in which the risk mitigation activities were explained to them and they were asked to vacate the ravine banks and clear the area in order to avoid overloading the slopes. The negotiations that ensued were difficult and time-consuming, and resulted in on-site mitigation measures being applied to 92 of the 146 dwellings identified as at risk. These mitigation measures consisted of clearing banks, removing minor structures, placing pipes to channel wastewater and rain run-off into the sewage system, rendering impermeable those areas considered highly vulnerable and placing wire fencing around private property. Where feasible, landscaping was done to maintain harmony with the restoration of the ravines.

The aim was to keep the need for family resettlement to a minimum, but in cases where on-site risk mitigation was not feasible, properties were expropriated in a process that involved three types of assessment: technical, social and legal.

- The technical evaluation assessed the environmental and physical vulnerabilities of structures located along the rims and banks of the ravines, and their exposure to those vulnerabilities.
- The social assessment considered various vulnerabilities that were summarized according to the Social Vulnerability Index.
- The legal assessment examined the type of ownership or tenancy title submitted by the inhabitants to determine whether the property could be expropriated.

For solid waste management, the municipal garbage collection company (EPMASEO) conducted a study of models appropriate for difficult slopes where access to dwellings is by stairs. The investigation found that approximately 40 per cent of domestic rubbish produced by people living on the hillsides was deposited in the ravines (Fulco, 2005). An initial participatory diagnosis was conducted, and several proposals were subsequently made, of which the following were implemented: modification of the garbage collection routes in the Phase 1 project zone; establishment of a street-sweeping and clean-up micro-enterprise, which was later absorbed by EPMASEO; and an education campaign in support of solid waste management.

Environmental Sanitation Programme, Phase II

Phase II of the Environmental Sanitation Programme was launched at the end of 2007 and continued through 2014. This phase maintains the objectives and approach pursued under Phase 1. Total investment in the programme is US$112.8 million, including US$67.1 million from the IDB and US$45.7 million local contribution. A total of US$12 million has been invested in slope management projects and the conditioning and restoration of public spaces (Figure 11.5).

Phase II includes the preparation of the Environmental and Social Management Plan, which contains indicators of performance of the Plan that are assessed twice a year. The plan calls for a municipal cooperation agreement with both EPMAPS and EPMASEO. During the construction of these works, 'social intervention methodology' has been applied as a means of ensuring that local population understands what is being done and how the works are progressing. This methodology also provides a way for managing and resolving disputes. Training has played a central role in the programme. First, with the intention of making the community aware of appropriate land use, the negative impact of changes in soil use on steep slopes, the enrichment of soil on the slopes, and the need for crop rotation, training was provided to 900 inhabitants of the Pichincha-Atacazo corridor, which is located in the transition zone between the urban and natural boundaries. Second, to improve waste recycling and to optimize the existing collection system, training was provided to 2,900 hillside residents. In this case, the support is greater in scope, and a baseline has been prepared for analysing the community's behaviour, perceptions and attitudes with respect to waste management and disposal. The intention is to propose a

Figure 11.5 Example of environmental conditioning, Quebrada Navarro

Source: PSA (2012)

model for environmental management of the slopes. The individuals who have received training are expected to act as environmental managers in the handling of solid waste and in the operation of temporary deposit sites. In August 2012, as part of a cooperation agreement with the University of California at Berkeley, a workshop was held, in Quito, to develop a proposal for overall improvement of the rim and banks of the Ortega Ravine. The event included professors and students from the University of California and two local universities as well as technical staff from EPMAPS. The proposal addresses (i) landscape upgrading; (ii) decontamination of water; (iii) mitigation of hydrodynamic risks; (iv) mitigation of geotechnical risks; and (v) active community participation. This pilot proposal takes into account PSA experience with the restoration of ravines, lessons learnt and the possibility of improving future interventions.

Institutionalizing the slope protection programmes

A report on municipal management for risk reduction and emergency preparedness (UNDP, 2007), in discussing the actions undertaken by EPMAPS, indicates that this entity has contributed to the prevention and mitigation of risks and thus to generating a comprehensive vision of risk management. Through its work on the Pichincha slopes, EPMAPS has played an important role in reducing vulnerabilities in an area that is home to at least 300,000 people. Since 2006, the challenge has been to institutionalize initiatives, particularly those not directly related to the EPMAPS mission and vision. Achievements in this respect include the following:

- In 2009, the Risk Evaluation Methodology (see the explanation below) was shared with the Safety and Governance Secretariat,[2] which adopted it and is now using it. An Environmental Responsibility Branch was established within EPMAPS, with responsibility for managing the slopes covered by the EPMAPS mission and vision.
- In 2010, the Public Metropolitan Habitat and Housing Company, which is part of the MDMQ, adopted an approach of working with and supporting families before, during and after resettlement.
- In 2012, there was an orderly handover by EPMAPS to the Secretariat of the Environment, in particular its Natural Heritage Division, of available information and studies conducted for putting planned activities into operation.

The Risk Evaluation Methodology prepared for and applied to slope protection projects under the PSA has been adopted by the Safety and Governance Secretariat of the metropolitan authority of Quito. Every district administration now has a risk management unit that uses it. Some 50 officials from the District Administrations have been trained in the methodology, and they have already applied it successfully in dealing with the problems that arose during the 2011 rainy season.

Conclusions

The eastern flank of the Pichincha-Atacazo massif in the city of Quito is a typical case where disaster risks have been heightened by the uncontrolled encroachment of settlements into hilly areas that are unstable from a hydro-geomorphological viewpoint. In this naturally unstable setting, the frequency of flooding and landslides is increased by chaotic land occupation, the construction of dwellings with no regard for the minimum building standards, and the infill of ravines to expand the area available for construction with works that reduce their water carrying capacity. The situation is exacerbated by the poor management of solid wastes, which are dumped into the ravines, and the disposal of wastewater into the streets or the ravines.

In 1996, MDMQ designated EPMAPS to take the necessary steps to protect and undertake environmental restoration of the slopes, with backing from three loans from the IDB. Financing from these loans enabled the construction of engineering works to mitigate the effects of floods and landslides; creation of the *Comisaría de Laderas*; strengthening of municipal management in aspects relating to urban and peripheral planning, land use and community development; relocation of families from high risk zones; restoration and replanting of ravines for incorporation into the urban green area; proposal of a Slope Management Agency and a plan for managing the physical environment; and research to improve the technical information available concerning the slopes.

During execution of the third loan, EPMAPS has devoted part of its efforts to institutionalizing the activities underway in order to put them on a permanent footing and guarantee the sustainability of the overall project. EPMAPS has retained full responsibility for these activities, consistent with its mission and vision. Thanks to these achievements, Quito has reduced its environmental risks.

Lessons learnt from the project include the following:

- To resolve problems associated with the urban hydrological cycle and to guarantee the cycle's sustainability, solutions must be planned and implemented with the full participation of all the entities involved in delivering urban services and of the beneficiary community.
- The construction of engineering works alone is not enough to solve problems. Such construction must lead to the incorporation of spaces that offer recreational opportunities to local residents who will then come to see themselves as custodians of the works and of the restored environments.
- Project execution is an ongoing task. Institutions responsible for the project must be strengthened in their capacity to maintain and expand it.

Notes

1 The *minga* is a traditional way of organizing community work that persists in the low-income neighbourhoods of Ecuador's cities. It originated in the practices of reciprocity and solidarity that allowed for the redistribution of wealth in indigenous communities. It constitutes a space for social interaction.

2 The *Secretarías* constitute the local government ministries. They have operating departments in the District Administrations, which are the decentralized municipal units for local administration.

References

Aguinagua, V. and Paucar, S. (2011) 'Propuesta de un plan de comunicación corporativa para la empresa pública metropolitana de agua potable y saneamiento (EPMAPS), con el fin de incrementar las visitas a la quebrada Jerusalén, ubicada en el sector centro de la ciudad de Quito', Thesis, Universidad de Las Américas, Quito.

Del Pino, I. and Marín de Terán, L. (2005) *Algunas reflexiones sobre el Ecuador prehispánico y la ciudad inca de Quito*, Consejería de Fomento y Vivienda, Andalucía, Spain.

EPMAPS (Empresa Pública Metropolitana de Agua Potable y Saneamiento) (2011) *En las faldas inmensas de un monte ...*, EPMAPS, Quito.

Fulco, L. (2005) 'Evaluación técnica, institucional, ambiental, social y económica de alternativas de modelos de gestión para el manejo de desechos sólidos en las áreas comprendidas entre las Quebradas el Tejar y la Raya Sur', unpublished report, Quito.

Jiménez, E. (2011) 'Concepciones hacia un sistema de gestión comunitaria del riesgo en las laderas suroccidentales del Distrito Metropolitano de Quito', MSc, thesis, École Politechnique Nationale, Université de Nice – Sophia Antipolis, Nice, France.

MDMQ (Municipio del Distrito Metropolitano de Quito) (2010) *Atlas de amenazas naturales en el Distrito Metropolitano de Quito*, Distrito Metropolitano de Quito, Quito, Ecuador

Peltre, P. (1989) *Estudios de Geografía Volumen 2: Riesgos Naturales en Quito – Lahares, aluviones y derrumbes del Pichincha y del Cotopaxi*, Corporación Editora Nacional, Quito, http://

sthv.quito.gob.ec/spirales/3_publicaciones/libros/Quebradas_y_riesgos_naturales.pdf, accessed 29 March 2014.

PSA (2006) *Archivo fotográfico del* año 2006, Programa de Saneamiento Ambiental, Quito.

PSA (Programa de Saneamiento Ambiental de Quito) (2012) *Archivo fotográfico del* año 2012, Programa de Saneamiento Ambiental de Quito.

Secretaría del Ambiente (2012) 'Agenda ambiental municipio del Distrito Metropolitano de Quito 2011–2016', http://institutodelaciudad.com.ec/attachments/article/55/agenda_ambiental_rev_final%208-02-2012.pdf, accessed 29 March 2014.

Unda, M. (1998) *Vida en las laderas*, Centro de Investigaciones CIUDAD, Quito, www.flacsoandes.org/biblio/catalog/resGet.php?resId=47867, accessed 29 March 2014.

UNDP (United Nations Development Programme) (2007) *Catálogo de instrumentos en gestión municipal para la reducción de riesgos y preparativos para emergencias*, UNDP, La Paz.

Zevallos, O. (1996) 'Ocupación de laderas: incremento del riesgo por degradación ambiental urbana en Quito, Ecuador', in M. A. Fernández (ed.), *Ciudades en riesgo – degradación ambiental, riesgos urbanos y desastres*, Red de Estudios Sociales en Prevención de Desastres en América Latina, Limón, Costa Rica, http://www.desenredando.org/public/libros/1996/cer/CER_Intro_ene-7-2003.pdf, accessed 29 March 2013.

Zevallos, O. (2009) 'Degradación, vulnerabilidad y riesgo hidrogeomorfoclimático en areas urbanas de laderas', in J. Erazo (ed.), *Inter/Secciones Urbanas: Origen y Contexto en América Latina*, Flacso Sede Ecuador y Ministerio de Cultura del Ecuador, http://works.bepress.com/cgi/viewcontent.cgi?article=1009&context=jaime_erazoespinosa, accessed 29 March 2014.

12 Water resources management and climate adaptation in the Río Grande Basin, Argentina

A case study introducing the Hydro-BID modelling system

Fernando Miralles-Wilhelm, Eugene Brantly, Marcelo Borsellino, Edgardo Castellano, Robert Dykes, Fekadu Moreda, Raúl Muñoz-Castillo, Jay Rineer, Bill Wheaton and Alan Wyatt

Introduction

Cities in Latin America have experienced an increase in the number and intensity of climate events such as El Niño Southern Oscillation over the past 30 years, causing record precipitation events in Venezuela (1999 and 2005); flooding in the Argentinean Pampas (2000 and 2002); droughts in the Amazon basin (2005), Caribbean (2009–2010), Mexico (2011–2012) and Argentina (2011); and back-to-back record hurricane seasons in the South Atlantic (2004, 2005). The impacts of climate change do and will continue to vary across the region, as summarized by the Intergovernmental Panel on Climate Change (IPCC) (Magrin *et al.*, 2007).

Across Latin America changing rainfall patterns will produce more intense rainfall in some areas and drought in others. More frequent and more violent storms will damage infrastructure and strain drainage systems at some locations, while reduced rainfall will increase water stress at other locations. Rising sea levels will threaten coastal aquifers with saltwater intrusion. For cities in mountainous regions, the loss of water storage in glaciers and *páramos* (Alpine tundra ecosystems typical of South American Andes) could cause long-term reductions in river flows and groundwater recharge, resulting in stress on water supplies for people, crops and ecosystems, as well as reduced capacity for generating hydroelectricity. Hydropower is the main source of electricity for most Latin American countries and is at great risk from changes in water flows and water availability.

The factors affecting people's vulnerability to expected changes in water resources are known for the most part. On a macro scale, the main drivers are population growth and migration, uncontrolled urban growth, and inadequate infrastructure and services. At a finer level of resolution, the vulnerability of specific areas and populations to changes in water resources is a function of several key factors, including local geological and hydrological characteristics, local geography, land use, erosion, vegetation cover, the presence or absence of groundwater, suitable and sufficient water storage sites and facilities, current levels of consumptive water demand relative to available supplies, water quality, policies and institutional frameworks governing water allocation and use, and inter-basin transfers.

The vulnerability of water supplies to climate change and other threats can vary widely within a river basin. Some areas will be highly vulnerable to small changes in precipitation or temperature, while other areas are 'buffered' and will be relatively immune to even large changes in precipitation or temperature. These differences are most highly correlated with the level of water use and socio-economic development within the sub-basin, the drainage area and position within the larger watershed, the degree of groundwater flux to and from surface waters, and the reliability of existing storage systems. Measures designed to help adapt to climate change will be more effective, in both performance and cost, if they are planned using an integrative process that takes full advantage of available knowledge.

The Inter-American Development Bank (IDB) provides financial and technical support for infrastructure projects in water and sanitation, irrigation, flood control, transport and energy. Many of these projects depend upon water resources and may be affected negatively by climate change and other developments that alter water availability. Assessing the potential for future changes in water availability is an important step toward ensuring that infrastructure projects meet their operational, financial and economic goals. It is also important to examine the implications of such projects for the future allocation of available surface water among competing users and uses to mitigate potential conflict and to ensure that such projects are consistent with long-term regional development plans and preservation of essential ecosystem services.

It is with this objective that an integrated water resources management simulation modelling system for Latin America, *Hydro-BID*, was developed. The central purpose of this integrated suite of watershed modelling tools is to help policy makers make informed choices among alternative policies (e.g., water allocation for different uses, planning for existing or new infrastructure, and consideration of different climate scenario outcomes) for water resources management at the regional, basin and sub-basin scales. *Hydro-BID* generates output in the form of a daily time series of flow estimates for the selected location and period. As an illustration of the system's inputs, operation and outputs, this chapter documents a case study addressing surface water allocation issues in the Río Grande basin in Argentina.

Analytical hydrography data set

The Hydro-BID modelling tool includes the following:

- An analytical hydrography data set (AHD) representing more than 230,000 catchments in the Latin American and the Caribbean region and their corresponding river and stream segments.
- A geographic information system-based navigation tool to browse AHD catchments and streams with the capability of navigating upstream and downstream.
- A user interface for specifying the area and time period to be modelled and the location at which water availability will be modelled.
- A climate data interface to obtain rainfall and temperature inputs for the area and time period of interest.
- A rainfall-runoff model based on the Generalized Watershed Loading Function (GWLF) model (Haith and Shoemaker, 1987; Haith *et al.*, 1992).
- A routing scheme for quantifying time of travel and accumulating flow estimates across downstream catchments.

Hydro-BID's AHD is a spatially explicit (i.e., geographic information system [GIS]-based) dataset of surface water. A digital elevation map (DEM) was constructed based on global Shuttle Radar Topography Mission (SRTM) elevation data for Latin America (at approximately 90m resolution), and this was used to delineate basins in Hydro-BID using spatial interpolation techniques. The AHD itself serves as a regional spatial data platform for integrating the disparate data needed to support regional hydrography models. It provides a basis for parameterizing models in a consistent way, provides the upstream/downstream flow connectivity necessary for the models, and provides the data necessary for displaying results in map form.

The AHD has four key features enabling the development and implementation of tools that analyse water in the context of its movement in a network:

- A nested polygon catchment structure with a unique catchment identification scheme.
- A corresponding line structure for the stream segments contained within the catchments.
- Descriptive attributes that create a linked network of upstream and downstream stream segments and catchments.
- Database structure for populating and attaching additional model-specific attributes.

The AHD contains data on each stream segment's maximum and minimum elevations, length, average slope, and other geometric characteristics. The database structure also provides placeholder fields for a wide range of additional descriptive attributes. Each catchment in the AHD can be attached to areal-based data that are relevant to the models being executed, for example, area, land cover and land use, average slope, predominant soil types, and climate data. Models will use data drawn from the AHD and will generate results that can be attached back to the AHD.

The AHD uses a vector-based data model in which features are represented as points, lines, and polygons rather than a raster, or gridded approach in which data are stored in a grid cell. The vector-based data model of the AHD allows for a more natural (as well as more numerically accurate) representation of vector quantities such as flow, velocity, and hydrographic basin connectivity compared with raster- or grid-based hydrographic approaches. The AHD is independent of any single water model. AHD's hydrological network provides a consistent, flexible, and, most important, scalable framework for use with a broad range of models, containing data for parameterizing model inputs as well as the results of model runs to support post-processing analysis and decision support. The AHD is different from many geospatial hydrological data sets because it includes the upstream/downstream flow relations as database attributes, enabling intelligent analysis, modelling, and simulation. It is also different from satellite imagery or static maps of hydrography for the same reason – the hydrological network connectivity is encoded into the data.

The AHD is also unique in that it has been derived from a single globally available elevation dataset. The stream segments and corresponding catchments that the AHD contains are developed by an automated algorithm using only these elevation data. This means that the stream segments in the AHD are derived and are not necessarily located at actual stream locations as would be found on a detailed map or satellite image. In contrast, the creation of other similar data sets such as the United States National Hydrography Dataset Plus (NHDPlus) involved many additional input layers, including detailed field surveys of the actual stream networks that had to be developed by the U.S. Geological Survey (USGS) over many years at a great cost.

For many hydrological and water resources modelling uses, the analytically derived features contained in the AHD are both appropriate and adequate, but as additional detail is needed, the AHD can be updated to include actual surveyed stream segments, and increasingly detailed data. The AHD provides the ability to hold increasingly detailed surface water delineations, where needed, while still providing a consistent data structure for all users in the region. In this sense, the developed AHD provides users with a flexible and highly scalable platform for water resource modelling, mapping, analysis and management.

Hydro-BID simulation model

Figure 12.1 presents a schematic representation of the integrated Hydro-BID system for quantitative simulation of hydrology and climate change. Hydro-BID utilizes the data structure and the catchment and stream network topologies of the AHD. It incorporates data on land uses, soil types, rainfall and temperature within the study area, as well as observed stream flows for use in calibration.

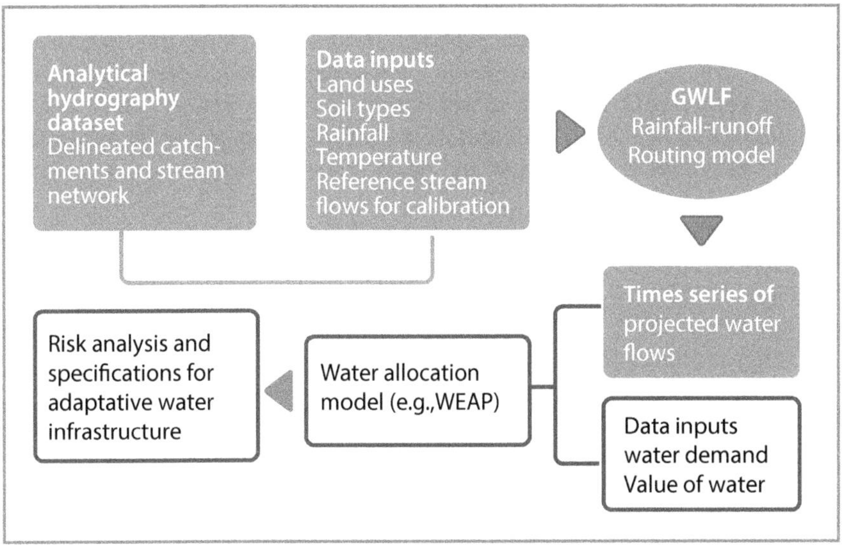

Figure 12.1 Hydro-BID simulation model software flow chart

Source: Authors

Hydro-BID includes a pre-processor interface to disaggregate monthly climate data into a daily time series of temperature and rainfall, which is the required form of climate data input. The interface renders daily projected climate data by combining a monthly time series of future projection data with a daily time series based on historical data. The method assumes that the daily variation reflected in historical data is a valid basis for projecting daily variation in future conditions, with proportional increases or decreases in future temperature and rainfall, as shown in the monthly time series of future projections. To apply this method, the Hydro-BID user will select a historical data set and period; select a future projection scenario from the list of available scenarios; and provide monthly future projections in a specified format. The system uses these inputs to produce daily climate data for the entire length of the projection period.

The system applies a standard GWLF model coupled with a novel lag-routing methodology initially described in Linsley *et al.* (1975). This methodology allows a more realistic representation of drainage over long channels in basin hydrology formulations like that in Hydro-BID, delaying water from an upstream channel to the downstream network by a lag time and allowing for the attenuation of the flow hydrographs. The computational implementation of this methodology is currently under patent pending status. The GWLF model formulation contains only a few parameters, which makes it easy to use and apply in settings with limited data. Model output is generated as a time series of projected water flows, at either a daily or a monthly scale. The system has a graphical user interface to accept model input, as well as to display graphical and tabular summary outputs. Output from the Hydro-BID system should be analysed further in conjunction with additional data and projections for water demand, cost, the economic value generated by water uses, and the expected performance of the infrastructure in question, in order to develop water allocation schemes, risk analyses and adaptive designs. Hydro-BID can be used with historical climate data or future climate projections to model basin-wide surface water flows, assess the potential impacts of climate change on water flows and infrastructure, and support the design of adaptive projects and strategies. The adaptation options include response to long-run declines in stream flows; an increase in flows; an increase in seasonal and inter-annual variation in flows; and extreme events such as drought and floods.

Hydro-BID has been developed to support:

- Water resources planning and management agencies.
- Drainage/flood control authorities.
- Irrigation authorities.
- Hydroelectric power generators.
- Water supply and sanitation utilities.
- Industrial water users

Hydro-BID case study application

The Río Bermejo basin in Northwest Argentina has been the subject of past projects in water supply, sanitation and water resources management by the IDB and other donors, and it is expected to benefit from more technical assistance and financing to address ongoing water resources management issues. This basin was selected for a first case study to illustrate the application of *Hydro-BID* in a collaboration project between the IDB and the Comisión Regional del Río Bermejo (COREBE). The selection process required several discussions for collaborators to understand the purpose, data requirements and potential outputs of the model.

After an agreement to apply the model to the upper part of the basin (Río Grande sub-basin), a mission to Argentina was made to accomplish the following:

- Explain the Hydro-BID model to local officials.
- Understand priority water resources issues in the case study area.
- Gather basic data on water resources, climate, and water demands.
- Establish linkages with other institutions in Argentina for further data collection.
- Determine climate modelling capabilities in Argentina.

These discussions led to the selection of a particular case study application:

> To model runoff on the Río Grande sub-basin and identify potential adaptation measures that could meet future demand for urban water supply for the city of San Salvador de Jujuy and future demand for irrigation water south of the city.

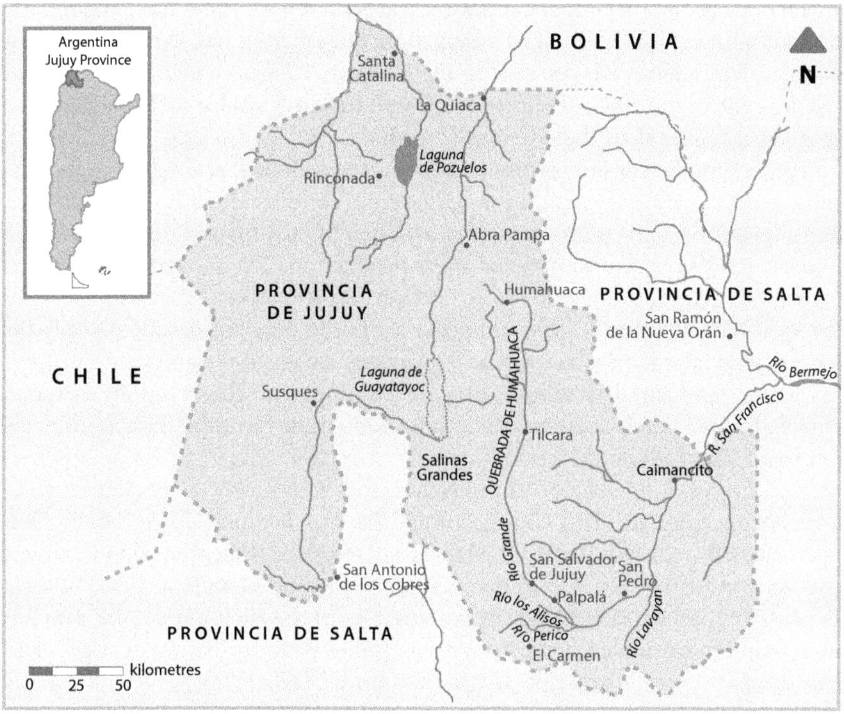

Figure 12.2 Jujuy Province, Argentina

Source: Courtesy of Instituto Geográfico Nacional

The province of Jujuy (Figure 12.2) is located in the extreme north-west of Argentina, bordering Bolivia to the north, Chile to the west, and the Argentinean province of Salta to the south and east. Key features include the large, dry mountain region (*altiplano*) to the west, with peaks over 4,200m in elevation, and the dramatic canyon Quebrada de Humahuaca. This valley forms the basin for the Río Grande, which flows south from the mountains, through the provincial capital of San Salvador de Jujuy (altitude 1,200m), and turns to the east to join with the Río Lavayan to form the Río San Francisco. That river flows north, through the wide eastern agricultural valley (altitude approximately 500m), eventually merging with the Río Bermejo.

The area has warm, wet summers and cool, dry winters, with very low rainfall in the high altitude areas. During the dry winter and spring months there are frequent water shortages in the provincial capital San Salvador de Jujuy, and restrictions on irrigation water in this agricultural province. Regional and provincial officials are concerned that climate change will make these periods of scarcity more frequent and more extreme.

The total provincial population, as of 2010, was approximately 673,000, with an average density of 12.7 inhabitants/km^2. The mountainous Puna region has a density of approximately 1.4 inhabitants/km^2, whereas the Valles area has more than 100inhabitants/km^2. San Salvador de Jujuy has a population of approximately 265,000. The population growth rate has slowed in recent decades, from about 2.0 per cent in the 1980s to 1.7 per cent in the 1990s and 1.0 per cent in the 2000s. Population growth in San Salvador de Jujuy has been somewhat higher than the province as a whole.

Today, Jujuy's economy is moderately underdeveloped, yet highly diversified. Jujuy's economy has significant agricultural activity, mainly in sugarcane. Sugarcane processing represents more than half of the province's local economic output and 30 per cent of national sugar production. The second major crop is tobacco. Other crops include beans, citrus fruits, tomatoes and other vegetables for local consumption. Cattle and goats are raised on a small scale, mainly for local dairies; and llamas, vicuñas and guanacos are used in significant numbers for wool. Industrial activities include mining for construction material, petroleum extraction at Caimancito, salt production from the Salinas Grandes salt basin, and paper production.

The city of San Salvador de Jujuy has only one source of potable water – a diversion from the Río Grande called the Los Molinos Dam, about 5km upstream from the city. Water for 34,000 ha of irrigated land, other smaller towns and agro-industrial use is also diverted from the Río Grande at Los Molinos. Because of high sediment loads in the water from the Río Grande, the dam has no remaining storage capacity, so water withdrawals for urban water supply and irrigation are provided by run-of-the-river flow (IDB, 2013).

During the late dry season, August to November, the river flow is very low. Urban water supply shortages and service cut-offs are common. Water diversion from the Río Grande for irrigation is also diminished during the dry season, despite a high demand for irrigation water. However, those using water for irrigation have additional water sources. Four smaller rivers feed water into four dams that supply water to irrigated lands and agro-industrial users, and potable water to two small towns, Palpalá and El Carmen, with a combined population of 155,000. Nonetheless, there has been competition among all types of users of water from Los Molinos Dam, each trying to secure an adequate share of the Río Grande river flow.

Given these considerations, the stakeholders in Jujuy (consisting of COREBE, municipal representatives and agricultural irrigation organizations) requested that the project team evaluate the vulnerability and sufficiency of water supply for urban San Salvador, while also considering supply for irrigation and other uses from the Los Molinos Dam, and taking into account future climate change. They asked the team to screen and suggest adaptation measures that should be considered by the stakeholders in the case study area. The stakeholders and the project team agreed that the case study would be conducted at the pre-feasibility level and would not include detailed engineering designs or cost studies. Figure 12.3 shows a map of water infrastructure – dams, channels, towns and irrigated areas – in the case study area.

Once the work began, the project team discovered that the analysis needed to be expanded beyond the Río Grande to include modelling of the adjacent water

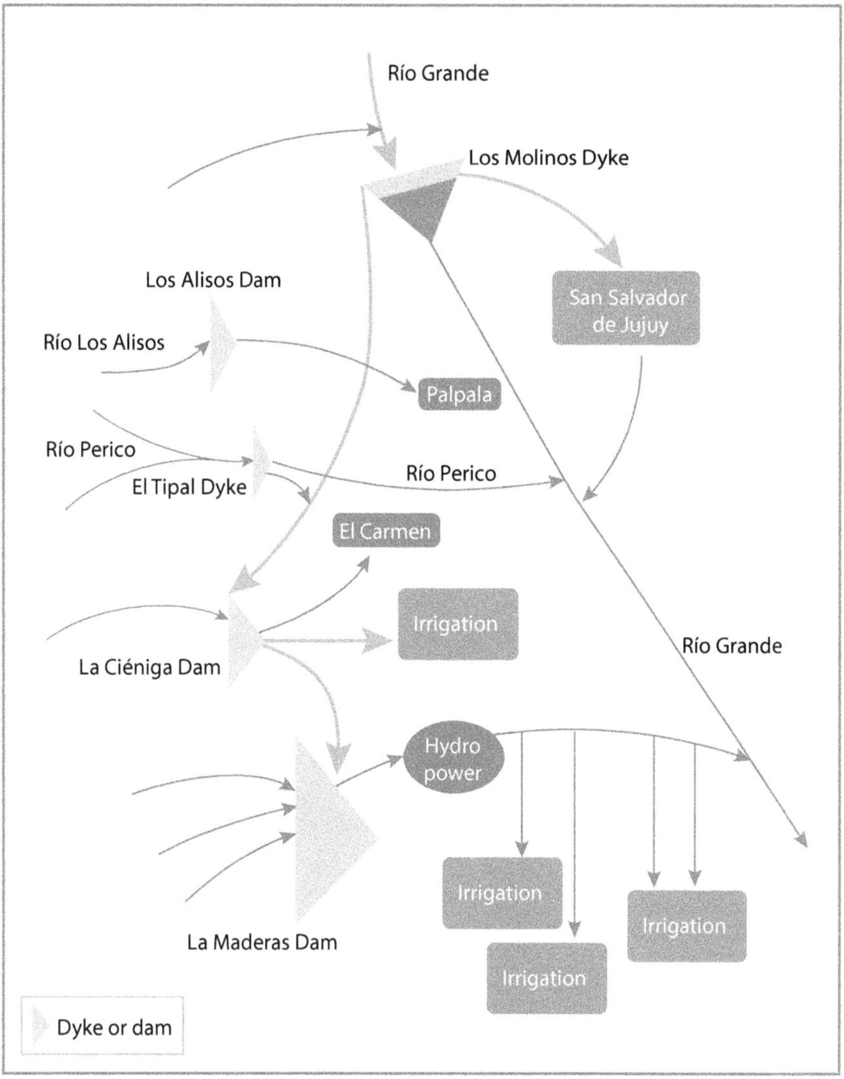

Figure 12.3 Diagram showing existing water infrastructure in the Río Grande and upper Río San Francisco basins

Source: Authors

sources and water uses in the southern part of the Río San Francisco, including the streams feeding the four dams, urban water supply in nearby smaller towns, and agro-industrial uses. The overall approach for the case study application of the Hydro-BID model to the water resource management issues in the Río Grande is outlined in the diagram in Figure 12.4.

After initial work with local stakeholders to define the study objectives and conduct basic data collection and assessment, the work proceeded along three

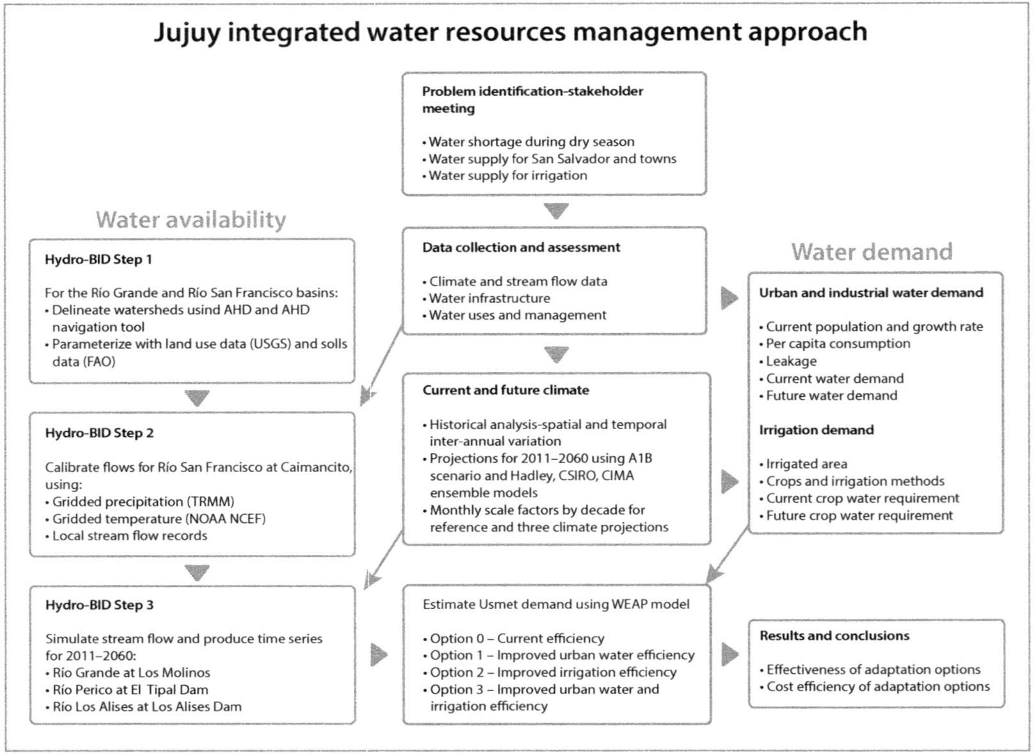

Figure 12.4 Jujuy integrated water resource management approach

Source: Authors

major areas: water availability, using the Hydro-BID model in three steps; current and future climate analysis; and water demand. These analyses provided the necessary information to compute the unmet water demand, assuming the following:

- Climates: a reference climate and three climate projections.
- Adaptation options: an option 0 (no improvements to water efficiency) and three adaptation options, each involving different water efficiency improvements.

From the unmet water demand results, each of the adaptation options was evaluated using two criteria (IDB, 2013). The first centres upon effectiveness which is understood as a reduction in unmet water demand. The second regards cost efficiency, considered as cost per unit reduction in unmet demand (IDB, 2013). The remainder of the case study project proceeded as follows:

Project start-up. This phase included discussion of priority water resources issues and formulation of case study objectives.

Data collection and assessment of natural conditions. In this phase data collection and assessment of natural conditions focused on stream flow, current and future climate, soils, and land use. Local and regional agencies and provincial government authorities provided useful data. The main failing was a lack of reliable, recent, stream flow data in the Río Grande basin. The nearest location with good data was the town of Caimancito, downstream on the Río San Francisco, for which data on daily stream flow were available for the period 1947–2011. This location was chosen as a base for Hydro-BID model calibration even though the area of the catchments upstream of Caimancito is far larger than the Río Grande basin. The project team enlisted the assistance of the Centro de Investigaciones del Mar y la Atmósfera (CIMA – Centre for Oceanic and Atmospheric Research) in Buenos Aires to prepare climate predictions for the case study area, based on their experience with different approaches. They prepared projections using the latest global climate modelling tools (IDB, 2013). The team also used downscaled future climate data from the Climate Wizard platform developed by the Nature Conservancy, the University of Washington and the University of Southern Mississippi as an alternate climate prediction tool to gauge sensitivity of climate assumptions.

Data collection and estimation of water demand. The characterization of existing water infrastructure and water management practices was straightforward, but estimating water demand for irrigation proved challenging due to a lack of reliable evapotranspiration data. Detailed calculations using CROPWAT and other methods recommended by the Food and Agriculture Organization of the United Nations (FAO) yielded results close to recent estimates by local water management personnel in COREBE using both analytical and empirical methods.

Modelling of current and future stream flows. This phase included calibration of the Hydro-BID model at Caimancito, and the running of the Hydro-BID model to predict future stream flows in the main channel and several sub-basins of the Río Grande basin. The calibration used international data sets for many parameters because daily values for rainfall and temperature were not available at the same locations as daily stream flow. Model calibration yielded acceptable results, with good accuracy in the dry season.

Comparison of water supply and demand. This phase included integration of projected future stream flows and water demand information in the model to calculate unmet water demand for several climate projections and several adaptation scenarios. The climate projections included (i) no change; (ii) a general circulation model (GCM) ensemble prepared by CIMA; and (iii) two downscaled projections from Climate Wizard. Adaptation options included (i) no change; (ii) improved urban and industrial water efficiency; (iii) improved irrigation application efficiency; and (iv) improved efficiency for urban and irrigation uses in combination. The unmet water demand by month and by decade was determined for each potential situation.

Assessment of adaptation options. Each adaption option was assessed using the CIMA GCM ensemble (which showed the highest climate change) by determining the effectiveness of the option, as measured by the reduction in unmet demand; and the cost efficiency of the option by comparing net cost per cubic metre saved, derived from the present value of all capital and operating costs associated with the option divided by the total reduction in unmet demand.

Hydro-BID model set-up and calibration

Model set-up

The Hydro-BID model was set up for the Río Grande/Upper Río San Francisco basin, which includes 249 AHD basins and a total catchment area of 22,070km^2. The basin was selected for calibration for the following reasons:

- Recent stream flow data is available.
- Stream flow is measured weekly (data could be converted to monthly averages for calibration).
- The human influence in the basin is relatively small at the catchment basin level; therefore, calibration could be performed for a general water balance.
- Calibration could not be performed for just the Río Grande because of a lack of reliable, recent flow records.

The observed data at sites throughout the basin are limited and also cover periods earlier than the 1980s. Therefore, the project team used the National Aeronautics and Space Administration (NASA) Tropical Rainfall Measurement Mission (TRMM) (Kummerow *et al.*, 1998) daily precipitation data for consistency. The AHD catchment overlaid the gridded daily precipitation map to obtain catchment average precipitation data.

For temperature, the National Oceanic and Atmospheric Administration (NOAA) National Centres for Environmental Prediction (NCEP) reanalysis of temperature covering the Río San Francisco basin was used (Kalnay *et al.* 1996). The minimum and maximum daily temperatures were used to compute the mean daily temperature map. The AHD catchment overlaid the gridded daily temperature map to obtain catchment average temperature data.

For each AHD catchment in the basin, we first generated the proportion of catchment area by land cover type. Second, we determined the dominant hydrological soil group within each portion of the AHD catchment. Third, we determined the curve numbers for each portion of the AHD catchment, based on the land cover type and hydrological soil group, using the curve number look-up table included in Hydro-BID.

Model calibration

Three parameters – curve number multiplier, runoff coefficient, and seepage parameters – were calibrated to obtain a reasonable overall volumetric error of less than 15 per cent. Figure 12.5 shows model results in the form of a comparison of observed and simulated monthly flows for the calibration period. Overall, the model simulates the water balance well but overestimates high summer flows. The difference in overall stream flow over the decade 2001–2010 was 14 per cent. The overall correlation between observed and simulated was 85 per cent.

By examining the seasons, we can see that the recession hydrograph and low flows are reproduced well. The low flow periods are most important for modelling water shortages. It should be noted that these results (and the current

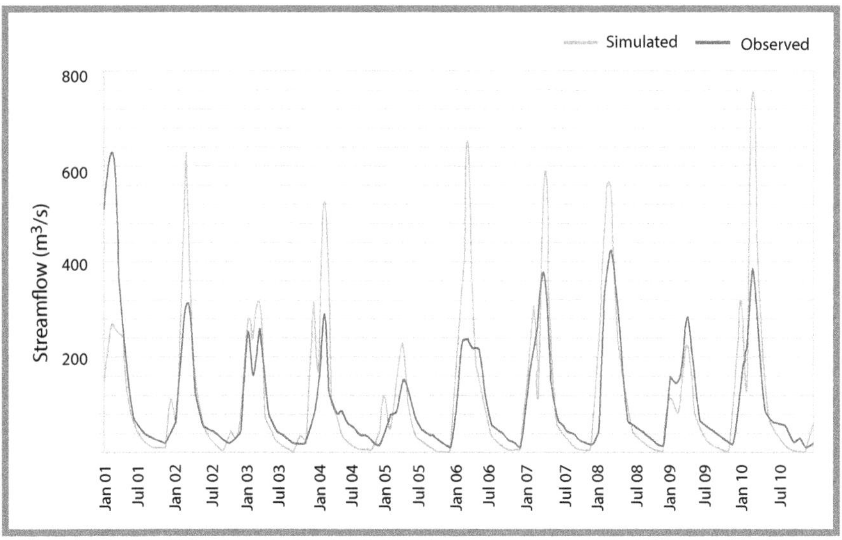

Figure 12.5 Observed and simulated flow, Río San Francisco at Caimancito

Source: Authors' calculations and elaboration

calibrating parameters) are sufficient to study the relative importance of climate change impacts on water resources management. However, improved modelling within the interior of the Río Grande/Río San Francisco basin will require better stream flow records for smaller catchments and model recalibration. The calibration parameters developed here will be a good starting point for refined modelling of the interior of the Río Grande/Río San Francisco basin.

Simulation results and evaluation of selected adaptation options

Simulation results

Given the complex geography of the river sub-basins and different water uses in this case study, we needed to simulate flows at the outlet of three sub-basins:

1 Río Grande upstream of the Los Molinos Dam (69 AHD catchments)
2 Río Los Alisos (1 catchment)
3 Río Perico (7 catchments).

The simulated monthly flows at these locations were compared to observed flow in previous years, showing similar seasonal flow trends. Given the moderate inter-annual flow variability in this basin, some difference of monthly mean flows for different periods is not unexpected. Despite the different periods, the simulated and observed flows correspond well in the dry winter months. The accuracy of model results in the dry winter season is required for assessing shortages, while differences in the wet summer months are far less relevant. However, we propose that in the future, with one or two years of complete data collection, better regional parameters can be obtained for future detailed water resource analyses of the basin at sub-catchment level. Nonetheless, we now have a daily stream flow model (present day), which describes how much water is flowing where and when. This can help in immediate water resources management and allocation issues. It can also help in other water resources issues, such as selecting sites for new flow stations.

The period 2000–2010 is considered a reference decade. Computations for future decades are based on future temperature and precipitation values from several climate projections. This is accomplished by scaling up/down the monthly temperature and monthly precipitation datasets by linear escalators derived from the Climate Wizard results, and gridded temperature and precipitation datasets from the climate modelling performed by CIMA. All of the climate model results are fed into the Hydro-BID model to compute new stream flow values for the Río Grande and Río San Francisco.

The simulated daily flows were averaged to create monthly time series for the period 2011–2060. Four monthly flow time series were generated for the Río Grande, Río Perico and Río Los Alisos rivers for the reference scenario and three climate projections:

1 Reference (the flow from 2001–2010)
2 UKMO Hadley HadCM3.1 (Pope *et al.*, 2000)
3 CSIRO Mk3 (Gordon *et al.*, 2002)
4 CIMA ensemble.

Evaluation of selected adaptation options

Using results from the Hydro-BID model simulations, we examined the unmet demand for 2011–2060 for the reference scenario, for the three selected climate projections, and for a variety of adaptation options. The key result of that work is that there will be a substantial unmet demand – in the order of 200 million m^3 per year – if water use efficiencies do not change. The key findings of the analysis can be summarized briefly below.

Option 0: Current efficiency

This option consists of no efficiency improvements, leakage in urban water supply systems remaining at 30 per cent and irrigation efficiency remaining at 37 per cent. The amount of irrigated land and crop distribution remains unchanged, and urban and industrial water demand rises strictly on the basis of population growth (1.5 per cent).

Option 1: Improved urban water efficiency

Under this option, improvements are made to water infrastructure. The distribution system leakage rate for the city of San Salvador de Jujuy and surrounding towns is reduced from 30 per cent of water supply to 10 per cent of water supply, saving a modest amount of water. As before, water consumption per person is constant, but a population growth rate of 1.5 per cent is assumed. In addition, water savings in the surrounding towns is 'transferred' to nearby irrigation use, reducing unmet demand for irrigation. Under this option there are interactions between urban and industrial water and irrigation water. More specifically, if water savings from improved efficiency in urban and industrial water usage produces surplus, that extra water can serve to diminish unmet irrigation demands.

Option 2: Improved irrigation efficiency

Under this option, improvements are made to the irrigation application efficiency. The overall irrigation efficiency is assumed to rise from 37 per cent to 55 per cent, which would be associated with no improvement in conveyance efficiency and an increase in application efficiency with a shift from basic furrow irrigation to surface drip irrigation. If water savings from improved irrigation produce zero unmet demand in irrigation, then any remaining water savings can contribute to reducing any unmet demand in urban water supply. However, this holds true only for the surrounding towns and not for San Salvador de Jujuy because San Salvador de Jujuy is upstream from the irrigated lands and does not have a storage reservoir.

Option 3: Improved urban and irrigation efficiency

This option is the combination of Options 1 and 2. Interactions between the two demands can take place in either direction.

Table 12.1 provides a summary comparison of unmet demand and the savings for the four climate projections, and the four adaptation measures, for the beginning and the end of the analysis period from 2011 to 2060, based on the details provided by the results of the model (but not shown here). Obviously, improved efficiency decreases unmet demand substantially, especially with irrigation efficiency, but some unmet demand remains under each scenario.

The key findings of these analyses can be summarized as follows:

- Under current circumstances, the unmet demand is approximately 34.3 million m^3/year.
- By 2060, without any improvement in water use efficiency by urban and irrigation users, unmet demand will rise to approximately 162–205 million m^3/year, depending on which climate projection applies.
- Improving the efficiency of urban water use (Option 1) will reduce unmet demand modestly over the study period, limiting its rise to approximately 141–183 million m^3/year.
- Improving the application efficiency of water use in irrigation (Option 2) has a substantial impact, limiting the increase in unmet demand to about 16–44 million m^3/year.
- Improving both urban and irrigation water efficiency (Option 3) provides an additional impact, limiting the increase in unmet demand to about 6–30 million m^3/year.

Thus, it is essential to improve the efficiency of water use in irrigation, just to keep unmet demand at approximately its current level over the next 50 years. In order to relieve unmet demand for water in the city of San Salvador de Jujuy, it will be necessary to take additional measures, such as securing an additional water source and developing water storage capacity.

We examined both the effectiveness and the cost efficiency of the three adaptation measures associated with improved efficiency. It is worth remembering that for purposes of this case study we define effectiveness and cost efficiency in the following manner:

1. Effectiveness of an option is the extent to which unmet demand is reduced by using that option. This is the most basic means of measuring effectiveness, but others are possible.
2. Cost efficiency is the present value of the cost of all the interventions (capital expenditures and maintenance) associated with that option, over

Table 12.1 Comparison of unmet demand and savings for four climate projections and four adaptation options

Total unmet water demand (mm³/year)	*2011–2020*				*2051–2060*			
Climate projection	*Reference*	*Hadley*	*CSIRO*	*CIMA*	*Reference*	*Hadley*	*CSIRO*	*CIMA*
Option 0 – Current efficiency	34.3	31.7	55.1	42.5	162	162	180	205
Option 1 – Improved urban water efficiency	26.3	24.3	45.0	34.1	141	141	159	183
Option 2 – Improved irrigation efficiency	0.53	0.45	1.45	0.75	15.5	15.5	27.2	44.2
Option 3 – Improved urban and irrigation efficiency	0.19	0.16	0.42	0.25	5.7	8.3	16.1	30.0
Savings in unmet demand (mm³/year)								
Savings from Option 1	8.1	7.3	10	8.3	21	21	21	22
Savings from Option 2	34	31	54	42	146	146	153	161
Savings from Option 3	34	32	55	42	156	153	164	176

Source: Authors' calculations and elaboration

the period 2011–2060, divided by the total reduction in unmet demand associated with that option.

For example, as shown in Table 12.1, the effectiveness of Option 1 in the reference climate, for the decade 2051–2060, is the reduction of unmet urban water demand from 162 million m^3/year to 141 million m^3/year, which is a reduction of 21 million m^3/year, or about 13 per cent. Under the CIMA ensemble projection, that effectiveness is a drop from 205 million m^3/year to 183 million m^3/year, which is a reduction of 22 million m^3/year, or about 11 per cent.

Another measure of effectiveness is the frequency of shortages. Under Option 0 and assuming the various climate change projections, the potable water users in San Salvador de Jujuy could expect at least 20 per cent of demand covered 75 per cent of the time in October. In other words, water will be quite constrained in October under climate change. By contrast, if Option 1 were applied, the citizens of San Salvador de Jujuy could expect at least 50 per cent of demand to be covered in October, which is a significant improvement. A new water source could help fill at least some of this gap.

The effectiveness of irrigation in reducing unmet demand is quite substantial. The effectiveness of Option 2 in the reference climate, for the decade 2051–2060, is the reduction of unmet demand from 162 million m^3/year to 15.5 million m^3/year, which is a reduction of 146.5 million m^3/year, or about 90 per cent. Under the CIMA ensemble projection, that effectiveness is a drop from 205 million m^3/year to 44.2 million m^3/year, which is a reduction of 160.8 million m^3/year, or about 78 per cent.

Turning to cost efficiency, we tabulated all the costs and water savings of the adaptation options, focusing on Option 1, and the cost effectiveness of efficiency improvements for sugar cane separately from tobacco. These results are summarized in Table 12.2. Improving water supply system efficiency results in an estimated 'cost of saved water' of about US$0.02 per m^3, which is highly attractive. The development of new sources of water and treatment facilities could easily cost ten times that amount. There are similar cost-efficiency values for improved irrigation efficiency for sugar cane and for tobacco, but the present value of investments and maintenance are much higher than for urban water efficiency improvements (Option 1). At the same time, water savings are also much higher. The net cost effectiveness of improving irrigation efficiency for both sugar cane and tobacco is higher than Option 1, but still very attractive compared to new sources of water.

Limitations of the approach

Hydro-BID has, as does any simulation model, a number of embedded assumptions that are important to note, as they may limit its application to several types of water resources management problems. These limitations are summarized here, together with avenues for further development and

Table 12.2 Cost efficiency of adaptation for urban water supply and irrigation

Option 1		*Option 2*					
Urban water supply		*Irrigation*					
			Crop	*Ha*	*Cap cost/ ha*	*Cap cost*	*Life, yrs*
Unit capital cost for leakage reduction	400$ per m^3/day saved	Capital cost for surface	Sugar	19,212	$1,200	$23,054,400	2
San Salvador and town water savings	24,253 initial m^3/ day saving	Trickle irrigation systems	Tobacco	14,238	$2,400	$34,171,200	10
Rehabilitation project capital cost	$9,701,200	Annual maintenance cost		14,238	$200	$2,847,600	
Annual maintenance cost	3%	Discount rate		10%			
Discount rate	10%						
Present value		*Present value*					
Cost	$7,602,008	Cost $, sugar	$81,900,805				
Reduction in unmet demand, m^3	337,000,000	Sugar water savings, m^3	2,005,421,667				
		Cost of sugar adaptation, $/$m^3$	$0.041				
Cost of urban adaptation, $/$m^3$	$0.023	Cost $, tobacco	$74,224,126				
		Tobacco water savings, m^3	987,745,000				
		Cost of tobacco adaptation	$0.075				

Source: Authors' calculations and elaboration

improvement of the system. First, surface water flow has been shown to be very sensitive to topography. The AHD component in Hydro-BID has been derived from radar remote sensing (SRTM), and thus merits calibration with ground-based topographic measurements to adjust for any significant inconsistencies. This was done for the case study documented in this chapter. In any event, modifying the AHD (which is open and available to any Hydro-BID user) with such adjustments, or even switching AHD topography for more detail (e.g. higher resolution) is a fairly straightforward task.

Second, although hydrological exchanges between surface water and groundwater are included in the GWLF water balance formulation, groundwater flow itself is not part of Hydro-BID. This is an important consideration for applying this model to water supply problems where groundwater is an important source of water. In its current form, Hydro-BID cannot be applied to problems such as groundwater well system design and operation, or groundwater contamination. This is an area of future development, and ongoing options being considered include linking Hydro-BID to well-known groundwater simulation models such as MODFLOW.

It is important to recognize that for water management problems, a hydrological model is frequently not sufficient to address issues such as water allocation (either rule or optimization-based), economic analysis and other policy-related issues. Hydro-BID has been designed to support the solution of such problems, but in its current version, should be complemented with other tools that focus on the management side of water resources. Ongoing research and development efforts are focused on adding these capabilities to the simulation system, precisely to provide a more inclusive user interface that is able to address broader water resources management issues.

Conclusions

The case study described in this chapter has provided valuable experience in use of the Hydro-BID modelling system and will inform the design of future applications and user training. The case study also demonstrates the value of using a modelling system with greater spatial resolution and geographic flexibility than is possible with other available hydrologic models. Hydro-BID is a useful tool for assessing water resources management challenges and options under current and future conditions, for any user-defined geographic area from the catchment to basin scale.

The Hydro-BID model was applied successfully in the case study, despite significant gaps in local data for Jujuy Province. Analysis of these gaps has highlighted the need for more extensive and intensive monitoring of water demand, climate, stream flow and other input parameters for use in future work. The model calibration was performed, therefore, primarily using international data sets and using stream flow data at the outlet of a large watershed. More recent reliable long-term stream flow data at interior points in the basin would improve the accuracy of model calibration for smaller watersheds.

Available data and climate models indicate that there will be a modest and steady increase in average temperature in the Río Grande/Upper Río San Francisco basin through to the year 2060. Average annual precipitation in the basin shows neither an increasing nor a decreasing trend over this period, although inter-annual variation will remain quite high. The direct impact of climate change on river flows appears to be small. The model confirms the current experience of water users in the basin. It yields results showing unmet demand (water shortages) during the dry months of August to November for the residents of San Salvador de Jujuy and downstream growers.

Water demand in the Río Grande/Upper San Francisco basin will increase dramatically over the next 50 years for two reasons: (i) increases in urban and industrial water demand, which reflect population growth in San Salvador de Jujuy and other towns in the basin; and (ii) increases in irrigation demand are driven primarily by increasing temperature, which increases evapotranspiration experienced by crops. Without adaptive actions, urban residents and growers who rely on irrigation will all experience increasingly severe water shortages over the study period.

Improving the efficiency of urban water systems will only partially alleviate unmet water demand among urban and industrial users. Reducing physical losses (leakage) from an assumed 30 per cent of system capacity to a target of 10 per cent would reduce unmet urban and industrial water demand in 2020 by 7.3–10.0 million m^3/year, depending on the climate projections, with all of the reduction occurring during the dry season. Because San Salvador de Jujuy does not have a reservoir, water that is saved by improving irrigation efficiency downstream cannot presently be used to alleviate unmet water demand in the city. Fully satisfying San Salvador de Jujuy's water demand will require additional investments in storage or new water sources.

The differences in overall results using different climate projections are significant but not enormous. This analysis predicted stream flow, water demand and unmet water demand for an 'average year', whether under the reference scenario or any of the three climate projections. Historically, there is considerable inter-annual variation in rainfall, temperature and stream flow. Under any future climate projection, there will be hot, dry years which could lead to increases in unmet demand and cool, wet years with decreases in unmet demand. The fundamental conclusion is that increasing water efficiency will provide a large benefit for adapting to anticipated changes in climate. Because of the substantial inter-annual variability, developing increased water storage and new sources of water will be of particular use to San Salvador de Jujuy.

There are many data gaps that limit the analysis of water resources management issues in Jujuy Province, for present and future conditions. The highest priority data needs are stream flow measurements in the Río Grande sub-basin; accurate and detailed data on current water usage; and new primary data on evapotranspiration, crop water requirements and crop yields in the irrigated areas.

In spite of its limitations, Hydro-BID is a very useful tool for predicting future river flows for user-defined areas from the catchment to basin scale. The model is flexible and does have a robust geographic/hydrological algorithm, data layering, and an ability to use either local or 'off-the-shelf' international input datasets. It can be used as the core of advanced and sophisticated work on integrated water resources management in river basins.

References

Gordon, H.B., Rotstayn, L.D., McGregor, J.L., Dix, M.R., Kowalczyk, E.A., O'Farrell, S.P., Waterman, L.J., Hirst, A.C., Wilson, S.G., Collier, M.A., Watterson, I.G. and Elliott, T.I. (2002) *The CSIRO Mk3 climate system model, CSIRO Atmospheric Research Technical Paper No. 60,* CSIRO Atmospheric Research, Victoria, Australia, http://www.cmar.csiro.au/e-print/open/gordon_2002a.pdf, accessed 17 May 2014.

Haith, D.A. and Shoemaker, L.L. (1987) 'Generalized watershed loading functions for stream flow nutrients', *Water Resources Bulletin*, vol. 23, no. 3, pp. 471–478.

Haith, D.A., Mandel, R. and Wu, R.S. (1992) *Generalized watershed loading functions: User's Manual*, Dept. of Agricultural and Biological Engineering Cornell University, Ithaca, NY.

IDB (Inter-American Development Bank) (2013) *Hydro-BID case study: A water resource model of the Río Grande Basin in Argentina*, for Inter-American Development Bank Contract, Washington.

Kalnay, E., Kanamitsu, M., Kistler, R., Collins, W., Deaven, D., Gandin, L., Iredell, M., Saha, S., White, G., Woollen, J., Zhu, Y., Leetmaa, A. and Reynolds, R. (1996) 'The NCEP/NCAR 40-Year Reanalysis Project', *Bulletin of the American Meteorological Society*, vol. 77, no. 3, pp. 437–471.

Kummerow, C., Barnes, W., Kozu, T., Shiue, J. and Simpson, J. (1998) *The TRMM Sensor Package*, http://dx.doi.org/10.1175/1520-0426(1998)015<0809:TTRMMT>2.0.CO;2, accessed 17 May 2014.

Linsley, R.K., Kohler, M.A. and Paulhus, J.L. (1975) *Hydrology for Engineers*, McGraw Hill Companies, New York.

Magrin, G., Gay García, C., Cruz Choque, D., Giménez, J.C., Moreno, A.R., Nagy, J.G. and Villamizar, A. (2007) 'Latin America. Climate change 2007: impacts, adaptation and vulnerability'. Contribution of Working Group II, in Parry, M.L., Canziani, O.F., Palutikof, J.P., van der Linden, P.J. and Hanson C.E. (eds), *IPCC Fourth Assessment Report Climate Change 2007*, Cambridge University Press, Cambridge.

Pope V.D., Gallani, M.L., Rowntree, P.R. and Stratton, R.A. (2000) 'The impact of new physical parametrizations in the Hadley Centre climate model: HadAM3', *Climate Dynamics*, vol. 16, pp. 123–146.

Part IV
Green infrastructure development

13 Green growth and water services in Latin American cities

Greg Browder

Introduction

Green growth is growth that is environmentally sustainable (World Bank, 2012). It is efficient in its use of natural resources, clean in that it minimizes pollution and environmental impacts, and resilient in that it accounts for natural hazards and the role of environmental management in preventing physical hazards. Green growth is now permanently fixed within the global agenda, and Latin America needs to ensure that its economic growth is efficient, inclusive, clean, and resilient.

Green growth is consistent with sustainable development, not a competing paradigm. It offers a development pathway that reconciles the urgent need for sustained growth with the imperative of avoiding lock-in to unsustainable growth patterns and irreversible environmental damage. Green growth is not anti-growth; it represents a change in how we manage economies for sustainability. Green growth policies must be carefully designed to maximize benefits and minimize costs for society at large. The improvement of environmental performance and the care for the less well-off and most vulnerable should be seriously considered.

Water supply and sanitation play an integral role in the green growth agenda of the Latin American and Caribbean (LAC) cities. It is impossible to imagine the continuing trend of urbanization within Latin America, without tying it to a greener future built on sustainability principles – without clean drinking water and sanitation for all, water for agriculture, energy and industry, protection against urban flooding, and vibrant rivers, lakes, wetlands and coastal areas. This vision of the water sector is achievable for most LAC countries within a generation – if they invest sufficiently and make sound decisions on sector reforms.

In the context of the green growth agenda, urban water sector objectives are:

- The provision of efficient and inclusive water services delivered at reasonable rates for everyone.
- The provision of water services that protect the environment, manage water as a scarce natural resource, and incorporate water into the urban fabric.
- The development of resiliency within water service provision that anticipates and responds to droughts, floods and other environmental events and conditions without significantly compromising service quality.

Efficient and inclusive water services and the green growth agenda

One of the targets of the Millennium Development Goals is to reduce by half the gap in basic water and sanitation coverage. Given the rapid urbanization and the economic shocks of the past two decades, the overall levels for access to improved water supply and sanitation in cities in Latin America is encouraging. Nevertheless, the general picture masks several disconcerting facts (WHO and UNICEF, 2013, pp. 34–35):

- About 36 million people – many in rural areas, but a significant number in cities – lack access to safe water.
- Some 107 million people lack access to improved sanitation.
- Only about 30 per cent of urban wastewater is currently treated.
- Many urban water utilities provide low levels of service in terms of continuity, pressure and water quality, and are struggling financially.

For urban areas in the region, access to water (the service coverage rate) is relatively high. According to the Joint Monitoring Programme (JMP), as many as 94 per cent of the population has piped connection to their premises (WHO and UNICEF, 2013, p. 35). The percentage of connection has increased over the past decade, which indicates that water utilities have managed to expand networks at a pace that exceeds urban population growth. Nevertheless, deficiencies persist. Those without municipal connection tend to live in peri-urban areas or urban slums. Of the 6 per cent of the urban population without a piped water connection, half resort to alternative sources classified as improved and half to those classified as unimproved (WHO and UNICEF, 2013) to meet their water needs. The extent to which these alternative water supplies are potable, however, is highly variable.

Sanitation coverage is lagging significantly behind water supply in Latin American cities, and the ensuing environmental and public health risks are substantial. JMP estimates that 87 per cent of urban households had improved sanitation in 2011 (WHO and UNICEF, 2013, p. 35), which includes connections to sewers and septic tanks as well as access to improved latrines and composting toilets. The remaining 13 per cent of the urban population, which has to resort to shared or substandard facilities, lives primarily in peri-urban areas and urban slums. Moreover, cities generate vast amounts of wastewater, which unless treated, threaten water quality, the environment and public health.

In rural areas, where just over 20 per cent of the population lives, the JMP estimates for 2011 that 82 per cent had access to safe water supply but only 64 per cent had access to piped water at their premises. Sanitation coverage rates are even lower: only 63 per cent of the rural population uses improved sanitation facilities (WHO and UNICEF, 2013, pp. 34–35). Many of the unserved live in remote areas and are members of indigenous communities.

Table 13.1 Regional average and selected water and sanitation hotspots in Latin America and the Caribbean, 2010

Region / country	*Improved sources Urban water (%)*	*Improved facilities Urban sanitation (%)*	*Improved sources Rural water (%)*	*Improved facilities Rural sanitation (%)*
Average of all countries in Latin America and the Caribbean	97	87	82	63
Bolivia	96	57	72	24
Nicaragua	98	63	68	37
Panama	97	77	86	54
Peru	91	81	66	38

Source: Adapted from WHO and UNICEF (2013, pp. 16–17 and 26–27)

Among the overall rates, there are localized hotspots of inadequate water and sanitation services coverage to be found. A number of Latin American countries have coverage levels well below the regional average (Table 13.1). And a number of sub-national regions have coverage levels dramatically below their national average. These include southern Mexico, the Amazon region of Colombia and the Northeast region of Brazil.

The JMP refers to the use of improved water sources and improved sanitation facilities. Yet even in the case of a piped water source provided by a water utility, potability of the water and reliability of the service is not guaranteed – in many cities, service is intermittent and the water is unsafe to drink. Likewise, a connection to a sewer does not guarantee that the wastewater will be properly treated and disposed of. In many cities the sewers are combined storm water and wastewater systems that flow to the nearest, usually contaminated, water body for disposal.

Low levels of service typically result from a combination of inefficient urban water utilities and inadequate investments. Although Latin America has some world-class urban water utilities, many are under-performing in operational and financial terms (e.g. non-revenue water, collection rates, overstaffing, operational efficiency and profitability). The causes vary from case to case, but many cities and countries are still in the process of crafting policies on water sector structure, regulation, public–private partnerships (PPPs) and financing that are appropriate for meeting the green growth agenda.

Green growth and environmental sustainability

As is the case in many parts of the world, large-scale water withdrawals, disruption of natural hydraulic conditions and flows, and substantial pollution have had severe ecological impacts on freshwater ecosystems in Latin America.

Many have experienced massive losses of biodiversity, and many have experienced or are close to collapse due to water quality or quantity issues. This is significant for the region's cities because of their reliance on freshwater ecosystems for their water service needs. Freshwater ecosystems are rated the most endangered ecosystems on the planet, surpassing even rainforests and oceans (IUCN, 2000).

The lack of adequate wastewater treatment in urban areas is a major contributor to the degradation of freshwater (and coastal) ecosystems, and poses one of the greatest challenges to the Latin American urban water agenda. Table 13.2 summarizes the estimated treatment rates for selected countries. Most of these countries are involved in large-scale wastewater investment programmes. In addition to construction of treatment plants, there are other challenges cities struggle with as they try to improve the environment. Connecting households, particularly poor ones, to a wastewater collection system is often problematic due to the cost, the extensive community outreach required, and – in contrast to water supply – the lack of a strong incentive for households to connect. Many treatment plants operate below capacity because of the inadequate development of wastewater collection systems. Infiltration of groundwater and inflow of storm water into collection systems are common problems that reduce the efficiency of treatment plants. Some treatment plants are not fully operational or properly maintained due to shortages of funds.

Wastewater management costs are typically higher than water supply costs. This is primarily due to the higher costs of gravity sewers compared with pressurized water pipes, as well the higher costs of treating polluted wastewater versus relatively clean water for water supply systems. In many Latin American cities, however, water utilities struggle with efficiency and financial challenges just supplying water. The inclusion of comprehensive wastewater services, with their added costs and performance requirements, will in many cases require significant financial and operational performance improvements by the utility.

Table 13.2 Estimated wastewater treatment rates for selected countries, 2011–2013

Country	*Wastewater treatment (%)*	*Year of data*
Brazil	30–35	2012
Colombia	25	2010
Chile	99	2012
Mexico	41	2011
Peru	33	2010
Costa Rica	4	2013

Source: For Brazil (Reportlinker.com, 2014); for Colombia (Lopera Gómez *et al.*, 2012); for Chile (SISS, 2012); for Mexico (SEMARNAT, 2012); Peru (WaterWorld, n.d.) and Costa Rica (Programa Estado de la Nación, 2013)

Building resiliency in Latin American cities

Floods and droughts threaten the reliability of water services throughout Latin America's urban areas. The number of major floods and droughts in the region appears to have increased in the past few decades, and there is increasing evidence that climate patterns are becoming more extreme in Latin America (Vergara *et al.*, 2013). Some of the major impacts forecast include:

- Reduction in glacier coverage in the Andes (glaciers provide critical dry season flows for many cities).
- Increased flooding and inundation of coastal zones.
- Destabilization of the hydrological cycle in major basins.
- Intensification of extreme weather events.

Extreme weather events, through water service disruptions, have a broad economic impact. Floods halt commercial activity and cause property and other damage. Dealing with water scarcity and managing seasonal droughts is now an imperative in most Latin American cities. Ensuring a reliable water supply is essential for sustaining urban growth, yet as cities grow and climatic extremes intensify, finding sufficient water to meet municipal and industrial needs is becoming harder. Fortunately, a new array of tools to deal with scarcity issues is emerging and becoming more economically viable. These range from standard conservation measures to developing alternative sources such as reclaimed wastewater, and rainwater harvested from rooftops of buildings, both of which are components of a green growth approach to urban water issues.

Coping with higher rainfall levels and increased urban flood risks requires more attention to urban storm water management. Rainfall in urban areas equates to huge quantities of water that must be managed properly to avoid loss of life, flooded streets, and consequent traffic flow problems, and building and property damage. Yet in most Latin American cities, the planning, financing and management of storm water drainage attract limited attention. Funding typically comes from the municipal budget, as opposed to user fees or a dedicated tax, and is often unreliable. Management of drainage systems is typically the responsibility of a general municipal public works agency, which may not manage the system as a public utility service.

An alternative to the direct channelling of storm water through networks of pipes and sewers to nearby watercourses is emerging and is called sustainable urban drainage systems (SUDS) (Sharma, 2008; British Geological Survey, n.d.). By mimicking natural drainage, SUDS seek to (i) reduce storm water run-off by improving infiltration into the soil; (ii) use storage areas such as ponds to assist with flood peaks; and (iii) reduce pollution by better controlling non-point sources of pollution. This is green growth at work. For SUDS programmes to be effective, however, they must be closely linked to urban development and management practices.

The cost of investing in green growth for Latin American cities

The cost of upgrading existing urban infrastructure is difficult to know with certainty. One estimate presented by the Corporación Andina de Fomento (CAF) – Development Bank of Latin America – put the total investment needs for 2010–2030 (to provide universal water and sanitation services in marginal Latin American urban areas) at nearly US$250 billion (CAF, 2013, pp. 48–49). This is equivalent to annual investments of some US$12.5 billion, or 0.31 per cent of the region's gross domestic product (GDP) (CAF, 2011; 2013).

Water policies and regulation: promoting the next generation of reforms

Since the 1990s, Latin America has been at the forefront of initiating policy reforms to boost investment and improve service quality in the water sector. These reforms have yielded real benefits, as evidenced by expansions in service coverage and improvements in service quality, yet the challenges for the future remain substantial and will require even deeper and more effective reforms, coupled with large-scale investment. Many governments in the region have formulated strategic visions for the future, which are generally achievable if these governments create the proper enabling frameworks. Chile's experience illustrates the feasibility of rapid and sustainable advances (Box 13.1).

The general thrust of reforms since 1995 has been on separating policy-making, regulatory and service entities, and ensuring that each entity has adequate capacity and authority to undertake its tasks. Many of these reforms are ongoing in some countries. Ultimately, the effectiveness of water sector policies – such as investment subsidies and tariffs, regulations, environmental standards, the use of green growth approaches – can be judged by whether water services are provided efficiently and inclusively, which is still not the case in many Latin American countries. Key reform efforts in some of the larger countries of the region follow.

Argentina has struggled with incorporating private sector participation in the water sector. In the 1990s, as part of a broader national privatization policy, many cities (notably Buenos Aires) signed long-term concession contracts with private water companies. By the early 2000s, more than 70 per cent of the urban population was receiving services from a private water company – mostly international companies. Following the economic crisis in the early 2000s and the loss of currency convertibility with the US dollar, many of the concession contracts collapsed, including the high-profile Buenos Aires concession. Nevertheless, some viable mixed-capital companies persist in the provinces, and there are at least ten water concessions serving around 20 per cent of the urban population (World Bank, 2009).

Brazil's water sector is characterized by large state-wide water companies and cautious experiments with PPPs. The Ministry of Cities is responsible for overall sector policy, in coordination with the National Water Agency, which is

Box 13.1 The Chilean experience with water sector reforms

Water supply and sanitation in Chile is characterized by high levels of access and good service quality. Chile's water sector is unique in the region in that all urban water companies are privately owned or operated. Before privatization, which started in 1998, the government had already created adequate economic and service regulatory agencies, and invested heavily in the sector. As a result, in 1990, 97 per cent of the urban population was already connected to water and 82 per cent to improved sanitation.

According to the Chilean regulatory agency Superintendent of Sanitary Services (SISS), access to water supply in urban areas of Chile was 100 per cent and improved sanitation 98 per cent in 2010. Service quality is generally good and water supply is continuous, both in urban areas and in concentrated rural areas. On the wastewater side, 83 per cent of Chilean wastewater was being treated in 2010.

Responsibility for water sector policy in Chile is vested primarily in the public works ministry, which grants concessions and promotes rural water supply and sanitation through its Department of Sanitation Programmes. Responsibility for regulation is shared between the SISS and the health ministry, which controls drinking water-quality standards in both urban and rural areas. To guarantee political independence, SISS is a decentralized organization with its own budget. It has the right to impose fines on service providers that violate norms; these fines flow directly into its budget. SISS also acts on user complaints.

Water supply and sanitation services in Chile's urban areas are provided by more than 50 entities, but only eight companies serve 85 per cent of the urban population, with two of these serving half the population. In rural areas, local water cooperatives and water committees provide water supply services to almost 1.7 million rural consumers. In concentrated rural areas, significant development has been achieved because of the national Rural Potable Water Programme.

Water tariffs in Chile differ substantially between regions, reflecting differences in the cost of supplying water and sanitation services. Tariffs in urban areas varied between US$0.80/m^3 and US$4.10/m^3 in 2010. In rural areas, tariffs cover only operation and maintenance costs. Urban water and sanitation systems do not receive direct subsidies and are financed through the capital market, and ultimately through user fees. But Chile has an innovative means-tested system that allows qualifying poor households to receive a subsidy administered by the municipality to cover some of their water and sanitation bills. The subsidy is meant to target only those subsistence-level households that are unable to pay for water and sanitation services. Rural water systems receive a partial investment subsidy defined in the *Ley del Subsidio al Agua Potable y Saneamiento* (Drinking Water and Sanitation Subsidy Act).

Source: Frei Ruiz-Tagle (2013)

responsible for resources management. In the past, states were responsible for water service, but since the 1988 Constitution this responsibility was devolved to municipalities, most of which continued with the state water companies. There are 27 state water companies and about 580 municipal water utilities. In addition, there are 14 state regulatory agencies, whose authority and capacity vary considerably. An increasing number of public water companies are entering into PPPs, particularly to provide wastewater services requiring large investments (Douglas-Watson, 2011).

With the huge backlog in sewerage investments, many water companies are entering into PPPs to boost investments and improve efficiency. A recent notable initiative involves the Pernambuco state water utility, COMPESA, which in 2013 signed a 4.5bn-real (US$2.3 billion) sanitation concession agreement with a private consortium (BN Americas, 2013). The concession covers 15 cities in the metropolitan region of the state capital Recife, has a duration of 35 years, and involves expanding wastewater services for around 4 million people. The private partner will provide 70 per cent of the required financing, and is required to boost wastewater collection and treatment from 30 per cent to 90 per cent within the first 12 years.

Chile is unique in the region in managing to create a strong policy and regulatory context that has enabled private companies to take over service responsibility (Box 13.1). The key to Chile's early success was creating a functioning policy and regulatory context that emphasized self-financing, accountability and government support when needed. Chile also has a well-targeted and well-structured tariff policy for low-income households, which does not undermine water utility's financial stability. The introduction of the private sector after 1998 helped attract new waves of investment to finance wastewater treatment and upgrade water and sewerage infrastructure. Chile's water sector structure is characterized by relatively large regional companies, which are able to provide economies of scale and simplify regulation (Frei Ruiz-Tagle, 2013).

In *Colombia*, water service is decentralized to the 1,119 municipalities. National sector policy and planning is provided by the Vice Ministry of Water in the Housing Ministry (World Bank, 2009). In 1994 a public service law was enacted that required municipalities to use specialized water companies, whether public, private or mixed-capital entities. The result has been a wide range of institutional models, including public water companies in Bogotá, Medellín and Cali, and PPP arrangements in Cartagena and Barranquilla. An increasing number of smaller and medium-sized cities are turning to private specialized operators, but the quality and coverage of services is generally lower than the major cities in the rest of the country. Colombia has a unique tiered water tariff system whereby upper-income households pay more for utility services, and thereby cross-subsidize lower-income households. Although this works well in most cities, in some of the smaller and poorer cities cross-subsidies are insufficient and municipal governments are unable or unwilling to provide the full subsidies to water companies, generating financial stress in many water companies.

Principles for sustaining reform agenda

As Latin American countries move forward with the water reform agenda in the context of the broader objective of green growth, the following principles are suggested:

Policy coherence and commitment: Water sector policy is composed of a number of different dimensions, referred hereto as 'the policy mix', and each dimension has its own set of alternatives. Some of these key dimensions are presented below. The goal of government is to ensure that there is a coherence within the policy mix to ensure the sector as a whole performs well. This is not an easy task, and each country will have its own unique set of policies that are appropriate for its specific context. The most important factor in success is sustained government commitment to adjust the policy mix over time as required to improve sector performance.

Industry structure and decentralization: Water services can be provided at the municipal, regional or national level. In some countries in Latin America, often mandated in constitutions, water service is the responsibility of municipal governments. Excessive fragmentation of water services, particularly in smaller municipalities, can generate problems of economy scale and complicate economic and service regulation. In some countries, such as Mexico and Colombia, the industry structure is relatively decentralized and many small cities struggle with providing efficient services. In contrast, Chile and Brazil are relatively centralized with regional companies providing most of the water services. Some smaller countries such as Uruguay (OSE) and Nicaragua (ENACAL) have national urban water utilities. Each country must search for an appropriate 'industry structure' within its 'overall policy matrix' to ensure efficiency and economy of scale.

Regulatory structure: Water and sanitation services are natural monopolies, and thus both the service quality and price need to be regulated in some manner to protect the public interest. Each country has tackled this issue in its own way, with varying degrees of success. Colombia has created separate economic regulator and service regulatory agencies to oversee municipal water companies.[1] Many Brazilian states have state level regulatory agencies for public services, including water and sanitation. In Mexico, oversight of municipal water companies is typically provided by state water commissions. Unfortunately, many independent regulatory agencies struggle with issues of real authority, political interference and inadequate resources. A regional organization of regulatory entities, ADERASA, has been established to share information on public service regulation, and over time countries are learning to perfect their regulatory regimes. In the absence of an effective national or regional regulatory regime, the quality of services can vary considerably and is often linked with the quality of municipal governance.

Public–private partnerships (PPPs): Privatization was the rallying cry of water sector reforms in LAC in the 1990s, but the experience has been less than satisfying – with only Chile achieving full privatization. Nevertheless, gradually and with accumulating experience, the Latin American region is developing innovative

PPP models that meet their specific circumstances. Cities in Colombia such as Barranquilla and Cartagena have turned to a PPP model where financing comes primarily from the public sector but management expertise is provided by private sector partners. In Mexico, the private sector is being leveraged to help finance and operate critically needed wastewater treatment plants, primarily through Build-Operate-Transfer (BOT) models. Some state water companies in Brazil are entering into 'sanitation concessions' with private companies to help achieve their environmental goals. In São Paulo, the state water company SABESP has issued shares in the New York and São Paulo stock exchanges to both raise capital as well as ensure governance discipline and transparency. Many publicly owned water utilities are looking for strategic alliances with private companies to enhance their efficiency.

Financial policy: Providing high-quality water services in a developing country context is so challenging precisely because, at least relative to higher income OECD countries, there are fewer resources available at the household level to pay tariffs or at the government level to finance infrastructure. Governments must then carefully develop appropriate financial policies taking into account what can be referred to as the '3-Ts'.[2]

The term '3-Ts' refers to three primary sources of funds for water services: *tariffs* from users; *taxes* – either dedicated or general, and typically at the municipal level; and *transfers* from national/state government or from foreign donations. The exact mix can and does vary, but if the sum of the '3-Ts' is insufficient to cover actual investment, operation and maintenance costs, water service cannot be sustainable over the long term. Efficient water utilities can reduce costs by making smart investment decisions and cutting operational expenses; likewise, an efficient water utility with a strong and stable cash flow can attract lower-cost financing.

In most middle-income Latin American cities, the majority of residents can afford to pay the full costs of water supply and sanitation services. Storm water service is typically paid by general municipal taxes, although in some cases a dedicated storm water tax may be added to the property tax. Nevertheless, water and sanitation tariffs are often kept below cost-recovery levels for political reasons, or out of social equity concerns linking affordability to accessibility. Many LAC cities need more intelligent tariff design and subsidy policies to achieve the dual goals of financial sustainably of the water service and protection of low-income households.

Linking green growth with water governance institutions

Water supply, sanitation, wastewater management, storm water drainage and solid wastes are components of a broader hydrological and institutional environment. The concept of *integrated urban water management* integrates those components with land use planning and economic development (Bahri, 2012). The concept helps capture a city's (and its water services) relationship with the broader water resources context, in close connection with urban development dynamics.

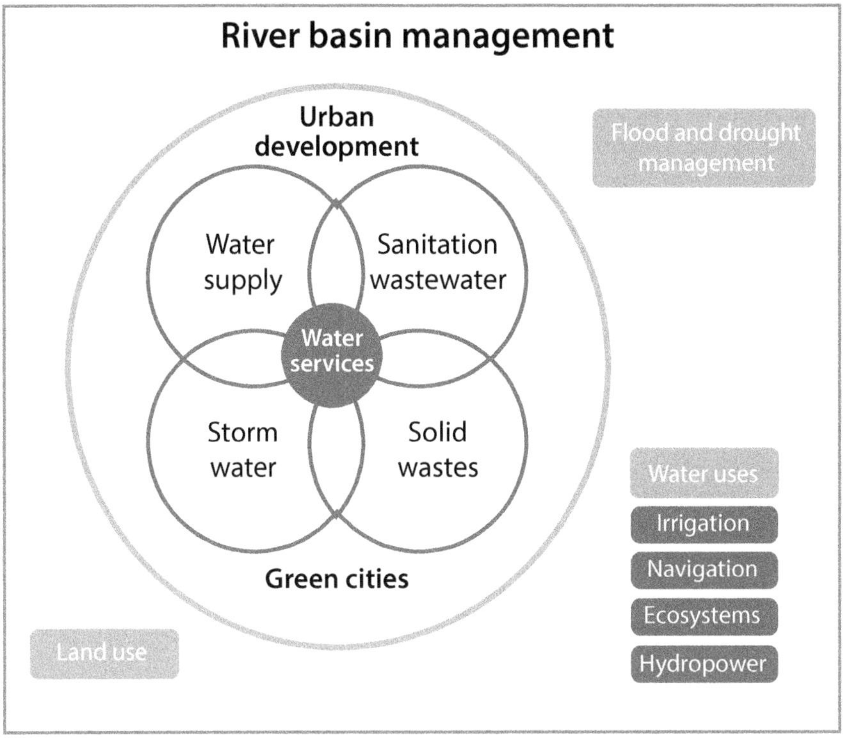

Figure 13.1 The concept of integrated urban water management

Source: World Bank (2012, Figure 2)

Successful adoption of the approach (Figure 13.1) will be central to achieving the green growth goals of efficient, inclusive, clean and resilient water services within overall economic and environmental sustainability.

The quality of a city's water services hinges on the broader water resource management framework: urban water supply in LAC typically constitutes 20–30 per cent of a country's water withdrawals, and water resource allocations and rights are fundamental to ensuring a reliable source of water.[3] Flooding from regional rivers typically requires a basin-wide management perspective. Drainage of storm water in urban areas can be hindered by high water levels in receiving rivers; similarly, rivers can overtop or burst through flood control embankments and cause catastrophic urban flooding. Typically, regional or national water resource management agencies are responsible for allocating water as well as regional hydraulic infrastructure, river management and integrated water management.

Intimately linked with water resources management is *environmental management*, which includes protecting aquatic ecosystems and setting ambient water-quality goals and wastewater effluent standards. The water-quality regulatory requirements typically determine the degree to which municipal wastewater, and sometimes storm water, needs to be treated and thus has a large impact on a water utility's costs. Water utilities and their municipal governments must coordinate with water resource and environmental management institutions, and be active participants in policy making to ensure that their interests are protected and water is managed in a sustainable manner.

Water services are embedded in the urban fabric and should be integrated into urban planning. Spatial development patterns have an important influence on water services. The density of development determines the extent of water distribution and collection networks. Latin American cities are typically characterized by high levels of sprawl requiring more kilometres of pipe and generally higher pumping costs, thus increasing investment and operating costs. The extent to which rainwater can naturally infiltrate into the soil, or be temporally stored in natural basins such as ponds or wetlands, determines the requirements for drainage pipes and pumping stations. Cities that preserve open spaces and/or take measures to temporarily store storm water peak flows can significantly reduce storm water management costs – while weaving water bodies as part of the urban landscape. Often, however, too little consideration is given to water issues in urban planning, leaving Latin American urban water utilities to simply respond to urban dynamics rather than proactively influence development to enhance green growth.

Conclusions

The green growth agenda is now permanently fixed on the global policy scene. Latin America needs to ensure that its impressive recent economic growth is efficient, inclusive, clean and resilient – the touchstones of green growth. The urban water sector will play a key, indeed a critical, role in influencing the overall trajectory of sustainable growth in the region.

There are still many challenges in Latin America. Although water supply service coverage over the decades has kept pace with rapid urbanization, peri-urban and rural areas are sometimes neglected. Service levels, in terms of continuity, pressure and drinking water quality, are often unacceptably low. And many middle-income cities are facing the dual challenges of wastewater and storm water management, where investment requirements are huge.

It is clear that green growth thinking and solutions need to be brought to bear on many urban water and sanitation issues in Latin America. This is not because green growth is second best when compared to the historical thinking and solutions that have been adopted for water issues in much of the developed world (where green growth is now strongly taking hold), but because green growth thinking and solutions are new, adaptive and arguably better than the approaches of yesterday, which did not consider the multiple dimensions to and complexities in designing truly sustainable solutions to environmental and economic growth issues.

Latin American cities, with the full and active support of national governments, should address water sector challenges on two fronts. First, in spite of two decades of policy initiatives, there is a need to promote the next generation of reforms. Each city has its own historical and political context and there are no predefined policies that will ensure success. The key is to continue to experiment until the right mix of policies is achieved with respect to green growth; water sector structure and decentralization; regulatory mechanisms; cost recovery; tariffs, taxes and transfers (the three Ts); and PPPs.

Second, stronger links and coordination with broader water resources and environmental management regimes should be established, and water should be integrated as deeply as possible into the urban planning framework. This integration also plays a defining role in green growth, whose ideal is the truly successful integration of sustainable environmental and economic considerations. It is becoming increasingly difficult, if not impossible, for the water sector to address independently its water, wastewater and storm water challenges without actively engaging in the formulation of environmental, energy, transportation and overall urban management policies. The principles and foundations underlying green growth provide a promising way to do so for the future.

Notes

1 The economic regulator is Comisión de Regulación de Agua Potable y Saneamiento Básico-CRA. The service regulator is Superintendencia de Servicio Público Domiciliarios-SSPD.
2 A more detailed explanation of these concepts are found in Chapter 3.
3 See 'AquaStat: General Summary Latin America and Caribbean' at: www.fao.org/nr/water/aquastat/regions/lac/index4.stm.

References

Bahri, A. (2012) *Integrated Urban Water Management,* Global Water Partnership Technical Committee, http://www.gwp.org/Global/Activities/News/August%202013/GWP%20Policy%20Brief_TEC%2016_FINAL.pdf), accessed 3 May 2014.

BN Americas (2013) Compesa, Odebrecht sign US$2.3bn sanitation PPP in northeast Brazil, http://www.bnamericas.com/news/waterandwaste/compesa-odebrecht-sign-us23bn-sanitation-ppp-in-northeast-brazil1, accessed 3 May 2014.

British Geological Survey (n.d.) *Sustainable Drainage Systems (SuDS)*, NERC Science of the Environment, http://www.bgs.ac.uk/suds/, accessed 3 May 2014.

CAF (Corporación Andina de Fomento) (2011) *La Infraestructura en el Desarrollo Integral de América Latina. Diagnóstico Estratégico y Propuestas para una Agenda Prioritaria,* CAF, Caracas, http://segib.org/actividades/files/2012/05/ideal2011.pdf, accessed 3 May 2014.

CAF (Corporación Andina de Fomento) (2013) *Equality and Social Inclusion in Latin America: Universal Access to Water and Sanitation,* Corporación Andina de Fomento, Caracas, http://publicaciones.caf.com/publicacion?id=1594, accessed 3 May 2014.

Douglas-Watson, J. (2011) *Environment and Water Opportunities in Brazil,* UK Trade and Investment, London, www.ukti.gov.uk, accessed 1 May 2014.

Frei Ruiz-Tagle, E. (2013) 'A government perspective: the sanitation services revolution in Chile', in CAF (ed.), *Equality and Social Inclusion in Latin America: Universal Access to Water and Sanitation,* CAF, Caracas.

IUCN (International Union for Conservation of Nature) (2000) 'Freshwater biodiversity in crisis', in *EarthTrends: Featured Topic,* World Resources Institute, Washington, DC.

Lopera Gómez, M. J., Campos, S. M. and Olarte, B. C. (2012) *Segundo Reporte Nacional: Producción de Aguas Servidas, Tratamiento y Uso en Colombia,* República de Colombia, Bogotá, Colombia, http://www.ais.unwater.org/ais/pluginfile.php/378/mod_page/content/144/COLOMBIA.pdf, accessed 6 May 2014.

Programa Estado de la Nación (2013) *Decimonoveno Informe Estado de la Nación en Desarrollo Humano Sostenible,* Programa Estado de la Nación, San José, Costa Rica, http://www.estadonacion.or.cr/estado-nacion/informe-actual, accessed 3 May 2014.

Reportlinker.com (2014) *Brazil Wastewater Treatment Plants Market Forecast and Opportunities, 2018* PR Newswire, New York, http://www.prnewswire.com/news-releases/brazil-wastewater-treatment-plants-market-forecast--opportunities-2018-240467461.html, accessed 3 May 2014.

SEMARNAT (2012) *Informe de la Situación del Medio Ambiente en México. Compendio de Estadísticas Ambientales. Indicadores Clave y de Desempeño Ambiental. Edición 2012,* Secretaría de Medio Ambiente y Recursos Naturales, Mexico City, http://app1.semarnat.gob.mx/dgeia/informe_12/pdf/Informe_2012.pdf, accessed 6 May 2014.

Sharma, D. (2008) 'Sustainable drainage system (SuDs) for storm water management: a technological and policy intervention to combat diffuse pollution', *11th International Conference on Urban Drainage*, International Association on Hydraulic Engineering and Research (IAHR) and International Water Association (IWA) Edinburgh, http://web.sbe.hw.ac.uk/staffprofiles/bdgsa/11th_International_Conference_on_Urban_Drainage_CD/ICUD08/pdfs/753.pdf, accessed 6 May 2014,

SISS (2012) *Informe Anual de Coberturas Urbanas de Servicios Sanitarios,* Superintendencia de Servicios Sanitarios, Gobierno de Chile, Santiago, http://www.siss.gob.cl/577/articles-9886_recurso_1.pdf, accessed 3 May 2014.

Vergara, W., Rios, A. R., Galindo Paliza, L. M., Gutman, P., Isbell, P., Sudding, P. H., Samaniego, J. and Pachauri, R. (2013) *The Climate and Development Challenge for Latin America and the Caribbean. Options for Climate-resilient, Low-carbon Development,* Inter-American Development Bank (IDB), with World Wide Fund for Nature (WWF) and the United Nations Economic Commission for Latin America (ECLAC), Washington, DC.

Water World (n.d.) *Peruvian Progress for Widespread Water Coverage,* PennWell Corporation, http://www.waterworld.com/articles/wwi/print/volume-27/issue-3/regional-spotlight/latin-america/peruvian-progress-for-widespread.html, accessed 3 May 2014.

WHO and UNICEF (2013) *Progress on Drinking Water and Sanitation, 2013 update,* WHO/UNICEF Joint Monitoring Programme for Water Supply and Sanitation, New York and Geneva, http://www.wssinfo.org/fileadmin/user_upload/resources/JMPreport2013.pdf, accessed 3 May 2014.

World Bank (2005) *Models of Aggregation for Water and Sanitation Provision, Technical Report,* The World Bank, Washington, DC.

World Bank (2009) *Public-Private Partnerships for Urban Water Utilities: A Review of Experiences in Developing Countries*, The World Bank Group, Washington, DC., http://elibrary.worldbank.org/doi/book/10.1596/978-0-8213-7956-1, accessed 6 May 2014.

World Bank (2012) *Integrated Urban Water Management. A Summary Note,* The World Bank Group, http://siteresources.worldbank.org/INTLAC/Resources/257803-1351801841279/1PrincipalIntegratedUrbanWaterManagementENG.pdf, accessed 6 May 2014.

14 Water funds as a tool for urban water provision and watershed conservation in Latin America

Fernando Veiga, Alejandro Calvache, Silvia Benítez, Jorge León and Aurelio Ramos

Introduction

During the last decade water funds initiatives have emerged throughout Latin America. Water funds are coordinated, cross-societal agreements with the purpose of offering an innovative payment mechanism for investment in green infrastructure and programmes such as watershed conservation, restoration of freshwater ecosystems and sustainable small businesses. Sustainable revenues are generated through contributions from water users and public and private institutions interested in watershed conservation and the protection of water sources. These sustainable revenues are in general derived from a combination of endowment funds interests and other flows of revenues based for example on user fees, municipal ordinances or private commitments.

The first water fund in Latin America was established in Quito, Ecuador, in 2000. Following its successful implementation, other similar initiatives have been established in other locations within the region. More recently, the Latin America Water Funds Partnership – a joint initiative from the Nature Conservancy, the FEMSA Foundation, the Inter-American Development Fund and the Global Environment Facility – has become another important driver for the establishment of water funds across the region. At the end of 2013, there were 17 water funds operating in the watersheds that supply water to approximately 34 million people living in several major Latin America cities, from Monterrey, Mexico, to São Paulo, Brazil. Currently another 15 water funds are in the process of design and negotiation, for example in Mexico City and Santiago, Chile. Through these processes more than 100 partners (relevant local and regional public and private institutions) have been involved.

These cities, like nearly all cities throughout the world, depend on natural freshwater ecosystems such as *páramos* (high mountain grasslands), rainforests, marshes or coastal lagoons for drinking water, industry, hydropower and other uses. These ecosystems are increasingly threatened by urban growth and climate change (Revenga and Mock, 1999). Processes of degradation and land use changes, including the construction of dams, water extraction, overfishing, mining, pollution, and conversion of forests to agriculture or ranching are causing irreversible transformations to natural water provision systems and compromising the hydrological services they provide (MEA, 2005; Célleri and Feyen, 2009).

A well-known example of a city with a green infrastructure approach is New York, which decided in 1997 to make investments of around US$1.5 billion in watershed management to protect water quality rather than investing in new water purification plants that would have cost the city US$6 billion to build and US$250 million annually to operate (Appleton, 2002). In a similar vein, the *Fondo para la Protección del Agua* – Fund for Water Protection (FONAG), created in 2000 to protect Quito's water supply sources, supports watershed conservation projects and programmes – mainly reforestation, park ranger programmes (surveillance of protected areas), and training and capacity building for local communities.

This chapter presents water funds as a tool for urban water provision and watershed conservation in Latin America. It first presents the rationale behind water funds. The chapter then shows how water funds work and addresses the current status of their implementation in Latin America. Selected case studies are presented to illustrate key aspects. Then lessons and challenges associated with the design and implementation of water funds are discussed. The chapter ends with the conclusions.

The rationale behind water funds

Ecosystem services

Water funds rely on the concept of ecosystem services applied to freshwater. An ecosystem is a dynamic complex of plant, animal and micro-organism communities and the non-living environment interacting as a functional unit. Ecosystem services are defined as the benefits people obtain from them (Daily, 1997; MEA, 2005). According to the Millennium Ecosystem Assessment (MEA), ecosystem services can be classified as *provisioning* (food, water, timber and fibre); *regulating* (climate, floods, disease, wastes and water quality); *cultural* (providing recreational, aesthetic and spiritual benefits); and *supporting* (soil formation, photosynthesis and nutrient cycling). The human species, while buffered against environmental changes by culture and technology, is fundamentally dependent on the flow of ecosystem services. In the case of freshwater, natural ecosystems provide direct benefits such as water quality control, hydrological regulation and sediment retention. High-Andean forests and *páramos*, for example, help maintain a constant, good-quality water supply. For many people, the loss of these ecosystems results in phenomena that range from flooding to desertification and soil aridity, causing a dramatic loss of water quality (Dudley and Stolton, 2005).

Forest ecosystems can slow the rate of run-off in a watershed by taking up water and delaying time to soil saturation (Johnson, 2000). Forest soils usually have a higher water storage capacity than non-forest soils, and the more complex structure of the forest ground surface and underlying soil allows more efficient infiltration and filtering of contaminants compared with a deforested watershed. Removal of forest cover can result in a crusting of the soil surface that reduces or prevents water infiltration and groundwater recharge. Forest soils also reduce soil erosion and sedimentation of waterways, and extensive root systems help hold soil more firmly in place and resist landslides compared with clear-cut or heavily disturbed watersheds. Forest canopies intercept rain and snowfall, and understory forest vegetation and leaf litter protect the soil from the harsh impact of rain that does fall through the canopy.

Pagiola (2002) makes the linkages between ecosystems and the economic benefits they provide for human societies clear, mentioning the following freshwater services: (i) reducing sediment loads in waterways to decrease sedimentation of reservoirs and the associated production and maintenance costs for irrigation systems, hydroelectric power plants, water supply systems and fisheries; (ii) regulating the timing of water flows so as to reduce flood risk in the wet season and the likelihood of water shortages in the dry season; (iii) increasing the volume of available water, either year-round or specifically in the dry season; and (iv) improving the quality of water available for domestic use. It is clear that in many circumstances ecosystems do play a role in the provision of water services, even if their precise nature and magnitude is not always understood due to data constraints.

Economic incentives for sustainable catchment management

Because the benefits of watersheds are often provided for free, businesses tend to take them for granted until water supplies become stressed or disappear. Social, demographic and economic trends suggest dramatically increasing demand-side stress on water resources, concomitant with detrimental land use changes in watershed ecosystems, which will amplify the situation of stress by increasing uncertainty and variability on the supply side. Scientists predict that, if humanity continues on its current trajectory, two-thirds of the world's population will face water shortages by 2025. These problems, unless addressed, will substantially diminish the benefits that future generations obtain from ecosystems (MEA, 2005; Tallis and Markham, 2012).

The most important point to make for freshwater ecosystem services is the causal relationship between land use at the headwaters of the watershed and downstream effects. Surface water flows generally originate in high altitude, natural ecosystems often surrounded by poor, rural ranching and farming communities that rely on the rich soils and abundant grasses in the *páramos* and forest ecosystems for grazing and for crops. If these natural ecosystems are converted or the already converted agricultural and cattle ranching lands are not well managed, the water quality, the regularity of water flows, and biodiversity can be compromised (Goldman *et al.*, 2010). Land users in and near the ecosystems typically do not receive any compensation for providing ecosystem services. On the contrary, providing such services often imposes costs on these land users through restricting their land use choices (Pagiola, 2002) or creates stresses on people living in those areas.

Good conservation management and incentives to discourage further encroachment on natural ecosystems can provide long-term protection for the watershed if the incentives are sufficient to compensate the people who live in the upper reaches (Goldman *et al.*, 2010). These incentives vary depending on the water fund through which they are being implemented, but generally they are direct cash payments, support for micro-enterprises such as nurseries, and assistance to implement best practices for agricultural and ranching management. The basic idea is that the economic incentives provided to land users (landowners, traditional communities) in the upper reaches of the watershed will be the drivers for land uses changes – or for the avoidance of undesirable land use changes – that benefit the provision of ecosystem water services for users downstream in the watershed. The upstream service providers receive incentives from the downstream end users of the water services. These users include public and private water utilities and hydroelectricity plants; irrigation districts and agricultural associations; and private companies producing beer, bottled water and other beverages as well as food (Goldman *et al.*, 2010).

From a business perspective water is a key ingredient in nearly every production process. It is used for cleaning, cooking, rinsing, cooling and as a conduit for waste and transportation. Furthermore, water quality impacts the real estate market, commercial fishing operations and the recreation sector. In so many ways, water impacts and is impacted on by business (Tallis and Markham, 2012). For the end users the main motivation to invest in conservation will be the greatest return of services they value, based on the understanding that it is less expensive to invest in watershed protection than to mitigate risks related to watershed degradation (Goldman *et al.*, 2010). Other motivations for business are corporate social responsibility (CSR) and, sometimes, environmental impact assessments (EIA).

In the past, water's availability for industrial processes has been typically considered certain and its supply easy and relatively cheap to secure. The recent number of new external drivers on water resources and their management, however, has made water use a much riskier proposition for industry (WWAP, 2012). Effective operation of an industry requires a sustainable supply of water in the right quantity, of the right quality, at the right place, at the right time and at the right price (Payne, 2007). Industry will find itself increasingly competing for limited water resources as water demands and consumption increase in all sectors, particularly agriculture with its substantial water needs. Water scarcity impact on industrial water supply security is viewed as an increasing business risk. Geographic and seasonal variations compound the situation, as do water allocations and competing water needs in a given region (e.g. agriculture versus drinking water or residential supply versus industry).

According to Tallis and Markham (2012), the financial value of natural capital becomes apparent when the costs of protecting an ecosystem for improved water quality or flood control are compared to investments in new or improved grey infrastructure. Companies are now finding that investments in natural capital can be worthwhile because wetlands, forests, grasslands and coastal habitats perform many of the same tasks as grey infrastructure often better and more cost-effectively. For large water users this advantage is reflected in avoided costs; for example, water of higher quality with a reduction in treatment cost for water companies in cities. Water treatment costs can be huge – and well beyond the investments necessary to maintain watershed ecosystems in good condition so that they can perform the same function. It can be more efficient and beneficial to invest in watershed conservation (green infrastructure) than covering annual costs for filters, energy for sediment removal, chemicals for water purification and new treatment plants (grey infrastructure) (Tallis and Markham, 2012) – or to combine the two approaches.

There are some good examples of the economic rationale for water funds. A study supported by Empresas Públicas de Medellín (EPM) found that 10 per cent loss in vegetation cover due to further deforestation in a water supply region would result in an average increase in monthly costs for purification of US$4.47 million at the primary water treatment facility (Mendieta, 2012). A study in the São Paulo metropolitan area showed that conservation investments made in priority ecosystem services areas can reduce the sediment load by 50 per cent, saving about US$4.9 million/year in the two water systems that provide water to this metropolitan region, based on avoided costs of water treatment and dredging (Souza, 2013). The feasibility studies developed for specific water funds have also generated evidence of economic benefits.

How water funds work

Water funds have three main components which when applied together have been proving to be a strong and innovative payment for ecosystem services (PES) mechanism. The first component is the building of a long-term sustainable funding source that will allow a long-term investment plan. The second is the organization of most of the relevant stakeholders in the watershed, private and public, who have an interest in the conservation of the watershed, in a body which makes decisions and supports project implementation on the ground. The third component is the robust science behind the process of prioritizing locations of interest and activities to be implemented.

Together, the three components promote sustainability, resources allocation efficiency and transaction costs reduction. Ultimately, the mechanism provides a long-term, transparent source of funding administered by a multi-institutional decision-making body that decides how to most effectively allocate the fund's revenue under competing demands (Goldman *et al.*, 2010). In some funds the contributions are voluntary while in others they are legally required, according to local policies. In Brazil the mechanism is slightly different from the water funds endowment model: under the Water Producer Program, a concept developed by the Brazilian National Water Agency, an annual distribution model is used where some of the water user fees (or other sources of funding) are collected and distributed each year rather than going into a trust.

Each water fund has its own location-dependent objectives and goals, but in general the funds are invested in conserving watersheds to improve or maintain water-related benefits and regulate water-related risks (Goldman-Benner *et al.*, 2012). Their objectives are to: (i) improve or maintain water quality and water quantity for downstream users; (ii) maintain regular flows of water throughout the year; (iii) reduce flooding events; (iv) maintain or enhance natural ecosystem biodiversity, both freshwater and terrestrial; and (v) improve or maintain human well-being and quality of life for upstream human communities. Some typical conservation activities supported by water funds are native trees planting, forest enrichment, *páramo* restoration, soil restoration, fencing, recovery of native species, contour farming and terracing, crop rotation, conservation tillage, pasture rotation, or surveillance of protected areas.

Prioritization of activities and selection of the sites where they will be implemented come from hydrological models which estimate the ecosystems services provision and evaluate the investments returns. These models allow water funds managers to develop scenarios that show responses on environmental services provision according to land use patterns and climate variations over time. These results are essential when drafting the water funds' environmental services goals and figuring out related financial costs to achieve these goals (The Nature Conservancy, 2012). Conservation plans developed with the use of such ecosystem services models enable the water funds stakeholders to identify where and how the investments should be allocated to generate the best ecosystem services returns. Discussion among the main water users informed by the models also contributes to the process.

Watershed activities are funded by the interest earned on endowment funds as well as additional revenues obtained from water users or external donors, and a portion of the capital itself may be used to pay for watershed management projects and operating costs. This proportion can vary from fund to fund and from investor to investor. In Brazil the mechanism is slightly different from the water funds endowment model: under the Water Producer Program, a concept developed by the Brazilian National Water Agency, an annual distribution model is used where some of the water user fees (or other sources of funding) are collected and distributed each year rather than going into a trust. Figure 14.1 shows the most common funding sources and how conservation activities and water fund operation are financed by the interest of the endowment.

Implementation of water funds in Latin America

Based on the pioneer example of FONAG in Quito, Ecuador, and the first experiences of the Brazilian model, TNC and its local partners are replicating and improving the water funds model in several countries in Latin America and the Caribbean, including Bolivia, Brazil, Chile, Colombia, Costa Rica, the Dominican Republic, Ecuador, Guatemala, Honduras, Panama, Mexico, Peru

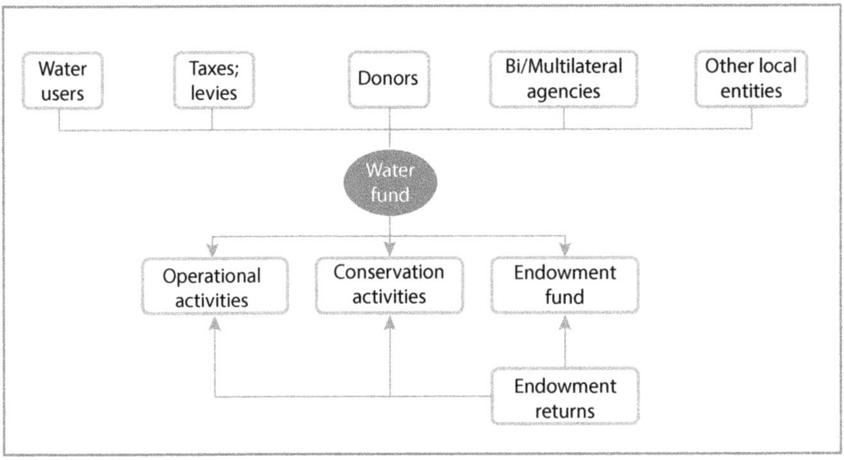

Figure 14.1 General water fund scheme

Source: The Nature Conservancy and Inter-American Development Bank (2010, p. 8)

and Venezuela. Water funds are recognized as the fastest growing model of watershed payments in Latin America (Bennett *et al.*, 2013).

A key milestone for water funds in Latin America was the creation of the Latin America Water Funds Partnership by TNC, the FEMSA Foundation, the Inter-American Development Bank (IDB) and the Global Environment Facility (GEF) in June 2011. The Partnership aims to support the establishment of 32 water funds across Latin America and the Caribbean by 2016. These 32 funds intend to raise US$27 million seed money to support their establishment as well as US$143 million investments from the main water users and US$500 million from other contributors. The funds will be used in conservation, restoration and support of best management practices in watersheds covering 28,000km^2, potentially benefiting 50 million people.

The Partnership, besides supporting the establishment of water funds across the region, has the following specific objectives: (i) identify and share best practices; (ii) develop projects that can benefit from a regional approach; (iii) support water, environmental and socio-economic monitoring; (iv) foster green infrastructure lending in the most important multilateral financial institutions; (v) support internalization of watershed conservation costs into water tariffs; and (vi) generate business cases that can reinforce the green infrastructure approach.

Implementation status

At the end of 2013, there were 17 water funds operating in some of the most important Latin American metropolitan areas (Figure 14.2). There are another 15 water funds in the design and negotiation stages (Table 14.1). The last six months of 2013 faced four water funds launches: Monterrey in Mexico, Medellín in Colombia, and Santo Domingo and Yaque del Norte in Dominican

Figure 14.2 Water Funds in Latin America and the Caribbean, 2013

Source: The Nature Conservancy

Republic. All of them experienced significant participation of public and private representatives. A noticeable example of this affirmative was the launch of the Monterrey Water Fund in September 2013 which had 62 public and private institutions joining the water fund as partners.

Together, water funds have raised US$66,056,089 of total contributions, which is destined to feed the endowment funds, to support their management

Table 14.1 Water fund location and status in Latin America, December 2013

Water fund name and country	*Stage*			
	Evaluate potential	*In design*	*Created and operating*	*Mature*
Patagonia Water Fund, Argentina	•			
La Paz Water Fund, Bolivia		•		
Doce Watershed Water Fund, Brazil	•			
Water Spring – São Paulo, Brazil	•			
Paraiba do Sul Watershed, Brazil		•		
Taquarussu Watershed – Palmas, Brazil		•		
Upper Tietê Watershed – São Paulo, Brazil		•		
Camboriu Watershed, Brazil			•	
Espirito Santo State Water Fund, Brazil			•	
Guandu Watershed – Rio de Janeiro, Brazil			•	
PCJ Watershed – São Paulo, Brazil			•	
Pipiripau Watershed – Brasilia, Brazil			•	
Santiago-Valparaiso Water Fund, Chile		•		
Barranquilla Water Fund, Colombia	•			
Cali Water Fund, Colombia	•			
Sierra Nevada Water Fund, Colombia	•			
Cartagena Water Fund, Colombia		•		
Agua por la vida y la sostenibilidad – Cauca Valley Water Fund, Colombia			•	
Agua Somos – Bogotá Water Fund, Colombia			•	
Cuenca Verde – Medellín Water Fund, Colombia			•	
San José Water Fund, Costa Rica		•		
Yaque del Norte Water Fund, Dominican Republic			•	
Santo Domingo Water Fund, Dominican Republic			•	
Sucumbios Water Fund, Ecuador	•			
Guayas-Guayaquil Water Fund, Ecuador		•		
Ayampe Puerto López Water Fund, Ecuador		•		
Fonapa Cuenca Azogues Water Fund, Ecuador			•	
Procuencas Zamora Water Fund, Ecuador			•	
Tungurahua Ambato Water Fund, Ecuador			•	
Fonag – Quito Water Fund, Ecuador				•
Foncagua – Guatemala City Water Fund, Guatemala		•		
Tegucigalpa Water Fund, Honduras	•			
San Pedro Sula Water Fund, Honduras	•			
Cancún, Quintana Roo Water Fund, Mexico		•		
Monterrey Water Fund, Mexico			•	
Semilla de Agua Chiapas Water Fund, Mexico			•	
Valle de México Water Fund, Mexico	•			
Panama Channel Water Fund, Panama		•		
Sixaola Water Fund, Panama		•		
Arequipa Water Fund, Peru	•			
San Martín Water Fund, Peru	•			
Piura Water Fund, Peru		•		
Aquafondo Lima Water Fund, Peru			•	
Mérida Water Fund, Venezuela		•		

Source: The Nature Conservancy

and technical operations, and implementation of conservation activities on the ground.[1] Of this amount, 66 per cent came from public funds, 28 per cent from private funds and 6 per cent from multilateral and bilateral institutions and non-governmental organizations. Water funds have been supporting activities on a total of 220,194 hectares of land. Of this total, 94,085 hectares belong to private landowners, and 126,109 hectares are public or communal lands. These areas are located in watersheds covering almost 1.45 million hectares and which benefit approximately 34 million people. The water funds have been generating positive incentives based on watershed conservation for 4,282 families living in the upper reaches of these watersheds.

Fencing riparian corridors, restoring degraded lands, using best practices in ranching, reforesting with native species, and training and capacity building have been the most frequent activities supported by water funds. This work is done directly with local communities, usually living in their small properties, communal lands or protected areas, who voluntarily decide to participate in the water fund. Funded activities bring direct benefits to them such as: better water supplies for their own use; additional income from more efficient farming and agricultural practices; support to micro-enterprises; and in the case of some funds, especially in Brazil, direct cash payments. In the following section the cases of Quito and Bogotá are presented, along with some brief notes on the Brazilian experience.

Case studies

Fondo para la Protección del Agua (Quito Water Fund)

FONAG, or Quito Water Fund, was the first water fund created in Latin America. The successful results of this model have led to its replication in the region and have generated interest in other continents. The idea for this mechanism was devised in 1997 as a response to concern for Quito's water supply and for the scarce funding for protected areas where water sources are located (Calvache *et al.*, 2012). About 80 per cent of the water for the city of Quito, Ecuador – a city of nearly 2 million people – comes from three protected areas and their buffer zones: Cayambe-Coca Reserve, Antisana Reserve and Cotopaxi National Park. These areas include the headwaters of more than 20 rivers and six large watersheds, and they provide a crucial habitat for many of Ecuador's bird and mammal species. Nevertheless, various activities threaten this biodiversity and the availability of clean water supply.

Much of the pressure on the land comes from the people living in the watershed because they depend on natural resources for their livelihoods. Available productive land is diminishing as soils lose nutrients, forcing families to move up the watershed into the natural ecosystems – a mixture of forest and *páramos* – that are the fundamental hydrological regulators (Goldman-Benner *et al.*, 2013). The development of economic incentives that can bring new sources of income for those communities who live upstream through activities that reduce the threats to these areas or increase their protection was considered to be the key to keeping the provision of these vital ecosystem services for Quito.

The financial approach to be used was evaluated by TNC and its partners and selected because it was deemed to be ecologically sustainable, politically viable, efficient and participatory (Krchnak, 2007; Goldman-Benner *et al.*, 2013). The *Empresa Pública Metropolitana de Agua Potable y Saneamiento* (Metropolitan Water and Sewage Company) (EMAAP-Q, today EPMAPS) and TNC boosted the creation of the trust fund which was established in 2000 as FONAG, an endowment fund with a planned life of 80 years. The fund began with seed funding of US$21,000. Participants progressively joined as constituents: the Quito Power Company (EEQ) in 2001, the Andean Brewery (now known as the National Brewery) in 2003, the Swiss Development Cooperation Agency (COSUDE) in 2005 and Tesalia Springs Co. (a bottled water company) in 2007. In 2010, the Training System for the Management of Natural Renewable Resources Consortium (CAMAREN) also became a member of the trust fund. One crucial step for the consolidation of FONAG was the approval in 2004 of the municipal ordinance dedicating 2 per cent of the water fees revenues collected by EPMAPS to the fund.

FONAG's mission is to restore, care for and protect the watersheds that supply water to the Metropolitan District of Quito and surrounding areas. Its vision is to be the mobilizing agent for all actors to exercise their civic responsibility on behalf of nature, especially water resources. And its objectives are to lead processes and consensus – through dialogue, proper decision making, strengthening research and the appropriate use of technology – to achieve integrated management of water resources in which active, responsible participation based on solidarity leads to sustainable water management.

FONAG has a Board of Directors formed by representatives of its constituents (the institutions who have contributed to the fund). This is the highest decision-making body of the fund and it is in charge of defining policies and strategies to guide the development and fulfilment of the fund's objectives. The technical secretariat is in charge of FONAG's operation and management, serving as the operational body of the fund and reporting to the Board. Financial resources are managed by the trust fund, which is also the legal representative of the fund. This mechanism ensures transparency in the decision-making process and in the use of resources.

FONAG has had significant financial growth. It started out with US$21,000 in 2000, as mentioned above; by December 2013 it had US$12,979,000. Regarding its investment in activities, FONAG has been very successful in attracting leverage funds. For each dollar spent by FONAG, at least three dollars have been obtained from external contributors. In 2008, for example, FONAG's budget was US$4.1 million, of which US$700,000 came from the endowment interest and US$3.4 million came from contributions by donors and partners (Calvache *et al.*, 2012). It invests 80 per cent of its resources in long-term programmes and activities and 20 per cent in specific projects.

FONAG works to guarantee sufficient good-quality water through co-financing actions aimed at protecting watersheds to achieve the natural regeneration of the resource. It carries out activities in the upper basins of the Guayllabamba, Oyacachi, Papallacta and Antisana rivers, covering a total area of 542,000 hectares. The water fund finances the following programmes: Recovery of Vegetation Cover, Environmental Education, Surveillance and Monitoring, Water Management, Communication, and Training.[2]

Agua Somos (Bogotá Water Fund)

Agua Somos (We are Water), or the Bogotá Water Fund, is not only one of the most important water funds established in Latin America because of its value for Bogotá itself – Colombia's capital city of more than 8 million people – but also because this fund helps protect one of the most impressive *páramos* areas in the continent. More than 80 per cent of the water consumed by Bogotá's residents is supplied by the Chingaza system, located to the east of the city. The system comprises the watersheds of the Chuza, Guatiquia and Teusaca rivers in addition to the Chuza dam, which has a capacity of 257 million m^3. Both the dam and the rivers' headwaters are located within Chingaza National Park, which was created in 1974. The Park covers 76,600 hectares and harbours high-Andean forests and *páramos* of vital importance: they provide the habitat for a large number of species and regulate water flow, control sediments and guarantee the supply of good-quality water for Bogotá.

Alterations in land use have significantly modified some areas of Chingaza National Park, particularly its buffer zones, where land used for livestock and agriculture has expanded concomitantly with human settlements. The park's budget is insufficient to meet the needs identified in its management plan; therefore, ecosystem conservation efforts and the possibility of offering alternative sustainable economic activities to local communities are also insufficient. This situation has exposed the long-term risk in the supply of good-quality water for Bogotá's residents from the Chingaza system, a risk that is equally valid for the Tibitoc (north of the city) and La Regadera (south of the city) systems, which supply water to the remaining 20 per cent of Bogotá's population.

Chingaza National park surrounding areas provide water for either the Chingaza system to Weisner plant, or the Tibitoc system. While the condition of páramo areas is generally good inside the national park, outside its boundaries the suburban expansion of Bogotá, cattle ranching pastures and potato crops predominate in the landscape. Farmers in the buffer area of the park are aware of the importance of these 'water mines' but without proper economic incentives for conservation uses, it is difficult for them to prevent cattle from entering these areas (see Figure 14.3).

In 2006, TNC began contacting public and private institutions to promote the idea of creating a water fund for Bogotá and its neighbouring municipalities. One of the entities approached was the Empresa de Acueducto y Alcantarillado

Figure 14.3 Cattle ranching in high *páramo* areas, inside the Chingaza system but just outside the national park

Source: The Nature Conservancy

de Bogotá (EAAB – Bogotá Water Supply and Sewerage Company), which spends a considerable amount of resources annually on water treatment, an amount that could increase in the future if measures necessary to stop ecosystem transformation and to protect watersheds are not taken. Another was the National Parks Service, the national government agency in charge of managing national parks, which had expressed an interest in participating in innovative schemes that would allow it to leverage funding resources to implement its programmes and projects. In 2007, a memorandum of understanding was signed between EAAB, the National Parks Service, Patrimonio Natural Foundation and TNC to combine technical and financial efforts to determine the feasibility of creating a watershed investment mechanism aimed at conserving ecosystems.

While feasibility studies were taking place, TNC made contact with the private sector to promote the participation of corporations in the Bogotá Water Fund. SABMiller Bavaria, the largest brewery in the region, committed resources to the fund. Its interest in the fund, beyond a simple social and environmental responsibility, coincides with one of the fund's objectives: to reduce sediment levels and prevent an additional sediment load in the long term (thus saving costs for the treatment of water to be used in beer production). The

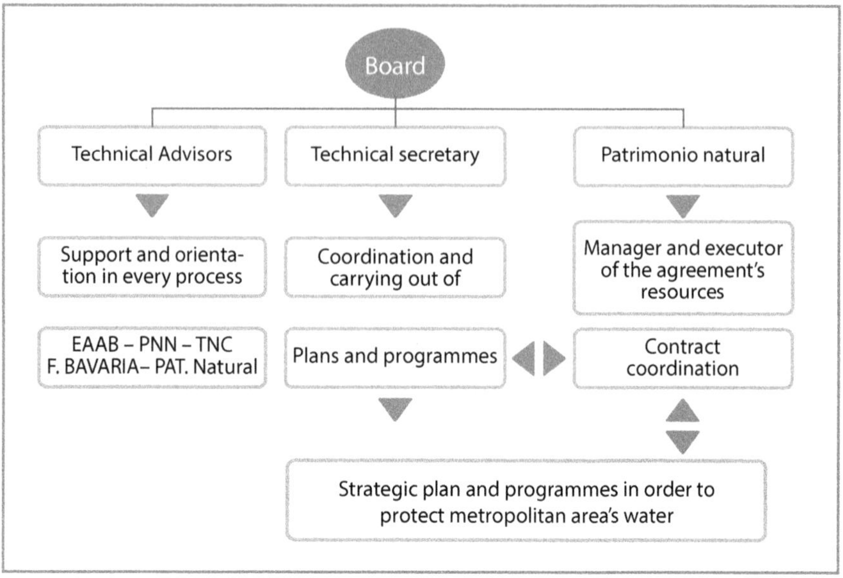

Figure 14.4 Governance arrangement of the Bogotá Water Fund

Source: The Nature Conservancy (2008, p. 16)

Bogotá Water Fund was launched publicly in May 2009. In October, a two-year cooperation agreement was signed in which EAAB, SABMiller Bavaria and TNC contributed an initial 1,300 million Colombian pesos (US$650,000) to kick-start the fund's operation. This money covered the costs of a full-time technical secretary to lead the water fund, the completion of legal studies necessary for its operation, a marketing campaign to promote the fund and leverage voluntary donations from citizens and the private sector, and the implementation of some restoration activities in the field. The governance arrangements of the fund are presented in Figure 14.4.

Between 2007 and 2009, feasibility studies with two main components were carried out. The first was the application of numerical models to determine the hydrological dynamic of the watersheds of the supplying system, which included quantifying water volume flows and yields, sediment levels and the importance of fog capture in the water balance for the area. This study was developed by the Centro Internacional de Agricultura Tropical (International Center for Tropical Agriculture – CIAT). The second study evaluated and identified legal alternatives for the establishment and structure of the Bogotá Water Fund, and analysed potential funding sources and resource flows.

The results of these studies indicate that the estimated investment of US$15 million over ten years would pay for interventions on 60,000 hectares of land in the watersheds that provide Bogotá's water supply. With these interventions it is projected that 2 million tonnes of sediments are prevented, representing savings in the order of US$35 million in treatment costs over

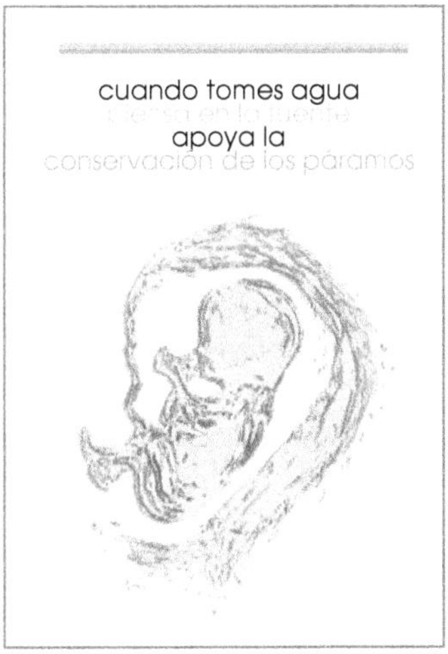

Figure 14.5 Logotype and slogan of the Bogotá Water Fund
Note: The slogan means: 'When you drink water, think about its source. Support the *páramos*' conservation'

Source: Yunis and Lizarazo (2013, p. 18)

the same time period (White *et al.*, 2007). The findings of the modelling work emphasize both the large returns possible from watershed management and the need to target investments to areas where land management changes will indeed lead to reductions in costs and increase in economic benefits.

The Bogotá Water Fund has developed a communication strategy with an awareness campaign aimed at reaching the primary stakeholders, those who can contribute to the water fund's objectives, as stated above. An essential element of the campaign was the development of a slogan that may stimulate a connection between the audience and the water fund's purpose and so lead to desired concrete actions. The slogan defined was 'When you drink water, think about its source. Support the *páramos*' conservation' (Yunis and Lizarazo, 2013) (Figure 14.5).

In addition, the water fund has started to support the implementation of conservation projects such as protection of riparian corridors, conservation of native forests and the building of a conservation plan taking into consideration a detailed characterization of erosion processes in the most important water supply systems. More recently the water fund started a new cycle of projects in partnership with Coca-Cola FEMSA in Colombia.

Water funding schemes in Brazil

The implementation of funding schemes in Brazil has followed several avenues based on governance models different from those used in other Latin American countries (Veiga and Gavaldão, 2011; Goldman-Benner *et al.*, 2013). These use an annual distribution model where part of the water user fees or other sources of funding are collected and distributed each year rather than going into a trust.

The first implementation method to be highlighted is a result of the National Water Policy, which established a water user fee and required the proceeds to go towards maintaining and increasing watershed health. In addition, the creation of watershed committees was enabled, comprising water users and representatives of governments and civil society, which have the legal power to decide the best way to spend the proceeds of the fund. Creating a water producer project from this policy arrangement involves several steps. The project's concept is presented to the watershed committee. Direct payments compensate watershed landowners for the positive externalities they generate when they restore and protect their lands. When the project is approved by the committee, the project partners, usually public and private stakeholders, create a management unit to implement and manage the project.

A second important way in which water funds in Brazil have been implemented requires municipal and state legal approval explicitly for developing PES schemes. These laws are important for two reasons: they can provide a legal framework for PES implementation, and they can enable and facilitate the use of public funds for PES schemes. An important example of this method is the water fund FUNDAGUA in Espíritu Santo state, which is primarily funded by a percentage of the oil royalties collected by the state and has a governance body mainly constituted by public agencies.

Lessons learnt

Over the past decade, especially after the launch of the Latin America Water Funds Partnership, this Partnership and its local partners have been working to implement and refine the innovative water funds concept. The pilots and replicates have allowed a clear model of implementation to be developed. An increasing range of local and regional partners have been part of the efforts, offering the opportunity to produce a large-scale tool that effectively connects main water users and ecosystem services providers. Thanks to the many countries in which water funds are being established and the diversity of situations, institutional schemes, stakeholders, ecosystems protected and ecosystem services generated, there is a huge opportunity for knowledge sharing and collaboration across the region.

Water funds cases mentioned in this chapter as well as other water funds implemented so far, offer relevant and timely lessons to be learnt and reveal some challenges to be addressed. At the beginning of the process, it is important to understand and identify the main water demand drivers and the potential water funds beneficiaries. The experience in Quito showed that the success of the water fund will rely on stakeholder engagement and water users' willingness to pay for the ecosystem services. Success at this stage depends to a large extent on the awareness of potential private and public partners about the water risks they face, and their belief that the green infrastructure approach is an acceptable solution to mitigate these risks. It is also very important that local public opinion leaders participate actively in that respect.

In the design and negotiation phase, the key is to get estimations about the potential return on the investments that will be done by future investors (mainly water users) based on ecosystem services models. This will allow an evaluation of the costs and benefits involved in supporting the water fund, as well as a clear definition of the quantitative goals in terms of the ecosystem services generated (tonnes of sediments reduced; per cent of base flow increased; per cent increase of the amount of water infiltrated, etc.) and the amount of money needed to reach them. A notable example of this approach is the Bogotá Water Fund. Studies have demonstrated how important it is to conserve Bogotá's water provision system, given anthropogenic pressures to natural parks and buffer zones where the water for the city comes from, and the fund was created using science tools to identify priority areas, and estimated financial and intervention areas goals. In the case of Bogotá, the conservation work is in its initial phase. Monitoring systems have been implemented in most of the water funds to track the targets' achievement in the long term.

There are some challenges to be addressed at the design phase, mostly related to the lack of data to feed the ecosystem services models and/or the lack of local technical capacity to operate these models. The Natural Capital Project, an association of Stanford University, University of Minnesota, TNC and WWF, has addressed these issues through the development of ecosystem services models able to run in several tiers of data requirements, as well capacity-building activities across the region.

There are many potential sources of public and private money that can contribute to the establishment and long-term consolidation of a water fund. On the public side, there are many regulatory options that can be important sources of resources such as: a) water user fees; b) legal frameworks for the payment of ecosystem services; c) the inclusion in water tariffs of watershed conservation costs; and d) the establishment of local regulations that earmark public revenues to be allocated to the water fund. In Brazil and Ecuador, the initiatives being implemented have been instrumental to prove the importance of these public sources of funds. The main challenge from now on is to build on these first examples as well as to pave the way for some new ones such as the inclusion in water tariffs of watershed conservation costs. Public awareness and its influence on decision makers can also prove an important driver for the expansion of existing water funds and the creation of new ones.

On the private side, besides the direct potential benefits of investing in the activities promoted by the water fund, companies are being driven by their evaluation of risks associated with increasing water scarcity. These risks can relate to the physical, regulatory and reputational dimensions and can prove significant drivers for corporate participation in a water funds scheme. This, especially when corporations have already realized the benefits of increased water efficiency inside their own plants and now understand that the conservation of the watersheds where their water comes from is critical for their business. There are many companies that are evaluating the water footprint of their direct operations and supply chains, trying to understand where they can reduce their impacts and/or compensate them in order to be partially or totally water neutral.

As a financial mechanism, a water fund's success largely depends on the adequate management of its resources, its independence, its legal and financial guarantees, and the self-sustainability of its financial resources. It is essential to determine the best way to administer these resources among the many available alternatives (e.g. a private market financial institution, an environmental trust) based on each country's conditions and the needs of each fund as one can see in the example of the Quito Water Fund and its consolidated growth.

Once the water fund is launched and the contributions and roles of their members are negotiated, there are important lessons to be learnt from the ones operating so far. The first one is to establish the monitoring baselines (hydrology, biodiversity and socio-economic) and start the monitoring activities from the early stages of the water fund. Monitoring is critical because the water funds stakeholders (investors, local communities, managers) require information about the efficacy of the models that are used to estimate the benefits, the effectiveness of the activities, the efficiency of implementation and the return on investments (Higgins and Zimmerling, 2013).

Water funds have been monitored at two levels. First, at the implementation level where a general board composed of water fund managers and coordinated by the Latin America Water Funds Partnership tracks three main aspects: i) the number of hectares of public and private land where conservation activities have been implemented; ii) the number of people benefited downstream (water users) and upstream (local communities); and iii) the amount of private and public funds collected and invested by the water funds.

The second level of monitoring focuses on two main topics: activity implementation and impacts. The monitoring of impacts (the level of accomplishment of the water fund in terms of water quantity/quality goals) has followed a general protocol developed in Higgins and Zimmerling (2013), adjusted to the local conditions of each fund, that include the hydrological, socio-economic and biodiversity impacts generated by their activities. This is already happening in Guandu (Rio de Janeiro), PCJ (São Paulo), Camboriú, Palmas, Cauca Valley, Quito and Lima water funds. Other water funds are in the process of implementing their monitoring protocols.

Another lesson points to the need to work on the ground through demonstrative projects from the beginning, even if the whole amount of the expected contributions to the new water fund has not yet been secured. This will generate local knowledge, local support and experience to inform the implementation of the planned activities and will facilitate the influx of new revenues to the fund. As with other payment schemes for ecosystem services, one of the key challenges for the implementation of conservation activities by water funds is the transaction costs associated with agreements with many local stakeholders such as smallholders and local indigenous communities. To overcome these bottlenecks, clear processes for calls for proposals and/or long-term programmes are being established, most of them with the participation of local associations and/or local NGOs providing support for those engaged in conservation activities.

Water funds, as expected, are focused on activities which promote water security to downstream users through nature-based solutions, but this focus does not mean that their activities and outcomes will be limited to this arena. For instance, there are activities linked to the Water, Sanitation and Hygiene (WASH) agenda which integrated health objectives with the watershed management approaches and focus on the rural and upper watershed communities. Water funds have enabled different water stakeholders to talk and plan together and facilitated information for better decision making.

Conclusions

Green infrastructure is a relatively new concept which has gained traction because it can convey in a simple way the concept of ecosystem services. As is the case with traditional infrastructure, the green one is also able to generate jobs and incomes. Balancing different interests, however, is far from easy. Water funds have shown they can assemble prominent local stakeholders to work together towards implementing concrete conservation plans. This is the experience of Quito and the other water funds established so far. It is worth noting that they build an environment of trust among their constituents, but also ensure that local communities are represented in the decision-making process.

Typically water funds resources come from a combination of public and private funds. Although the examples established so far have shown that there is an increasing awareness about the importance of green infrastructure investments, there is still much to do in order to translate this awareness into concrete investment in watershed protection. Securing long-term funding through tariffs or other legal mechanisms can take time and is still a challenge for several water funds.

Monitoring systems are paramount. Even though their design is based on the best science available, uncertainties remain regarding the models. The monitoring programmes installed so far in many water funds have started to generate actual data in terms of hydrological, socio-economic and biodiversity impacts, but as most of them are in the initial phase of data collection and some of them only in the design phase, there is a clear need to keep them up and running.

The inclusion of monitoring costs in the operation costs of water funds has proved challenging. It must be noted that demonstrating the success of conservation efforts through the monitoring of ecosystem services processes may take time. For example, *páramo* ecosystems restoration is slow, and changes in water-related indicators will take several years to respond to conservation actions. This is important to keep in mind in order to set realistic expectations among the key stakeholders.

The pace of creation of new water funds has been remarkable in Latin America; however, there remains the challenge of fostering local capacities along each step of the process behind the establishment, implementation and evaluation of water funds. Packages of toolkits, technical assistance for implementers, and capacity-building programmes are being developed and will play an important role in the near future. This process – supported by an increasing amount of experiences and the exchange of lessons learnt among water funds – will be essential for maximizing the positive impacts of water funds. The accomplishments reached so far show that, in spite of their limitations, water funds can play an important role in both preserving the environment and aiding water supply to Latin American cities.

Notes

1 The source for this comes from an internal dashboard where the information from all the water funds is collected on a quarterly basis by TNC water fund managers.
2 For more detailed information of the programmes see: http://www.fonag.org.ec.

References

Appleton, A. (2002) 'How New York City Used an Ecosystem Services Carried out Through an Urban–Rural Partnership to Preserve the Pristine Quality of Its Drinking Water and Save Billions of Dollars', paper presented on *Forest Trends Meeting*, Tokyo, November 2002.

Bennett, G., Carroll, N. and Hamilton, K. (2013) *Charting New Waters: State of Watershed Payments 2012*, http://www.forest-trends.org/documents/files/doc_3308.pdf, accessed 20 April 2014.

Calvache, A., Benítez, S. and Ramos, A. (2012) *Fondos de Agua: Conservando la Infraestructura Verde. Guía de Diseño, Creación y Operación*, Alianza Latinoamericana de Fondos de Agua, Bogotá.

Célleri, R. and Feyen, J. (2009) 'The Hydrology of Tropical Andean Ecosystems: Importance, Knowledge Status, and Perspectives', *Mountain Research and Development*, vol. 29, no. 4, pp. 350–355.

Daily, G.C. (1997) *Nature's Services: Societal Dependence on Natural Ecosystems*, Island Press, Washington, DC.

Dudley, N. and Stolton, S. (2005) 'The Role of Forest Protected Areas in Supplying Drinking Water to the World's Biggest Cities', in T. Trzyna (ed.), *The Urban Imperative*, California Institute of Public Affairs, Sacramento, California.

Goldman, R. L., Benítez, S., Calvache, A. and Ramos, A. (2010) *Water Funds: Protecting Watersheds for Nature and People*, The Nature Conservancy, Arlington, VA.

Goldman-Benner, R. L., Benítez, S., Calvache, A., Ramos, A. and Veiga, F. (2013) 'Water Funds: A New Ecosystem Service and Biodiversity Conservation Strategy', in S.A. Levin (ed.), *Encyclopedia of Biodiversity*, Academic Press, Waltham, MA.

Goldman-Benner, R. L., Benítez, S., Boucher, T., Calvache, A., Daily, G., Kareiva, P., Kroeger, T. and Ramos, A. (2012) 'Water Funds and Payments for Ecosystem Services: Practice Learns from Theory and Theory can Learn from Practice', *Oryx*, vol. 46, no. 1, pp. 55–63.

Higgins, J. V. and Zimmerling, A. (2013) *A Primer for Monitoring Water Funds*, The Nature Conservancy, Arlington, VA.

Johnson, N. (2000) 'Strategies to Build Sustainable Links between Forests, Water, and People', paper presented at the Environmental Markets and Forests Workshop, Katoomba, Australia, 11–12 April.

Krchnak, K. M. (2007) *Watershed Valuation as a Tool for Biodiversity Conservation*, The Nature Conservancy, Arlington, VA.

MEA (Millennium Ecosystem Assessment) (2005) *Ecosystems and Human Well-being: Synthesis*, Island Press, Washington, DC.

Mendieta, J. C. (2012) *Valoración Económica Ambiental de los Servicios Ecosistémicos Hidrológicos para el Mecanismo Financiero de Conservación de las Cuencas Abastecedoras de los Embalses Ríogrande II y La Fe que Abastecen de Agua a la Ciudad de Medellín y los Municipios del Valle del Aburrá en el Departamento de Antioquia*, The Nature Conservancy, Bogotá.

Pagiola, S. (2002) 'Paying for Water Services in Central America: Learning from Costa Rica', in S. Pagiola, J. Bishop, N. Landell-Mills (eds), *Selling Forest Environmental Services: Market-based Mechanisms for Conservation and Development*, Earthscan, London.

Payne, J. G. (2007) 'Matching Water Quality to Use Requirements', technical report for *UNIDO Technology Foresight Summit*, Budapest, 27–29 September 2007.

Revenga, C. and Mock, G. (1999) *Pilot Analysis of Global Ecosystems: Freshwater Systems and World Resources 1998–1999*, World Resources Institute, Washington, DC.

Souza, W.C. (2013) *Pagamento por Serviços Ecossistêmicos: Sedimentos e Nutrientes – Sistemas de Abastecimento da Região Metropolitana de São Paulo, São Paulo*, The Nature Conservancy, Curitiba, Brazil.

Tallis, H. and Markham, A. (2012) *Water Funds Business Case: Conservation as a Source of Competitive Advantage*, The Nature Conservancy, Arlington, Virginia.

The Nature Conservancy (2008) *Establishment of a Fund to Protect Watersheds that Provide Water to the City of Bogotá. Proposal*, The Nature Conservancy, Bogotá.

The Nature Conservancy and Inter-American Development Bank (2010) *Earth Fund Platform Identification Form (EFPIF)*, The Nature Conservancy, Cartagena.

The Nature Conservancy (2012) *Operations Manual – TNC Regional Platform for Water Resource Management GRT/CF-12631-RG*, The Nature Conservancy, Curitiba, Brazil.

Veiga, F. and Gavaldão, M. (2011) 'Iniciativas de PSA de Conservação dos Recursos Hídricos na Mata Atlântica', in F. Becker and S. Edda (eds), *Pagamentos por Serviços Ambientais na Mata Atlântica: Lições aprendidas e desafios*, Ministério do Meio Ambiente, Brasília.

White, D., Rubiano, J., Andersson, M., García, J., Saenz, L. and Jarvis, A. (2007) *Análisis de Oportunidades de Inversión en Conservación por Ahorros en Tratamiento de Aguas*, Centro Internacional de Agricultura Tropical – CIAT, Cali, Colombia.

WWAP (World Water Assessment Programme) (2012) *The United Nations World Water Development Report 4: Managing Water under Uncertainty and Risk*, UNESCO, Paris.

Yunis, J. and Lizarazo, A. (2013) 'Campaña "capital" para el Fondo de Agua de Bogotá', in C. Ortega, A. Ramos, S. Benítez, C. A. Hurtado, S. Ortiz and A. Guzmán (eds), *Reconocimiento de los Servicios Ambientales Hidricos en LatinoAmerica: Intercambio de experiencias y mejores prácticas*, The Nature Conservancy, Quito.

15 Green infrastructure and watershed protection in US cities

Insights for Latin America

Jonathan C. Kaledin

Introduction

The long history of urbanization in Latin America, coupled with continued urban growth, means that Latin American cities face the challenge of expanding water services while replacing older and ageing systems. Will Latin America continue along with a strong focus on large linear water systems, or will it begin looking more and more into decentralized and 'green' solutions to water issues as well? This chapter looks into how New York City, Boston, Seattle, Philadelphia and Milwaukee have managed their watersheds and undertaken other green infrastructure developments in novel ways that harmonize with natural systems, and draw insights of relevance for Latin American cities.

While the quality of water and sewerage services provided in the United States is very high, an increasing number of the country's urban water systems are approaching, or are already beyond, their useful lives. Many cities now face serious challenges in maintaining and upgrading their old systems (WeiserMazars, 2012). In September 2002, the Environmental Protection Agency (EPA) issued *The Clean Water and Drinking Water Gap Analysis* looking at capital, operations and maintenance financial needs to the year 2020 under both the Clean Water Act and the Safe Drinking Water Act. The Report noted that 'a significant funding gap could develop if the nation's clean water and drinking water systems maintain current spending and operations practices' (US EPA, 2002).

In spite of this awareness, improvements in urban water systems continue at a pace that has total annual investment being much less than that needed (ASCE, 2011). And much of the investment is not voluntary, but results from enforcement actions brought against cities under the Clean Water Act or the Safe Drinking Water Act by the federal or state governments.[1] Despite, perhaps, starting from a higher level of water and sanitation service provision, any number of US cities now face water system issues – financial, administrative, quantity and quality – that a large number of Latin American cities face too. Growing cities must invest more to address both regulatory requirements and increasing demand, and often must do so in the face of water supply constraints, which are likely to increase or at least become more unpredictable because of climate change effects (Bolger *et al.*, 2009).

Many cities are falling behind on the critical task of keeping their water systems up to date and effective. Reviewing the current state of, and possible future developments for, water systems in the US, the Johnson Foundation found that 'our water infrastructure, designed in the 19th and early 20th centuries, no longer meets today's needs and challenges' (Johnson Foundation, 2011). And with up to 25 per cent of drinking water in the United States being lost to leaks in decaying pipes instead of being delivered to customers, the American Society of Civil Engineers (ASCE) gave drinking water systems almost the worst 'school grade' possible – a D minus – in its 2009 overall assessment of the state of US infrastructure (ASCE, 2009). If one accepts the findings of the ASCE, the Johnson Foundation, and others, there can be no question that serious problems may be on the horizon for urban water systems in the United States.

Green growth and water systems in the United States

Despite the somewhat gloomy picture just presented, and as the upcoming case studies show, a new era in water system infrastructure development is beginning, with re-thinking of many aspects of infrastructure design and development. An example of the new generation of work being undertaken on US urban water issues is the new Engineering Research Center programme (an academic consortium funded by the US National Science Foundation) named 'Re-inventing the Nation's Urban Water Infrastructure (ReNUWIt)' and whose goal is to create 'fundamental, systemic, and far-reaching changes in the United States' ageing urban water infrastructure' (Brody Guy, 2012).

One element of the new thinking is the recognition that large-scale, centralized approaches to infrastructure – the 'pipes and concrete' solution – are not always appropriate, even in densely developed urban areas. The health and environmental benefits of centralized pipes and concrete water systems are notable and highly laudable; indeed, such systems have been absolutely critical to the flourishing of US cities throughout the twentieth century. But they have their limitations. The retrofitting of urban water system infrastructure can be enormously expensive, disruptive to service distribution and local economies, and even technically infeasible. Emerging is a growing recognition of green growth and its related green infrastructure: the efficiency and power of what the use and integration of natural systems can provide for urban water systems (Aspen Institute, 2009).

US cities leading the way on green growth approaches to urban water system issues – Milwaukee, for example – understand several things. Watersheds left in their natural condition, or restored to close to what was their natural condition, can provide many of the functions that advanced engineered filtration facilities provide. Green open spaces within city boundaries can serve as flood protection as well as provide highly valued recreational functions to city dwellers. Rain gardens and rainwater tanks, no matter how small, collectively make a difference in programmes designed to control urban storm water runoff, while also educating and involving city residents in water infrastructure issues.

These cities also understand that adaptive thinking and approaches must now be built into water systems, and that sometimes adaptability cannot be built into large pipe and concrete infrastructure solutions. Whether green or conventional, limits to adaptability exist for all forms of infrastructure, but in general, smaller and less centralized systems provide a greater degree of flexibility, leaving the door open for whatever the future holds. A new generation of analytical assessment is focusing on the quantifiable benefits that come from green infrastructure approaches in water systems, and cost comparisons between green and conventional approaches to water infrastructure (Bolger *et al.*, 2009). Now that rigorous cost-to-cost and cost-to-benefit analyses can be made, the benefits of green infrastructure are known to be more than anecdotal (van Ast *et al.*, 2013). Often, the economies and efficiencies of green approaches to water systems are seen as extremely attractive.

Protection of watersheds

Well before green growth principles matured, and the term ‘green infrastructure’ came into being, a number of cities actively pursued the acquisition or control of land surrounding water bodies used for water supply and other crucial purposes. Such protection of watersheds, while neither a new nor a technologically advanced concept, has proven quite successful.[2] While most surface water used for water supply in the US has some buffering land around it, relatively few cities have attempted to protect completely their watershed lands for the purpose of protecting the water supply. Cities that have undertaken comprehensive watershed protection activities have, in some instances, avoided the need to build legally required water filtration plants, or have been able to build less sophisticated plants than might otherwise be required. Considering that water filtration plants are extremely expensive, successful watershed protection is almost invariably a good way of ensuring the future integrity of a water supply. The following case studies on watershed protection show that several US cities have been quite successful at accomplishing watershed protection as part of the water system infrastructures.

New York City

New York City is the best known, and perhaps the best, example of a US city that has pursued the protection of watershed lands as the primary means of maintaining the quality of its water supply. New York’s water supply system was built over 150 years, starting with the portion east of the Hudson River, known as the Croton system. As New York grew during the late nineteenth and early to mid-twentieth centuries, it added the Catskills and Delaware systems on the west side of the Hudson River to its water supply portfolio. The upper reaches of the system are approximately 240km away from the city (Figure 15.1). Almost the entire system uses gravity (under normal circumstances) to deliver water to the city, thus using very little energy.

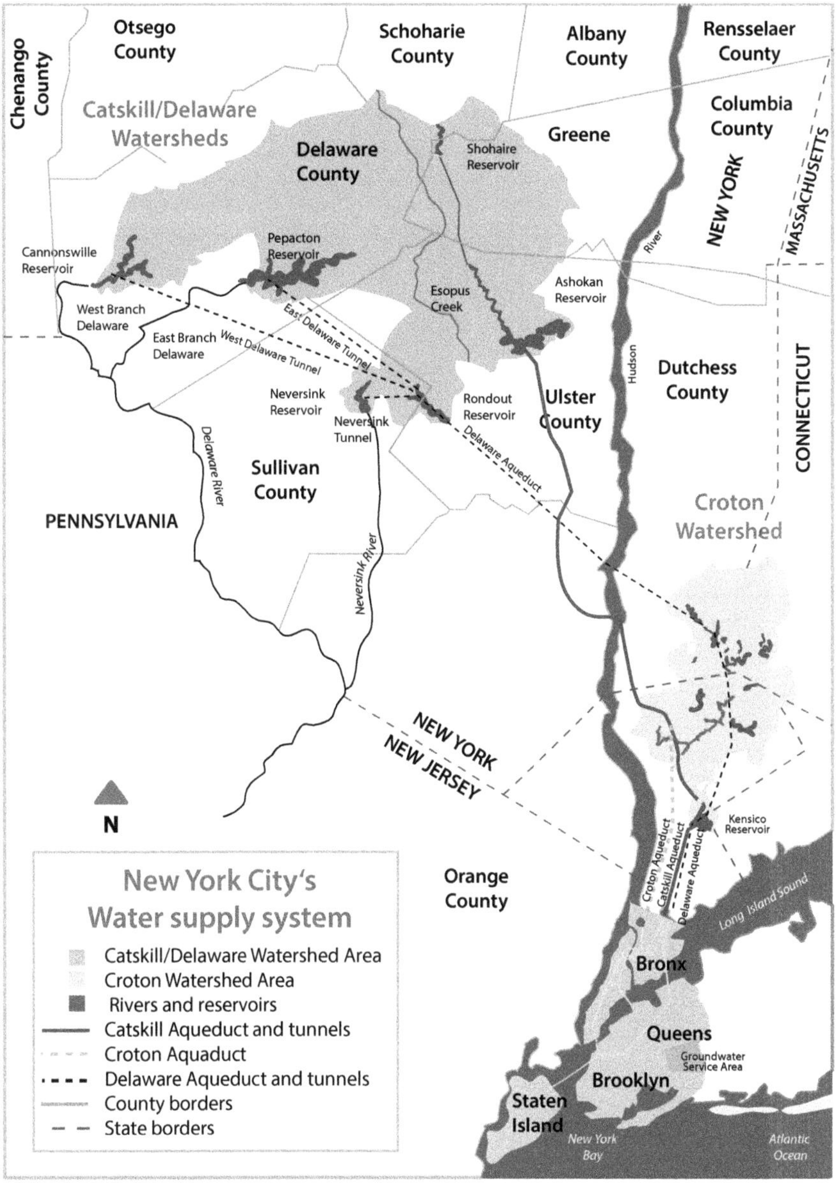

Figure 15.1 New York City's water supply system

Source: NYC Department of Environmental Protection, 2014

New York's water supply system is the largest predominantly unfiltered system in the United States (Osborne, 2007). The first-rate quality of the water stems from two factors: first, it originates far away from development and other activities that could potentially degrade it; and second, New York City has taken action through the years to protect the watershed lands around all of its reservoirs – built from the mid-nineteenth century to the 1960s. Today there are 19 reservoirs and three controlled lakes within the Croton, Catskills and Delaware systems, serving approximately 9 million people. Water is moved around and within the three systems to give the city maximum flexibility as to where to draw water from and to allow it to maximize the quality of its unfiltered water sources.

As the SDWA made US drinking water standards more stringent in recent decades, New York was required to meet filtration standards. But in 2007, the City requested and received its second ten-year Filtration Avoidance Determination (FAD) under the SDWA. US federal water supply regulations (40 CFR §141.71, §141.72 and §141.171) set forth most of the requirements for getting a FAD waiver. A public water supplier like New York must meet the stringent conditions established in these regulations in order to avoid building water filtration facilities. New York avoided $8 billion in water filtration infrastructure costs by meeting the filtration avoidance requirements in 2007. A condition for receiving its FAD waiver required New York to implement a comprehensive watershed protection programme. The programme first identified the most important lands to protect so as to preserve the water quality of the reservoirs within the Delaware, Catskills and Croton systems, and then implemented aggressive protection activities in a methodical manner.

Because investing in the protection of watersheds around its reservoirs has been significantly cheaper than building water filtration plants, New York has eagerly invested in such activities through what is known as the Lands Acquisition Programme (LAP). As the City states in its website about its water supply system, 'LAP is a key component of [our] efforts to continue to provide high quality drinking water without filtration of the Catskill-Delaware (Cat-Del) System, which provides water to over 9 million residents of the City' (NYC Environmental Protection, 2014). The LAP has protected more than 40,500 hectares of land in the 400,000 hectare Catskills and Delaware systems, and more than 800 hectares in the much smaller Croton system. When these figures are combined with lands protected by New York State and regional non-profit land conservation organizations such as the Open Space Institute, the percentage of permanently protected land in New York City's three water supply systems is 34 per cent (NYC Environmental Protection, 2014).

An important aspect of its land protection approach is that the city has been granted broad police authority by the New York State Legislature over its reservoirs and watersheds. As these water bodies and lands are located well outside the City's jurisdictional boundaries, without such a law there would be confusion about and challenges to the City's ability to control activities occurring on them. Today these lands and water bodies are closed to the public, although there are now discussions about whether allowing certain activities (e.g. hiking, camping, non-motorized boating and bird watching) could be compatible with the City's management of its reservoirs as unfiltered water supply sources and the surrounding watersheds as buffers to such systems.

Boston

Boston is another city that, through foresight and fairly aggressive protection of the lands around its water supply sources, is able to meet the stringent health requirements for drinking water imposed by the SDWA with reduced filtration. Much like New York, Boston draws on a water supply located some distance – about 105km west – from metropolitan Boston (Figure 15.2). The Quabbin Reservoir was constructed during the 1930s and 1940s, and was once the largest engineered water supply reservoir in the world. Building it involved the diversion of two rivers, the Swift and the Ware, and the government 'taking' and subsequent submerging of four rural towns in Massachusetts.[3] The goal in creating the massive Quabbin Reservoir, which today serves more than 3 million people, was to create a water supply system for metropolitan Boston that could be delivered mostly by gravity and did not require filtration. Water on its way east from the Quabbin Reservoir goes first into the Wachusett Reservoir, about 56km west of metropolitan Boston. The Wachusett Reservoir acts as a temporary holding tank or 'rest area' for water heading to Boston to be used.

Through decades of planning, management and land acquisition, the State of Massachusetts and the Massachusetts Water Resources Authority (MWRA) – the wholesale distributor of water throughout the Boston metropolitan area – have seen that more than 85 per cent of the watershed lands surrounding the Quabbin and Wachusett reservoirs are covered in forest and wetlands, and that about 75 per cent of the total watershed lands cannot be built on (MWRA, n.d.). By keeping so much watershed land in its natural condition, Massachusetts and its water authority ensure that metropolitan Boston's water supply is kept clean and requires less filtration than it might otherwise need (Barten *et al.*, 1998). Both the State and the water authority also patrol the lands, which are owned by the State, and enforce fairly strict land use limitations so that permitted uses do not degrade the water quality in the reservoirs.

Seattle

Seattle has also taken important steps, over a period spanning decades, to protect watershed lands in order to safeguard its water supply; the city has much to offer the rest of the world about how to do so. Seattle's water supply comes from two sources: the Cedar River watershed and the Tolt River watershed (Figure 15.3). Seattle started protecting the Cedar River watershed in the 1920s, after much of it had already been deforested. Careful forest and riparian management since then has produced a healthy and mature second growth forest that functions well to protect the quality of the watershed (Barten *et al.*, 1998). In turn, this

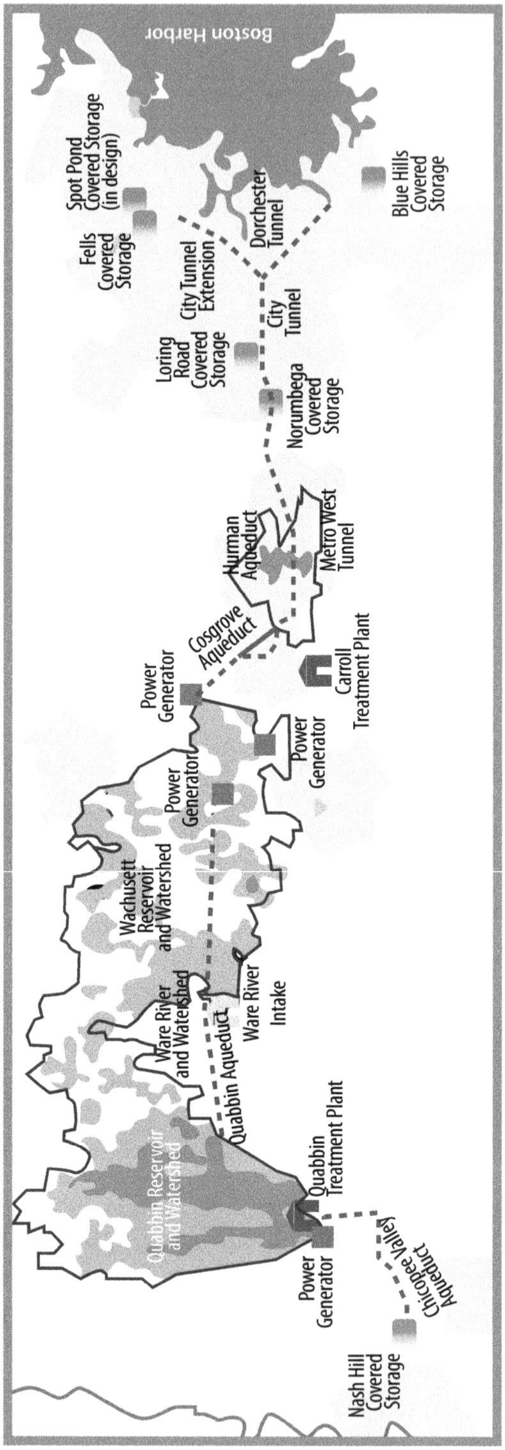

Figure 15.2 Boston's water supply system

Source: MWRA, n.d.

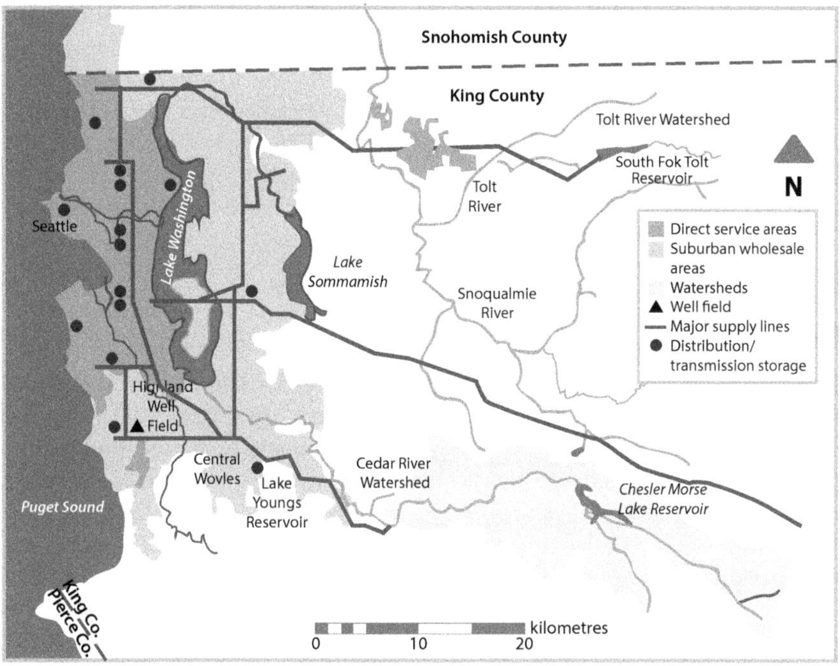

Figure 15.3 Seattle's water supply system

Source: Seattle Public Utilities, 2014

ensures its water bodies can help supply clean drinking water to 1.4 million people in the greater Seattle area, as well as provide critical downstream water flows for salmon, lakes and locks. Seattle owns almost the entire Cedar River watershed, more than 36,400 hectares, having started its land acquisitions almost 100 years ago. In 1996, the US Forest Service, which owns substantial tracts of land (particularly in the west US) gave the land that it owned in the watershed to Seattle. This allowed the city to increase its Cedar River watershed holdings.

The Tolt River (South Fork) is a smaller watershed, but is nevertheless essential to the overall system: it supplies metropolitan Seattle with about 30 per cent of its water. This part of Seattle's water supply system is relatively new, having been built in 1964. After a land swap in 1997 with a US forest products company, Seattle came to own about 70 per cent of the Tolt River watershed (Seattle Public Utilities, n.d.). The remaining 30 per cent lies in the Mt Baker-Snoqualmie National Forest, which is owned by the US Government, so the entire watershed is protected from certain types of development and use.[4] Seattle actively manages the use of both the Cedar River and the Tolt River watersheds, but because the water from these two watersheds is filtered as it reaches metropolitan Seattle, the use and management of these watersheds can be somewhat more relaxed than those of New York and Boston, where strict watershed land management allows for largely unfiltered sources of water to be used for drinking water purposes.

Seattle is doing much more than watershed protection in developing green infrastructure for water. The city is at the forefront of the movement to revisit the 'one size fits all' approach to the quality of water provided to different customers. Seattle is experimenting with maintenance and construction activities based on desired service levels of water, in regard to both quantity and quality. It is also involving its citizens in water system issues through programmes similar to those in Philadelphia and Milwaukee, which are discussed later in this chapter.

Green infrastructure within city boundaries

Philadelphia

Philadelphia, with a metropolitan population of 6 million (US Bureau of Census, 2014), is one of the largest cities in the United States. The city has embarked on an ambitious and progressive green infrastructure programme that, when implemented, will transform the way in which wastewater and storm water issues are addressed, and will raise awareness among Philadelphians as to how nature can help solve urban water problems. In 2011, Philadelphia's Water Department received permission from the State of Pennsylvania Department of Environmental Protection to use a wide range of locally implemented green infrastructure techniques to substantially reduce the city's combined sewer overflow discharges. Over the next 20–25 years, through its Green City/Clean Waters Programme, Philadelphia will invest up to US$2 billion in both green and traditional infrastructure to help keep its waterways unpolluted (Philadelphia Water Department, 2011a; Nutter, 2012; Philadelphia Water Department, 2011b).

Combined sewer overflows (CSOs) are a source of water pollution that plague many US cities, especially older ones in the northeast and Midwest (US EPA, 2001). These older cities have wastewater pipes that combine sewage and storm water in one transmission system (US EPA, 2001). The problem with CSO systems is that during storm events, large volumes of storm water cause the systems to reach capacity; they then 'overflow' and discharge and release both raw sewage and polluted storm water into local water bodies. And note that severe storms are likely to occur more frequently in many parts of the US as a result of climate change, so water pollution issues caused by CSOs may increase (Bolger *et al.*, 2009).

There are a number of conventional solutions to CSOs (Moffa, 1997). One is to replace the pipe system carrying the combined wastewater with two parallel pipe systems – one for sewage and one for storm water – but this requires tearing up streets and neighbourhoods to lay the new pipes. Another conventional solution is to construct huge underground tunnels and connect them to CSOs. During severe storms, rather than discharging upwards out of CSOs, these systems discharge downwards into their tunnels. Later, during dry weather, the sewage and storm water mix is pumped out of the tunnels and sent to wastewater treatment facilities. Chicago is an example of a US city that has applied this solution to the CSO.

Conventional infrastructure solutions to CSOs can be extremely expensive and disruptive, which explains in part why CSOs remain a significant source of water pollution, and why they have yet to be addressed in many places (US EPA, 2001). Historically, when faced with hard cost–benefit analyses, US cities have decided that the health and environmental benefits gained are not worth the tens or even hundreds of millions of dollars of costs in designing and constructing conventional infrastructure solutions to CSO discharges that pollute water bodies – especially when these polluting discharges happen only a few times a year.

To help solve Philadelphia's CSO pollution problems, the Green City/Clean Waters programme intends to transform the city's urban landscape, from one consisting almost exclusively of human-made surfaces that push and centralize storm water into the city's combined pipe system, into a green landscape that captures naturally as much of the storm water it receives as possible, and infiltrates and stores it through natural means. These green infrastructure 'technical devices' include storm water tree trenches, vegetated bumpouts, porous asphalt, rain gardens and sidewalk planters that will soak up the rain 'as Mother Nature once did' (Nutter, 2012). Over the next 20–25 years, Philadelphia expects to see a significant share of its landscape transformed (Philadelphia Water Department, 2011b).

A critical question for Philadelphia, and the governmental entities that make sure the city complies fully with federal and state water pollution control, is this. Will an approach that fully integrates green infrastructure with conventional infrastructure *succeed* in solving the city's remaining water pollution issues? The progress to date has been encouraging. Before publicly announcing the Green City/Clean Waters programme, the Philadelphia Water Department intensively developed green infrastructure demonstration projects that best served the Philadelphia landscape, and provided the city and its collaborators with the opportunity, not only to monitor the effectiveness of green infrastructure approaches, but also to improve projects after evaluating their initial performance. The projects also assisted in raising residents' awareness of the multi-dimensional benefits of the programme when it was announced.

It is clear from Philadelphia's initial work under the Green City/Clean Waters programme that there are multi-dimensional benefits of green infrastructure. The aesthetic pleasures of a green city – even if the paramount purpose is storm water control – attract city residents and encourage them to reinvest emotionally in their city. Green open spaces built to control storm water can be used for other purposes much of the time – for sports, picnicking and outdoor concerts. In addition, schoolchildren like to build things, and while they cannot participate in conventional infrastructure construction, they can certainly participate in planting trees, designing and planting gardens, and many other activities for the creation of green infrastructure. All of these factors played into the Philadelphia Water Department 'determining that a green storm water infrastructure-based approach would provide maximum return in environmental, economic, and social benefits within the most efficient timeframe, making it the best approach for the City of Philadelphia' (Philadelphia Water Department, 2011a).

Milwaukee

Milwaukee, Wisconsin, while perhaps not as well known internationally as the other cities discussed in this chapter, has a great deal to share with the international community regarding green infrastructure for water. The work that Milwaukee has undertaken over the past ten years places it at the forefront of global efforts to develop integrated, sustainable green infrastructure approaches to water pollution control issues. As is the case with Boston, New York, Philadelphia and Seattle, Milwaukee suffers no shortage of water.[5] Rather than protecting upstream sources, Milwaukee's green infrastructure efforts are aimed primarily at controlling water pollution – specifically to protect Lake Michigan, one of the largest freshwater lakes in the world.

The Milwaukee Metropolitan Sewerage District (MMSD) has, since the early 2000s, developed green infrastructure programmes that are fully integrated with conventional infrastructure for handling wastewater and storm water. What is important from a knowledge sharing perspective is how comprehensive the efforts have been to integrate new approaches for storm water control with more traditional ones. MMSD uses natural approaches to support and protect the hard infrastructure that it has invested in over the decades. The MMSD has also articulated why it has implemented and continued to run several green infrastructure water pollution control programmes – first and foremost because of the economics involved:

> By storing and draining water into the ground naturally, [the Greenseams programme] helps prevent future flooding while supporting and protecting MMSD's structural flood management projects – infrastructure investments worth hundreds of millions of dollars.
>
> (Milwaukee Metropolitan Sewerage District, n.d.)

The Greenseams programme is a flood management programme run by the MMSD that acquires outright ownership of, or conservation easements, over important land around Lake Michigan and its tributaries. The MMSD identifies strategically important land from a flood control perspective – in much the same way as New York identifies strategically important land from a water supply protection perspective – which may be undeveloped, privately owned flood control properties where growth is expected, wetlands, and riparian corridors of creeks and streams leading into Lake Michigan.

In promoting Greenseams to its ratepayers, the MMSD highlights the multi-dimensional benefits that come from green infrastructure approaches to water systems:

> All land acquired [under the Greenseams programme] will remain undeveloped, protecting water and providing the ability to store rain and melting snow. Wetlands maintenance and restoration at these sites will provide further water storage. Greenseams also preserves wildlife habitat

> and creates recreational opportunities for people living in the region. Where applicable, the properties can be used for hiking, bird watching and other passive recreation.
>
> (Milwaukee Metropolitan Sewerage District, n.d.)

An interesting aspect of Greenseams is that, after setting the goals of the programme and identifying the lands sought for acquisition, MMSD engaged the Conservation Fund, one of the US's major land conservation organizations, to run the programme (Milwaukee Metropolitan Sewerage District, n.d.). The Conservation Fund has the expertise and resources to implement a sophisticated and sustained land conservation effort like Greenseams, so MMSD through this partnership saved itself the effort of developing such expertise in-house.

Another important aspect of Milwaukee's approach to green infrastructure is its insertion within the city's overall approach to the next generation of its water infrastructure. Not only does Milwaukee use land acquisition and protection efforts along the lines of those in New York, Boston and Seattle, it also has several significant green infrastructure programmes within the city limits, along the lines of those in Philadelphia. For example, it has engaged its citizenry in the MMSD's Every Drop Counts programme, which appeals to customers through environmental protection, economic benefits and social responsibility:

> You can reduce water pollution, lower your water costs, and help preserve Lake Michigan – all by capturing water from your roof's downspout into a rain barrel. Rain barrels are fun, and easy to install!
>
> (St Croix County, n.d.)

Rain Barrels (water butts) and Rain Gardens are other programmes aimed at involving MMSD's customers in the city's overall green infrastructure efforts while keeping polluted storm water out of Lake Michigan and its tributaries. As with Philadelphia's efforts to create a more absorbing overall environment for storm water, MMSD asks customers to help with storm water control by planting certain types of gardens:

> Rain gardens soak up water from your roof, driveway, sidewalk or yard. Slightly different from your typical flower garden, rain gardens have wild flowers and deep-rooted native vegetation that help the ground soak up more water …
>
> (Fresh Coast, n.d.)

Rain gardens help reduce the risk of water pollution by keeping excess water out of sewers and from becoming polluted runoff. According to the Wisconsin Department of Natural Resources, rain gardens can absorb 30 per cent more water than a conventional, well-manicured, lawn (Wisconsin DNR, 2012). The MMSD makes it as easy as possible for its customers to participate in the Rain Barrels and Rain Gardens programmes. It advises customers where they can buy the rain barrels, and every year sells plants for a rain garden at a discounted price as an incentive to its customers. It stresses the community building aspects of participation by using inclusive language (e.g. 'all in our community' can participate) to describe the programmes. This will be a key message for green infrastructure programmes in the future, since many will rely on decentralized, local efforts to succeed.

Financing green infrastructure in the water sector

The growth of green infrastructure in the water sector is of course dependent, to a large degree, on the availability of funding. Whether a pipe and concrete or a green solution is preferred, no infrastructure project ever succeeds without the necessary capital. While there have been impediments in the past to allowing cities to use funding for green infrastructure, over the past ten years, as interest in and momentum for developing green infrastructure has grown, progress has been made on opening up financing mechanisms for green infrastructure in US cities (Bolger *et al.*, 2009).

It is important to give a brief synopsis of how water systems in US cities have historically been, and are now, financed so that Latin American counterparts can evaluate financing methods that may be available to them to pay for green infrastructure (see also Chapters 3 and 13). US cities now pay for the development of water systems through either cash financing or debt financing, but for decades (through the mid- to late 1980s) they received large amounts of federal and state grant funding for their water systems. Because conventional infrastructure systems are so expensive, they can rarely be paid for by cash financing. Therefore, US cities now almost always need to finance water systems through debt financing, and usually borrow money by issuing municipal bonds, which are granted favourable federal and state tax treatment for investors.

As green infrastructure becomes more widely accepted and prevalent in US cities, its financing becomes easier to achieve. Although the EPA notes that 'lack of funding is consistently cited as a barrier to the implementation of green infrastructure' (US EPA, n.d.), a host of funding mechanisms exist for US cities pursuing green infrastructure as part of their twenty-first-century water systems, and new mechanisms are getting more and more attention. The EPA is in fact a champion of such funding, and leads those municipalities willing to search for green infrastructure funding to useful sources.

What would be required next in the US is for traditional water system funding mechanisms, such as the municipal bond market, to 'open up' to provide funding to decentralized diverse systems that do not fit the traditional definition of water system utilities:

> the bond markets are traditionally used to finance development of centralized systems that are wholly owned by the issuing entity and secured by the revenues and physical assets of the system. When a utility wants

to finance work with its customer base [such as a rain barrel programme] to develop a citywide network of green infrastructure on private land, the bond market may no longer be a viable option.

(Johnson Foundation, 2011)

Insights from US urban watershed protection and green infrastructure efforts

What insight and knowledge do US cities have to offer Latin American counterparts regarding watershed protection and the use of green infrastructure for water systems? Common characteristics of the examples above can be identified, and these characteristics certainly can be used as the basis for preliminary thinking and planning by Latin American cities considering undertaking large-scale watershed protection as a means of protecting water supply systems. There has been much work in the United States in recent years on the restoration of watersheds – including forests, riparian corridors, and littoral zones within watersheds – and the transfer of this knowledge to Latin America can and should occur:

- The earlier that watershed protection activity occurs in the overall management of a water system, the greater the likelihood of its success. New York, Boston and Seattle all identified the protection of lands around their water supply reservoirs as critical features of the management of their water supply systems as they were building their surface water reservoirs, not years or decades afterwards.
- The watershed to be protected needs to be relatively pristine, or restoration of watershed lands, riparian corridors and littoral zones along water bodies should at least be relatively easy, for the interventions to have a good likelihood of being successful.
- The extent to which watershed protection activities can actually take the place of engineered filtration facilities also need to be determined. Ideally, the green infrastructure activities are the functional equivalent of, or better than, hard infrastructure – built at great capital cost and often with high operation and maintenance costs.
- Funding sources must be identified and secured for watershed protection activities. As is the case with traditional hard infrastructure, there needs to be both capital asset funding (buying or leasing land) and operations funding (managing the land or paying for behaviour changes in third parties) for watershed protection approaches.
- Educational and awareness-raising efforts about watershed protection must be undertaken. These activities will ensure that water users understand the connection between the water supply for their residences and businesses and the watershed activities they pay for, especially since watershed protection activities often occur at some distance from where water users are located.
- Broad authority to control activities on watershed lands must be put in place. This can be established through statutes that create such authority

for entities that manage water, or through contractual authority that is valid and binding upon third parties.

As Chapter 14 indicates, important work to protect watersheds as a means of protecting water supplies is already under way in parts of Latin America. Quito (Ecuador), Bogotá (Colombia) and other cities throughout the region are participating in 'water fund' programmes that raise money from urban water users and other sources to pay for the protection of watersheds through land acquisitions as well as paying landowners, farmers, foresters and villagers within upstream watersheds to alter their activities and behaviour so as to protect the quality of upstream water sources.

Regarding inner city green infrastructure water, insight and knowledge from the United States can also be of considerable value to Latin America's cities. This insight and knowledge includes the following:

- Water system infrastructure needs to be considered as an integrated system consisting of both natural and engineered components. With this in mind, to the extent that Latin American cities have not invested in, or otherwise developed, the engineered components of their water systems, they may want to start with the natural components, or make decisions along the lines of Philadelphia's green storm water infrastructure-based approach.
- The centralized versus decentralized needs of urban water systems need to be carefully considered. Is a centralized solution the best answer to a problem such as storm water control? Or can decentralized, green approaches be joined to an overall programme to control storm water? In general, decentralized multi-local systems have greater flexibility, and may be more adequate responses in the era of climate-change-induced uncertainties. Further, decentralized components of water systems can be scaled up, and may be more resource and energy efficient, while disaggregating large central systems, after they are built, into smaller components is much more difficult.
- The multiple benefits that come from green infrastructure approaches to water systems need to be determined and quantified. These are now quantifiable, and rigorous analytical work can and should be done when deciding what approach is best for a particular city.
- Water system infrastructure planning may start with the approach that, to the maximum extent feasible, systems that provide the right water for the right need should be created. This means, for example, that it is not necessary to provide potable water for watering outdoor areas. Water systems that integrate green infrastructure approaches so as to allow customers to differentiate and choose between different water offerings can be of great benefit in getting water users to appreciate the value of water, potentially using market mechanisms to establish the price of water of different qualities and enabling this to be understood and accepted by consumers.

Conclusions

Watershed protection and inner city green infrastructure approaches must be considered as the planning and commitments for future water systems in Latin America's cities take place. There are any number of important insights to be drawn from US cities regarding watershed protection and green infrastructure activities. Uniting citizens in urban areas with non-urban land users in the effort to have urban water systems emulate natural systems through 'green' infrastructure is indeed feasible. Cities with infrastructures of varying ages and status of repair all face the uncertainties brought about by the challenges of climate change; and there can be no doubt that what will emerge as our future urban water systems will be significantly different from past and existing systems.

Green growth solutions in both Latin America and US cities will play an increasingly vital role for urban areas and their water needs. Whether it is Quito or Philadelphia, the work currently being done to advance green growth concepts, design and implementation – sophisticated, long-term watershed protection in Quito, or sophisticated, long-term green infrastructure storm water solutions in Philadelphia – is rapidly moving from the fairly marginalized 'green stream' into the 'main stream' as viable solutions to any number of urban water issues.

The possibility of planning and building urban water systems that allow for differentiation and prioritization – both hard infrastructure and green infrastructure – is one of the two most promising aspects of green growth development. The other is its ability to connect humans to the natural world, especially in cities, where disassociation with the natural environment is increasing and is troubling for its ramifications for developing sustainable urban environments.

From a knowledge sharing and global communications perspective, one of the great aspects of the electronic age is that it allows for the instantaneous communication of information from one part of the world to another. Remarkable opportunities exist for US and Latin American cities to share insight and knowledge about the new generation of water system infrastructure that is being conceptualized, designed and built today – the water systems of the twenty-first century, and especially on new and interesting green infrastructure development issues. It is critical that work be continued at building partnerships between cities in Latin America, the United States and other parts of the world, so as to build a pool of common knowledge and practices from which all urban dwellers can ultimately benefit.

Notes

1 Under the Clean Water Act (CWA) and Safe Drinking Water Act (SDWA), states can have delegated authority that the federal government – the EPA – would otherwise hold.

2 The protection of watersheds in the United States can involve buying land outright, putting a conservation easement on land to restrict development and other uses,

long-term leasing of land, or using other land conservation mechanisms to ensure the success of protecting water bodies within watershed lands.

3 In the United States, when the federal government (or a state or municipal government) confiscates or 'takes' the land of individuals through the process known as eminent domain, it must compensate them at the fair market value of the property. To this day, land continues to be taken around water bodies for water supply, flood control and hydroelectric generation, and these takings (as was the case with Massachusetts' Quabbin Reservoir) often involve the permanent displacement of thousands if not millions of people.

4 Lands within national forests in the United States can be managed for timber and other purposes, such as ski areas, so from the perspective of managing a watershed, just because part of the watershed is within a national forest does not necessarily mean that it is adequately protected.

5 Although Boston and New York are located in the northeast of the United States, a part of the country that receives substantial annual precipitation, there are occasional drought warnings and water use restrictions put in place in these cities when their reservoir systems fall close to safe yield levels during dry years. Philadelphia draws its water from two major rivers, the Schuylkill and the Delaware, and thus does not need to worry about water supply from a quantity perspective. Milwaukee is in the best situation of all, having one of the largest lakes in the world to draw upon for water supply.

References

ASCE (2009) '2009 Report Card for America's Infrastructure', http://www.asce.org/Infrastructure/Report-Card/2009-Report-Card-for-America-s-Infrastructure/, accessed 14 February 2014

ASCE (2011) *Failure to Act: The Economic Impact of Current Investment Trends in Water and Wastewater Treatment Infrastructure,* Economic Development Research Group, Inc., in association with Downstream Strategies for the American Society of Civil Engineers, Reston, VA.

Aspen Institute (2009) *Sustainable Water Systems: Step One – Redefining the Nation's Infrastructure Challenge*, The Aspen Institute, Washington, DC.

Barten, P. K., Kyker-Snowman, P., Thomas, M., O'Connor, R. and Spencer, B. A. (1998) 'Massachusetts: Managing a Watershed Protection Forest', *Journal of Forestry,* vol. 96, no. 8, pp. 10–15.

Bolger, R., Monsma, D. and Nelson, R. (2009) *Sustainable Water Systems: Step One – Redefining the Nation's Infrastructure Challenge. A Report of the Aspen Institute's Dialogue on Sustainable Water Infrastructure in the U.S.,* The Aspen Institute, Energy and Environment Program, Washington, DC.

Brody Guy, A. (2012) *Remaking America's Urban Water Systems, Breakthroughs*, The Magazine of the UC Berkeley College of Natural Resources, CA.

Fresh Coast (n.d.) 'MMSD Rain Gardens Project', http://h2ocapture.com/learn/funding-programs/rain-gardens-project, accessed 10 June 2014.

Johnson Foundation (2011) *Convening Report: Financing Sustainable Water Infrastructure,* American Rivers, Ceres, and the Johnson Foundation at Wingspread, Waters, C. N., Racine, WI.

Milwaukee Metropolitan Sewerage District (n.d.) 'Greenseams', http://www.mmsd.com/floodmanagement/greenseams, accessed 29 May 2014.

Moffa, Peter E. (1997) *Control and Treatment of Combined Sewer Overflows*, Van Nostrand Reinhold Publishers, New York.

MWRA-Massachusetts Water Resources Authority (n.d.) 'How the MWRA Water System Works', http://www.mwra.com/04water/html/watsys.htm, accessed 29 May 2014.

Nutter, M. A. (2012) *Philadelphia's Green City Clean Waters Plan: Blending Tradition, Innovation to Renew Our Water Resources*, The United States Conference of Mayors / US Mayor Newspaper , Washington, DC.

NYC Environmental Protection (2014) 'Extended New York City Watershed Land Acquisition Program', http://www.nyc.gov/html/dep/html/environmental_reviews/lap.shtml, accessed 29 May 2014.

Osborne, R. (2007) 'Five Largest Unlfiltered Water Systems', http://watercrunch.com/2007/04/5-largest-unfiltered-water-systems, accessed 14 February 2014.

Philadelphia Water Department (2011a) *Amended Green City, Clean Waters. The City of Philadelphia's Program for Combined Sewer Overflow Control,* Program Summary, Philadelphia Water Department, Philadelphia.

Philadelphia Water Department (2011b) *PWD and PADEP Sign Historic Agreement to Officially Launch Green City, Clean Waters,* Philadelphia PWD Water Department, Philadelphia.

Seattle Public Utilities (n.d.) 'Tolt River Watershed', http://www.seattle.gov/Util/EnvironmentConservation/OurWatersheds/Tolt_River_Watershed/, accessed 9 June 2014.

St. Croix County (n.d.) 'Rain Barrel Program', http://www.co.saint-croix.wi.us/index.asp?SEC=%7B7634A9FD-BE2F-4C52-8C9E-F6381CA01A3D%7D, accessed 10 June 2014.

US Bureau of Census (2014) 'Top Metropolitan Areas', http://www.ntc-dfw.org/northtexas/poplargestmetro.html.

US EPA – United States Environmental Protection Agency (2001) 'Combined Sewer Overflow', http://www.epa.gov/nrmrl/wswrd/wq/stormwater/cso.pdf, accessed 9 May 2014.

US EPA – United States Environmental Protection Agency (2002) *The Clean Water and Drinking Water Infrastructure Gap Analysis,* Office of Water, United States.

US EPA – United States Environmental Protection Agency (n.d.) 'Funding Opportunities', http://water.epa.gov/infrastructure/greeninfrastructure/gi_funding.cfm, accessed 29 May 2014.

van Ast, L., Maclean, R. and Sireyol, A. (2013) *White Paper: Valuing Water to Drive more Effective Decisions,* Yarra Valley Water/Trucost, Australia.

WeiserMazars (2012) *U.S. Water Industry Outlook,* WeiserMazars LLP, New York.

Wisconsin Department of Natural Resources (2012) 'Raingardens: A Beautiful Way for You to Reduce Runoff Pollution', http://dnr.wi.gov/topic/stormwater/raingarden, accessed 17 May 2014.

Index

Bold page numbers indicate figures, *italic* numbers indicate tables

abundance, scarcity and 95–8
access to services: in Latin America 222–3, *223*; São Paulo, Brazil 77, *77*
affordability of services 51
Agência Reguladora de Saneamento e Energia do Estado de São Paulo (ARSESP) 78–9
agglomeration economies 21, 37
agriculture, wastewater reuse for 115–17
Agua para Todos (AOPT - Water for All) programme, Lima, Peru 94–5
Agua Somos (Bogotá Water Fund) 246–9, **247, 248, 249**
Alex, Hurricane. *see* Hurricane Alex in Mexico
allocation of water in São Paulo, Brazil 79–80
Andean Finance Corporation 41
Argentina: Jujuy province **203,** 203–4, **205**; privatization 226; reforms in water sector 226. *See also Hydro-BID* modelling system
Atotonilco Wastewater Treatment Plant 118–19

Bahri, A. 28
Bakker, K. 104
bank loans 54–5
barriadas 92–3
Bartram, J. 48–9
benchmarking 48, 60
Bird, R.M. 24
blended grants/repayable financing 57–8
Bogotá: Bogotá Water Fund 246–9, **247, 248, 249**. *See also* Río Bogotá
bond finance 55–6
Boston, USA 261, **262**
Brazil: costs recovery 51; reforms in water sector 226, 228; water funds 249–50; wells, unregulated 133. *See also* São Paulo, Brazil

capital costs, approaches to covering 50–1
Chaplin, S.E. 105
charges for WSS services 49–51, **50,** 78, 94, 227, 230
Chile: costs recovery 51; reforms in water sector 227, 228; responsibility for urban water services 39
cities: benefits of 21; comprehensive development perspective 24–5; costs of 21; economic importance of 21; Green Cities/Clean Water programme 265; green infrastructure within 264–8; growth in, challenges of 35–7; informal occupation 36–7; links with water 34–5; poverty and income distribution 25; resiliency, building 225. *See also* planning, urban
Clean Stream Programme, São Paulo, Brazil 85
Clean Wave Programme, São Paulo, Brazil 85
climate change: extreme weather 27; as global risk 16, *16*; and groundwater 129–30; proactive approach to 28–9; regional/local scale 27; and risk management

26–7; supplies of water, vulnerability of 198. *See also Hydro-BID* modelling system
Coastal Programme, São Paulo, Brazil 83
Colombia: costs recovery 51; Financiera de Desarrollo Territorial S.A. (FINDETER) 57; reforms in water sector 228; responsibility for urban water services 39. *See also* Río Bogotá
combined sewage overflows (CSOs) 264–5
community systems in Lima, Peru 94, 97
Companhia de Saneamento Básico do Estado de São Paulo (SANESP) 76, 80–6
Companhia de Technologia de Saneamento Ambiental (CETESB), São Paulo, Brazil 80
condominial sewerage systems in Lima, Peru 95–6
contamination of groundwater 139
Corporate Programme to Reduce Water Losses, São Paulo, Brazil 83
costs recovery 49–51, **50**
credit ratings 60

dams in Mexico City Metropolitan Area 114–15
debt finance 54–6
decentralization 229
demand for water: control measures as ineffective 39; control of in São Paulo, Brazil 82; Mexico City Metropolitan Area 112, 113
development, sustainable: comprehensive perspective 24–5; connections with other global risks 17–18
drainage systems: as component of urban water services 39; Mexico City Metropolitan Area 113–15, *114*; Quito 182; responsibility for 39; storm water systems 225; sustainable urban drainage systems (SUDS) 225

economics: development and public/private sectors 29; income distribution 28; urban sustainability, importance of 21. *See also* finance
ecosystem services 236–7
Ecuador. *see* Quito, Ecuador
environmental management 231–2
environmental restoration: Hydraulic Upgrade and Environmental Recuperation Project, Río Bogotá 175, **175**; Quito and Pichincha volcano 187–94, *189,* **190, 193**
environmental sustainability: connections with other global risks 17; and green growth 223–4, **224**; protection and preservation 40–1
equity finance 56–7
equity raising 60
Every Drop Counts programme 267
extreme weather: as global risk 16, *16*; hurricanes 27; prevention and mitigation 27; resiliency in cities 225

finance: affordability of services 51; bank loans 54–5; benchmarking 60; blended grants/repayable financing 57–8; bond 55–6; capital costs, approaches to covering 50–1; closing the water infrastructure gap 41–3, *42*; Companhia de Saneamento Básico do Estado de São Paulo (SANESP) 86; constraints on 46–7; costs recovery 49–51, **50**; credit ratings 60; debt 54–6; equity 56–7; equity raising 60; full range of approaches, use of 67; green growth 226; green infrastructure 268–9; groundwater 135; grouped 59; guarantees and insurance 59; innovative mechanisms 57–61, *58*; international transfers 53–4; lack of sustainable, impact of 47; location-specific factors 66; Mexico City Metropolitan Area 120–1; microfinance 55, 58–9; needed for MDGs 46; OECD and BRIC countries 46; official development assistance (ODA) **53,** 53–4; operational inefficiencies, improving 48–9; policy 230; private sector, role of 61–6, *64,* **65**; project 56; project preparation facilities 61; public/private mix of financing 65–6; Quito 188, 191, 192; realistic objectives for planning 66; repayable 54–7; São Paulo, Brazil 78, 86; strategic planning needed 67; sub-sovereign lending

59–60; subsidies, government 135; tariffs 49–51, **50,** 94, 227, 230; taxes 51–2, 230; transfers 52–4, **53,** 230; 3Ts as sources of 49–54, **50, 53,** 230; water infrastructure 23–4. *See also* water funds

Financiera de Desarrollo Territorial S.A. (FINDETER), Colombia 57

flooding: history of in Monterrey metropolitan area (MMA) 149; from Hurricane Alex in Mexico **150,** 150–3, **151, 152,** *153,* **153**; hydraulic capacity, lack of in Monterrey 156–60; Mexico City Metropolitan Area 110–11, 114–15; preventative engineering 160; resiliency in cities 225; response capacity following disasters 161–2; Río Bogotá 170–1, **171,** 171–2, **173**; urban design and hydrological risk 161

Fondo para la Protección del Agua (FONAG)(Quito Water Fund) 244–6

forest ecosystems 236–7

freshwater ecosystems 223–4

freshwater services 236–7

Gassner, K. 63

geography and urbanization and water services 23

Global Competitiveness Report (World Economic Forum) 17

Global Risks 2014 16, 16–17

Green Cities/Clean Water programme 265

green growth: costs of 226; decentralization 229; environmental management 231–2; environmental sustainability 223–4, **224**; financial policy 230; governance institutions 230–2, **231**; integrated urban water management 230–1, **231**; planning, urban 231–2; policy coherence and commitment 229; principles for sustaining reforms 229–30; public-private partnerships (PPPs) 229–30; regulatory structure 229; resiliency in cities 225; structure of the industry 229; United States of America (USA) 257–8; and water services 222–3, *223*

green infrastructure: Boston 261, **262**; within city boundaries 264–8; Every Drop Counts programme 267; financing 268–9; Greenseams programme 266–7; insights from USA for Latin America 270; Milwaukee 266–8; New York City 258, **259,** 260–1; Philadelphia 264–5; rain gardens 267–8; Seattle 261, **263,** 263–4; United States of America (USA) 257–8; watershed protection in USA 258–64, **259, 262, 263**. *See also* water funds

Green Plan, Mexico 136

Greenseams programme 266–7

groundwater: abstraction 127–8; adverse effects of extraction 135; aquifer over-exploitation 134, **134**; and climate change 129–30; complementarity with surface water 130, *131*; consumption and use 130–2, **132**; contamination 139; contribution to supply 126; coordination and integration of management 141; data on urban, lack of 128; deterioration of quality 48; development of 48; enforcement of regulations 139–40; environmental/ecological perspective 127; as favoured over surface water 126; government subsidies 135; *Hydro-BID* modelling system 216; institutional framework 136; long-term management, need for 133–4; Mexico City Metropolitan Area 113, 119, **134,** 134–6, 140–1; overuse of resources 134–5; political will, lack of 133; as poorly understood 127; production wells 135; protection and management programmes 133; recharge 129–30, 135, 138; restriction on use, problems from 139; salinity encroachment 129; São Paulo, Brazil **137,** 137–41; as strategic resource 127; unsustainable use of 128–9; wells, unregulated 129, 133, 135, 138, 139–40, 140–1

grouped finance 59

guarantees and insurance 59

hardware and operational inefficiencies 48–9

Holloway, A. 24–5

housing 40

Huaycán, Lima, Peru 97–8

Hurricane Alex in Mexico: causes of disaster triggered by 156–60, **158,**

159; damage caused by 153–6, **154,** *155, 156,* **157**; flooding from **150,** 150–3, **151, 152,** *153,* **153**; lessons from 160–2
hurricanes 27
Hutton, G. 48–9
Hydraulic Upgrade and Environmental Recuperation Project, Río Bogotá: concerns leading to 167; environmental restoration 175, **175**; location of projects **173**; objectives 172; structure 174; wastewater treatment 174
Hydro-BID modelling system: adaptation options, assessment of 208; analytical hydrography data set (AHD) 199–200, 216; calibration of model **209,** 209–10; case study 202–8, **203, 205, 206**; current/future stream flows, modelling of 207–8; data collection 207; demand for water, estimation of 207; effectiveness/cost efficiency defined 212, 214; evaluation of adaptation options 211–12, 214, *215*; features of 199–200; and groundwater 216; irrigation efficiency, improving 211–12, 212; limitations of approach 214, 216; natural conditions, assessment of 207; objective of 198; results from simulation 210–11; set-up 208–9; simulation of hydrology and climate change 200–2; supply and demand comparison 208; water efficiency, improving 211

income distribution 25, *26,* 28
inefficiency in operations, improving 48–9
inequality: Mexico City Metropolitan Area 120; and water and urban policies 25, *26*
informal occupation in cities 36–7; *barriadas* 92–3; Lima, Peru 92–3
information, collection and dissemination 60–1
infrastructure, water: complexity of development 39, **40**; financing 23–4; gap in, closing 41–3, *42*; improvement in, need for 36; Mexico City Metropolitan Area 115; São Paulo, investment in 81–2. *See also* green infrastructure; water funds
institutional frameworks: decentralization 229; Lima, Peru 96; Mexico City Metropolitan Area **111,** 111–12, *119,* 119–21; Monterrey metropolitan area (MMA) 163; Quito 193–4; São Paulo, Brazil 78–80, 137–8; structure 229
insurance 59
integrated urban water management 230–1, **231**
international transfers 53–4
investment in water infrastructure: low level of 23–4; Mexico City Metropolitan Area 121; needed to close gap 41–3, *42*. *See also* finance
irrigation, wastewater reuse for 115–17

Jiménez, E. 185
Jujuy province, Argentina **203,** 203–4, **205**. *See also Hydro-BID* modelling system

Kariuki, M. 57

Lagos, R. 25
land: control of for watershed protection 41; urban, changes in use of 35–6
Lands Acquisition Programme (LAP) 260
legal framework in São Paulo, Brazil 78–80
Lima, Peru: *Agua para Todos* (AOPT - Water for All) programme 94–5; *barriadas* 92–3; challenges of water supply system 89–90; community systems 94, 97; condominial sewerage systems 95–6; development of water services 93–5; as dual city 95; expansion of 92, 93; growth, problems associated with 96–8; Huaycán 97–8; informal city 92–3; institutional fragility 96; map **91**; micro-credit scheme 97; ontological basis of scarcity 90–1; Pachacútec 97; privatization, proposed 93; right, water as a 96; scarcity and abundance 95–8; scarcity of water 94, 95; SEDAPAL (Servicio de

Agua Potable y Alcantarillado de Lima) 93–4, 95–6; social and spatial discrimination 98–9; start of pubic water servcies 92; tariffs 94; Villa El Salvador 96–7
Llamas, R. 128
loans, bank 54–5
Ludwig, F. 27

Marin, P. 62, 63
market-based repayable finance 54–7
Markham, A. 238
Martínez-Santos, L. 128
Mehta, M. 58
Metropolitan Water Programme, São Paulo, Brazil 82–3
Mexico City Metropolitan Area: adverse effects of groundwater extraction 135; agriculture, wastewater reuse for 115–17; aquifer over-exploitation 134, **134**; Atotonilco Wastewater Treatment Plant 118–19; Cutzamala system **112,** 112–13; dams 114–15; demand for water 112, 113; drainage systems 113–15, *114,* 120; Eastern Discharge Tunnel project 114; Emisor Central pipeline 115; financial issues 120–1; flood control 114–15; flooding 110–11; government strategies for improvement 117–19; government subsidies 135; Great Channel 113, 114; Green Plan 136; groundwater **134,** 134–6, 140–1; groundwater development 113; groundwater withdrawals 119; historical water issues 110–11; inequity issues 120; institutional framework **111,** 111–12, 136; investment needs 121; jurisdictional issues 120; Mezquital Valley 116; non-point contamination sources 118; population 112, 113; solid waste management 118; storm water systems 120; supply of water 110, **111,** 111–13, **112,** 119–20; transfers 120; wastewater infrastructure 115, 120; wastewater treatment 118–19; water governance *119,* 119–21; Water Sustainability Programme for the Valley of Mexico 135–6; wells, unregulated 135, 140–1
Mezquital Valley 116
microfinance 55, 58–9, 97
Millennium Development Goals (MDGs): finance for 46; reducing costs for meeting 49
Milwaukee 266–8
Monterrey metropolitan area (MMA): causes of disaster triggered by hurricane 156–60, **158, 159**; Corral de Palmas dam 160; damage caused by Hurricane Alex 153–6, **154,** *155, 156,* **157**; flooding from Hurricane Alex **150,** 150–3, **151, 152,** *153,* **153**; geographic location 149; history of flooding 149; housing development patterns 157–8; hydraulic capacity, lack of 156–60, **158, 159**; institutional framework 163; lessons from Hurricane Alex 160–2; preventative engineering 160; reactive model of risk adaptation 162–3; response capacity following disasters 161–2; structures on riverbed 159–60; urban design and hydrological risk 161
municipal bonds 55

National Water Agency (ANA), São Paulo, Brazil 79–80
New Life Programme, São Paulo, Brazil 85–6
New York City 258, **259,** 260–1
non-point contamination sources 118

official development assistance (ODA) **53,** 53–4, 58–9
operational inefficiencies, improving 48–9

Pachacútec, Lima, Peru 97
Pacheco-Vega, R. 105
Pagiola, S. 237
Peltre, P. 183, 184
Peru. *see* Lima, Peru
Philadelphia 264–5
Pichincha volcano: anthropic threats 185; drainage from 182; Environmental Sanitation Phase I *189,* 190–2; Environmental Sanitation Phase II *189,* 192–4, **193**; geography of 182; historical context 184; institutionalization of initiatives 193–4; landslides 185; protection and environmental

restoration 187–94, *189,* **190, 193**; risk of disaster 183; settlements on slopes of 182–6, **185**; Slope Protection for Mount Pichincha Programme 188, *189,* 190; water and sewage services 186
planning, urban: green growth 231–2; links with water management as insufficient 35
planning of WSS sector: realistic objectives for 66; São Paulo, Brazil 76
Plano Nacional do Saneamento (PLANASA), São Paulo, Brazil 86–7
policy-making: coherence and commitment 229; context, importance of for 29; reforms in water sector 226–8; São Paulo, Brazil 76–80
political resource, water as 104, 105
pollution: agriculture, wastewater reuse for 115–17; contamination of groundwater 139; groundwater, deterioration of 48; non-point contamination sources 118; Río Bogotá 171–2, **173**; São Paulo, Brazil 77, 82
population of cities, growth in 35–7
poverty and water and urban policies 25, *26*
private sector: financial institutions 61; financing role of 61–6, *64,* **65**; public/private mix of financing 65–6; public-private partnerships (PPPs) 229–30; São Paulo, Brazil 78; service operators 61; water users 61
privatization: Argentina 226; variation in performance 105; wastewater governance 103–4
production wells 135
Programme for Rational Use of Clean and Reclaimed Water, São Paulo, Brazil 83
project finance 56
project preparation facilities 61
public-private partnerships (PPPs) 229–30
public sector, financing mix with private sector 65–6

quality management: groundwater, deterioration of 48; Río Bogotá 168–70; São Paulo, Brazil 77, 80
Quito, Ecuador: anthropic threats 185; climatic threats 184; disorderly urban sprawl 186; drainage system 182; Environmental Sanitation Phase I *189,* 190–2; Environmental Sanitation Phase II *189,* 192–4, **193**; financing of programmes 188, 191, 192; geography of 181–2; geomorphological threats 184; historical context 184; institutionalization of initiatives 193–4; landslides 185; protection and environmental restoration 187–94, *189,* **190, 193**; rainfall 182; risk of disaster 183; settlements on slopes of Pichincha volcano 182–6, **185**; Slope Protection for Mount Pichincha Programme 188, *189,* 190; water and sewage services 186; Water Fund 244–6

rain gardens 267–8
reclaimed water in São Paulo, Brazil 83
reforms: decentralization 229; financial policy 230; policy coherence and commitment 229; principles for sustaining 229–30; public-private partnerships (PPPs) 229–30; regulatory structure 229; in water sector 226–8
regulation of public utilities: reforms in water sector 226–8; São Paulo, Brazil 78–80; structure 229
repayable finance 54–7
responsibility for urban water services 39
right, water as a 96
Río Bogotá: challenges and recommendations 176–8; continuous amendments to projects 177; as 'dead river' 166; flooding 170–1, **171**; Hydraulic Upgrade and Environmental Recuperation Project 167, 172, **173,** 174–5, **175**; industrial discharges 177; interagency cooperation, obstacles to 176; lower basin 168; middle basin 168; pollution and flood control 171–2, **173**; population growth around 166; quality of water in 168–70; sewage treatment costs 177–8; upper basin 168; urban development/environmental

policies coordination 176–7; watershed **167,** 167–8
risk management, climate change and 26–7
risks, global: interconnections between 17; water as *16,* 16–17
Rojas, E. 29
Rosales, O. 25

salinity encroachment 129
sanitation services: access to 222–3, *223*; challenges facing 103; as component of urban water services **38,** 39; countervailing forces 103; factors affecting access to 102–3; political resource, water as 104, 105; privatization 103–4; research agenda for 104–5; responsibility for 39; scholarship on 103; social science perspective 103; technology, looking beyond 104; universal coverage in São Paulo, Brazil 84
São Paulo, Brazil: access to services 77, 77; achievements 86; Agência Reguladora de Saneamento e Energia do Estado de São Paulo (ARSESP) 78–9; allocation of water 79–80; challenges for 76; Clean Stream Programme 85; Clean Wave Programme 85; Coastal Programme 83; Companhia de Saneamento Básico do Estado de São Paulo (SANESP) 76, 80–6; Companhia de Technologia de Saneamento Ambiental (CETESB) 80; consumption of water 137; Corporate Programme to Reduce Water Losses 83; demand control 82; Federal Constitution 1988 78; finance 86; goals for coming years 87; groundwater **137,** 137–41; institutional framework 78–80, 137–8; investment in infrastructure 81–2; legal framework 78–80; Metropolitan Water Programme 82–3; National Water Agency (ANA) 79–80; New Life Programme 85–6; new water sources 82; Plano Nacional do Saneamento (PLANASA) 86–7; policy in 76–80; pollution control 77, 82; population 73; Programme for Rational Use of Clean and Reclaimed Water 83; quality management 77, 80; regulation of public utilities 78–80; security of water supplies 82–6; sewage collection 81; stress, Water 74, **75,** 76; supplies of water 74, **75,** 76, 137; Tietê River Decontamination Programme 84–5; universal services 76–7, 81, 84, 87; Upper Tietê Basin **137,** 137–8, 141; *Vida Nova/Mananciais* programme 82; water basin management units **137,** 137–8; Water Law 1997 79–80; wells, unregulated 138, 140
scarcity: and abundance 95–8; Lima, Peru 94; ontological basis of 90–1
Schwartz, J. 57
Seattle 261, **263,** 263–4
security of water supplies in São Paulo, Brazil 82–6
SEDAPAL (Servicio de Agua Potable y Alcantarillado de Lima), Lima, Peru. *see* Servicio de Agua Potable y Alcantarillado de Lima (SEDAPAL), Lima, Peru
service ladder 48–9
Servicio de Agua Potable y Alcantarillado de Lima (SEDAPAL), Lima, Peru 93–4, 95–6
sewage/sewerage: agriculture, wastewater reuse for 115–17; combined sewage overflows (CSOs) 264–5; Mexico City Metropolitan Area 115; Quito 186; Río Bogotá 177–8; São Paulo, Brazil 81
social sustainability and other global risks 17–18
solid waste management in Mexico City Metropolitan Area 118
Solomon, L.D. 105
sources, water, as insufficient 39
Stein, E. 29
Stiglitz, J. 17
storm water systems: drainage 225; Mexico City Metropolitan Area 113–15, *114,* 120
stress, water, in São Paulo, Brazil 74, **75,** 76
sub-sovereign lending 59–60
subsidies, government 135
supplies of water: access to 222–3, *223*; future changes in availability 198; industrial use 238; São Paulo,

Brazil 137; vulnerability to climate change 198. *See also Hydro-BID* modelling system
supply of water: complexity of infrastructure development 39, **40**; as component of urban water services 39; groundwater contribution 126; as insufficient 39; Mexico City Metropolitan Area 110, **111,** 111–13, **112,** 119–20; responsibility for 39; São Paulo, Brazil 82–6
sustainability: connections with other global risks 17–18; urban, importance of 21
sustainable urban drainage systems (SUDS) 225

Tallis, H. 238
tariffs 49–51, **50,** 230; Chile 227; Lima, Peru 94; São Paulo, Brazil 78
taxes 51–2, 230
technology and operational inefficiencies 48–9
Tietê River Decontamination Programme, São Paulo, Brazil 84–5
Tommasi, M. 29
transfers 230; intergovernmental tax 52; international **53,** 53–4; Mexico City Metropolitan Area 120; São Paulo, Brazil 78

UN-Habitat 24, 28, 35
United Nations Economic Commission for Latin America and the Caribbean (UNECLAC) 17–18
United Nations Environment Programme (UNEP) 17–18
United States of America (USA): adaptive thinking 257–8; bond finance 55–6; Boston 261, **262**; combined sewage overflows (CSOs) 264–5; Every Drop Counts programme 267; finance for green infrastructure 268–9; Green Cities/Clean Water programme 265; green growth 257–8; green infrastructure within city boundaries 264–8; Greenseams programme 266–7; insights for Latin America from 269–70; Lands Acquisition Programme (LAP) 260; Milwaukee 266–8; New York City 258, **259,** 260–1; old systems 256; Philadelphia 264–5; privatization 105; Seattle 261, **263,** 263–4; water system issues 256–7; watershed protection 258–64, **259, 262, 263,** 269–70
universal services 76–7, 81, 84, 87
Upper Tietê Basin **137,** 137–8, 141
urban water services: components of **38,** 38–9; as multidimensional 37–8; responsibility for 39; single sector perspective as ineffective 39–40
urbanization: benefits of 21; connections with other global risks 17; costs of 21; economic importance of 21; and geography 23; growth in, challenges of 35–7; poverty and income distribution 25; sustainability, importance of 21; as top development theme 17; trends 18–19, *20,* 21, *22–3. See also* planning, urban

Villa El Salvador, Lima, Peru 96–7

wastewater: agriculture, wastewater reuse for 115–17; challenges facing 103; combined sewage overflows (CSOs) 264–5; countervailing forces 103; factors affecting access to sanitation services 102–3; Hydraulic Upgrade and Environmental Recuperation Project, Río Bogotá 174; Mexico City Metropolitan Area 113–15, *114,* 115, 118–19, 120; political resource, water as 104, 105; privatization 103–4; research agenda for 104–5; scholarship on 103; social science perspective 103; technology, looking beyond 104; treatment, lack of adequate 224, *224*
water: comprehensive development perspective 24–5; and geography 23; as global risk 16, *16*; holistic approach needed 28; interconnection with other global risks 17–18; links with cities and urban planning 34–5; as political resource 104, 105; poverty and income distribution 25; and urbanization 18–19, *20,* 21, *22–3,* 23–4
water funds: Bogotá Water Fund 246–9, **247, 248, 249**; Brazil 249–50;

components of 239–40; defined 235; design and negotiation phase 251; economic incentives for 237–9; ecosystem services 236–7; hydrological models, use of 240; implementation of 240–4, **241, 242,** *243*; in Latin America 235; Latin American Water Funds Partnership 241; lessons learnt 250–3; local community involvement 244, 252–3; monitoring 252; objectives 240; Quito Water Fund 244–6; rationale behind 236–9; sources of public/private money 251–2; stakeholder involvement 250–1; watershed activities 240

Water Law 1997 (Brazil) 79–80
Water Sustainability Programme for the Valley of Mexico 135–6
watershed protection: Boston 261, **262**; land, control of 41; New York City 258, **259,** 260–1; Seattle 261, **263,** 263–4; United States of America (USA) 269–70; USA 258–64, **259, 262, 263**; water funds 240
wells: production 135; unregulated 129, 133, 135, 138, 139–40, 140–1
World Economic Forum 17
World Urban Forum 28
World Water Week 2011 15–16

Zevallos, O. 184

For Product Safety Concerns and Information please contact our EU representative GPSR@taylorandfrancis.com
Taylor & Francis Verlag GmbH, Kaufingerstraße 24, 80331 München, Germany

www.ingramcontent.com/pod-product-compliance
Ingram Content Group UK Ltd.
Pitfield, Milton Keynes, MK11 3LW, UK
UKHW022013180926
475701UK00003B/347

* 9 7 8 1 1 3 8 3 6 4 2 3 3 *